Birkhäuser

Static & Dynamic Game Theory: Foundations & Applications

For further volumes:
http://www.springer.com/series/10200

Pierre Bernhard
Jacob C. Engwerda
Berend Roorda, J.M. Schumacher
Vassili Kolokoltsov
Patrick Saint-Pierre, Jean-Pierre Aubin

The Interval Market Model in Mathematical Finance: Game-Theoretic Methods

Birkhäuser

Pierre Bernhard
INRIA Sophia Antipolis-Méditerranée
Sophia Antipolis, France

Berend Roorda
Department of Industrial Engineering
and Business Information Systems
School of Management and Governance
University of Twente
Enschede, The Netherlands

Vassili Kolokoltsov
Department of Statistics
University of Warwick
Coventry, United Kingdom

Jean-Pierre Aubin
VIMADES (Viability, Markets,
Automatics, Decisions)
Paris, France

Jacob C. Engwerda
CentER, Department of Econometrics
and Operations Research
Tilburg School of Economics
and Management
Tilburg University
Tilburg, The Netherlands

J.M. Schumacher
CentER, Department of Econometrics
and Operations Research
Tilburg School of Economics
and Management
Tilburg University
Tilburg, The Netherlands

Patrick Saint-Pierre
Université Paris Dauphine
Paris, France

ISBN 978-0-8176-8387-0 ISBN 978-0-8176-8388-7 (eBook)
DOI 10.1007/978-0-8176-8388-7
Springer New York Heidelberg Dordrecht London

Library of Congress Control Number: 2012951648

Mathematics Subject Classification (2010): 49K21, 49L20, 49N90, 91A23, 91A25, 91G20, 91G60, 91G80

Printed on acid-free paper

Springer is part of Springer Science+Business Media (www.birkhauser-science.com)

Preface

Mathematical finance was probably founded by Louis Bachelier in 1900 [19]. In his thesis and subsequent contributions, he constructed a stochastic model of stock price processes, essentially inventing the random walk or Brownian motion. But this was five years before Einstein investigated Brownian motion and long before Kolmogorov refounded probability on sound mathematical grounds; some basic probabilistic tools were missing. Bachelier's contribution was considered nonrigorous and, consequently, not recognized for its true pioneering value.

In contrast, few works in mathematical finance have enjoyed the fame and had the impact of Fischer Black and Myron Scholes' seminal paper [46]. In a bold move, it took the subjective concept of risk aversion out of the rationale for pricing financial derivatives, grounding such pricing on purely objective considerations.

"Objective," though, does not mean that no arbitrariness remains. In line with Bachelier, the Black–Scholes theory is based on an arbitrary choice of mathematical, stochastic model for underlying stock prices, which we will call "Samuelson's model," although some authors trace it back to earlier works.

Samuelson's model is called a "geometric diffusion," or "lognormal distribution." In that model, the price process $S(t)$ is assumed to obey the following Itô stochastic equation:

$$\frac{\mathrm{d}S}{S} = \mu \mathrm{d}t + \sigma \mathrm{d}B, \tag{0.1}$$

where μ and σ are known, deterministic parameters, or time functions, called "drift" and "volatility," respectively, and $B(\cdot)$ is a standard Brownian motion (or Wiener process).

Following these prestigious forerunners, most of the literature in mathematical finance relies on Samuelson's model, although notable exceptions have existed ever since, for example, [56, 57, 77, 87, 109, 117, 122, 124, 133].

The aim of this volume is to report several accomplishments using another class of models that we call, after [132], interval models. In these models, if n stocks are considered, it is assumed that a compact convex set of $\mathbb{R}^n$ is known that always contains the vector of relative stock price velocities (in a continuous-time setting) or one-step relative price changes (in a discrete-time setting). In the scalar case, corresponding to the classic Black–Scholes problem, and in discrete time, this means that we know two constants $\mathbf{d} < 1$ and $\mathbf{u} > 1$ – the notations used here are in reference to [57] – such that for a given $\delta t > 0$ and for all possible price trajectories

$$S(t+\delta t) \in [\mathbf{d}S(t), \mathbf{u}S(t)],$$

a line segment. In contrast, Cox et al. [57] assume that

$$S(t+\delta t) \in \{\mathbf{d}S(t), \mathbf{u}S(t)\},$$

the end points of a line segment, of course, a huge difference in terms of realism, and also of mathematics, even if in some cases we recover some of their results. More generally, in higher-dimensional problems, whether discrete or continuous time, this results in a tube of possible trajectories, or a so-called trajectory tube model.

These *interval models* were introduced independently, and almost simultaneously, by the authors of this volume. We only cite here some earlier papers as a matter of historical record. A common feature of these works is that, far remote from the mainstream finance literature, they suffered long delays between the date when they were written and their eventual publication, usually not in finance journals. Beyond Roorda et al. [132] already cited, whose preprint dates back to 2000, we mention here a 1998 paper by Vassili Kolokoltsov [95] and a paper from 2003 that only appeared in 2007 [86], a thesis supervised by Jean-Pierre Aubin defended in 2000 [128] – but a published version [17] had to wait till 2005 – and a conference paper by Pierre Bernhard, also in 2000 [37], an earlier form of which [35] did not appear in print until 2003.

If probabilities are the *lingua franca* of classic mathematical finance, it could be said that, although probabilities are certainly not ruled out, the most pervasive tool of the theories developed in this volume is some form of dynamic game theory. Most developments to be reported here belong to the realm of robust control, i.e., minimax approaches to decision making in the presence of uncertainty. These take several forms: the discrete Isaacs equation, Isaacs and Breakwell's geometric analysis of extremal fields, Aubin's viability approach, Crandall and Lions' viscosity solutions as extended to differential games by Evans and Souganidis, Bardi, and others, Frankowska's nonsmooth analysis approach to viscosity solutions, and geometric properties of risk-neutral probability laws and positively complete sets.

As a consequence, we will not attempt to give here a general introduction to dynamic game theory, as different parts of the book use different approaches. We will, however, strive to make each part self-contained. Nor will we try to unify the notation, although some of these works deal with closely related topics. As a matter of fact, the developments we report here have evolved, relatively independently, over more than a decade. As a result, they have developed independent, consistent notation systems. Merging them at this late stage would have been close to impossible. We will provide a concise "dictionary" between the notations of Parts II–V.

Part I is simply an introduction that aims to review, for the sake of reference, two of the most classic results of dynamic portfolio management: Merton's optimal portfolio and Black and Scholes' pricing theory, each with a flavor more typical of this volume than classic textbooks. The Cox–Ross–Rubinstein model will be presented in detail in Part II, together with the interval model.

Parts II and III mostly deal with the classic problem of hedging one option with an underlying asset. Part II tackles the problem of incompleteness of the interval model, introducing the fair price interval, and an original problem of maximizing the best-case profit with a bound on worst-case loss. Part III only deals with the seller's price – the upper bound of the fair price interval – but adding transaction costs, continuous and discrete trading schemes, and the convergence of the latter to the former, for both plain vanilla and digital options. Both parts deal in some respect with the robustness of the interval model to errors in the estimation of price volatility. Both use a detailed mathematical analysis of the problems at hand: portfolio optimization under a robust risk constraint in Part II, classic option pricing in Part III, to provide a "fast algorithm" that solves with two recursions on functions of one variable a problem whose natural dynamic programming algorithm would deal with one function of two variables.

It is known that in the approach of Cox, Ross, and Rubinstein, the risk-neutral probability associated with the option pricing problem spontaneously appears in a rather implicit fashion. Part V elucidates the deep links between the minimax approach and risk-neutral probability and exploits this relationship to solve the problem of pricing so-called rainbow options and credit derivatives such as credit default swaps.

Part V uses the tools of viability theory and, more specifically, the guaranteed capture basin algorithm to solve the pricing problem for complex options. A remarkable fact is that, as opposed to the fast algorithm of Part III, which is specifically tailored to the problem of pricing a classic option, the algorithm used here is general enough that, with some variations, it solves this large set of problems.

There obviously is no claim of unconditional superiority of one model over others or of our theories over the classic ones. Yet, we claim that these theories do bring new insight into the problems investigated. On the one hand, they are less isolated now than they used to be in the early 2000s, as a large body of literature has appeared since then applying robust control methods to various fields including finance, a strong hint that each may have a niche where it is better suited than more entrenched approaches. On the other hand, and more importantly, we share the belief that uniform thinking is not amicable to good science. In some sense, two different – sensible – approaches to the same problem are more than twice as good as one, as they may enlighten each other, be it by their similarities or by their contradictions.

France Jean-Pierre Aubin
France Pierre Bernhard
The Netherlands Jacob C. Engwerda
United Kingdom Vassili Kolokoltsov
The Netherlands Berend Roorda
The Netherlands J.M. Schumacher
France Patrick Saint-Pierre

Notation Dictionary

II	III	IV	V	Part number
T	T	T	T	Exercise time
X	$\mathscr{K}$	K	K	Exercise price
F	M	f	U	Terminal payment
0	$C^{\pm}, c^{\pm}$	β	δ	Transaction cost rates
	$S_0 = RS_0(T)$		S_0	Riskless bond price
		$j \in \{1, \dots, J\}$		Asset (upper) index
		Continuous time		
		Constants		
	μ_0		r_0	Riskless return rate
	$\tau^- + \mu_0$		$r^\flat$	Min risky asset return
	$\tau^+ + \mu_0$		$r^\sharp$	Max risky asset return
		Time functions		
$t \in [0, T]$	$t \in [0, T]$	$t \in [0, T]$	$t \in [0, T]$	Current time
	$R = S_0/S_0(T)$			End-time discount rate
S	$S = Ru$	S	S	Risky asset price
	$\tau + \mu_0$		r	Risky asset return rate
	$XS = Rv$		E	Portfolio exposure
	$v = \varphi^\star(t, u)$		$E = E^\heartsuit(t, W)$	Optimal hedging strategy
	Y		p_0	Number of bonds in portfolio
	X		p	Number of shares of risky stock
	Rw		W	Portfolio worth
	RW		$W^\heartsuit$	Optimal portfolio worth
	(Control)	Impulses	(Triggered)	
	t_k		t^n	Impulse times
	ξ_k		$\psi(x) - x$	Impulse amplitudes
		Discrete time		
		Constants		
h	h	τ	ρ	Time step
n	K	n	N	Total number of steps
	$e^{\mu_0 h}$	$\rho = 1 + r\tau$	$1 + \rho r_0$	One-step riskless ratio
d	$1 + \tau_h^-$	d^j	$1 + \rho r_{\mathrm{d}}$	Min one-step S ratio
u	$1 + \tau_h^+$	u^j	$1 + \rho r_u$	Max one-step S ratio
		Time functions		
$t_j = jh$	$t_k = kh$	m	$t_n = n\rho$	Current time
S_j	$S_k = R_k u_k$	S_m^j	S^n	Risky asset price
v	$1 + \tau_k$	ξ^j	$1 + r_\rho^n$	One-step S ratio
γ_j	X_k	γ_m^j		Risky shares in portfolio
$\gamma_j = g_j(S_j)$	$v_k = \varphi_k(u_k)$			Hedging strategy
	$R_k w_k$	X_m	W^n	Portfolio worth

Contents

Part I Revisiting Two Classic Results in Dynamic Portfolio Management

1 Merton's Optimal Dynamic Portfolio Revisited 3
1.1 Merton's Optimal Portfolio Problem 3
1.1.1 Problem and Notation 3
1.1.2 Solution 6
1.1.3 Logarithmic Utility Functions 8
1.2 A Discrete-Time Model 9
1.2.1 Problem and Notation 9
1.2.2 Solution 11
1.2.3 Market Models 15

2 Option Pricing: Classic Results 17
2.1 Introduction 17
2.2 Problem Formulation 18
2.3 Stop-Loss Strategy 20
2.4 Black–Scholes Theory 21
2.4.1 Black–Scholes Equation 21
2.4.2 Proof of Lemma 22
2.5 Digital Options 25

Part II Hedging in Interval Models

3 Introduction 31
3.1 Why Hedge? 31
3.2 A Simplistic Hedging Scheme: The Stop-Loss Strategy 33
3.3 Risk-Free Hedging in the Binomial Tree Model 34
3.4 Relationship with the Continuous-Time Black–Scholes–Merton Model 38

3.5 Risk Assessment Models 41
3.5.1 Current Models 41
3.5.2 Interval Model 42

4 Fair Price Intervals 45
4.1 Fair Price Interval of an Option: The General Discrete-Time Case 45
4.2 Fair Price Intervals in Interval Models 47
4.2.1 Fair Price Interval 47
4.2.2 Characterization of the Fair Price Interval in Terms of Strategies 48
4.2.3 Example 51
4.2.4 Characterization of Fair Price Interval in Terms of Martingale Measures 52
4.3 Computation of the Fair Price Interval for Path-Independent Strategies 54
4.4 Worst-Case Analysis 57
4.4.1 Introduction 57
4.4.2 A Nonextremal Path with Worst-Case Cost 59
4.4.3 Worst Cases in Interval Models Versus Tree Models 60

5 Optimal Hedging Under Robust-Cost Constraints 65
5.1 Introduction 65
5.2 Effect of Cost Constraints on Admissible Strategies 66
5.3 Calculating Maximum Profit Under a Cost Constraint 67
5.4 Extensions 72
5.4.1 Loss/Profit Ratio 73
5.4.2 Maximum Expected Profit Under a Cost Constraint 76
5.5 Summary 77

6 Appendix: Proofs 79

Part III Robust-Control Approach to Option Pricing

7 Continuous and Discrete-Time Option Pricing and Interval Market Model 91
7.1 Introduction 91
7.1.1 A New Theory of Option Pricing? 92
7.1.2 Related Contributions 93
7.2 Modeling 94
7.2.1 Contingent Claims 94
7.2.2 Market 95
7.2.3 Portfolio 97
7.2.4 Hedging 99
7.2.5 Conclusion: A Minimax Dynamic Game 101

7.3 Extensions ... 103
7.3.1 American Options ... 104
7.3.2 Delayed Information ... 105

8 Vanilla Options ... 107
8.1 Introduction and Main Results ... 107
8.1.1 Vanilla Options ... 107
8.1.2 Terminal Payment ... 108
8.1.3 Main Results ... 110
8.1.4 Joshua Transformation and the DQVI ... 114
8.2 Geometric Approach ... 118
8.2.1 Geometric Formulation ... 119
8.2.2 Primary Field and Dispersal Manifold $\mathscr{D}$ ... 121
8.2.3 Equivocal Manifold $\mathscr{E}$ ... 129
8.2.4 Focal Manifold $\mathscr{F}$ ... 133
8.2.5 Synthesis: Representation Formula ... 140
8.3 Viscosity Solution ... 141
8.3.1 Uniqueness ... 142
8.3.2 Verification of Representation Formula ... 153
8.4 Discrete Trading and Fast Algorithm ... 156
8.4.1 Dynamic Programming and Algorithms ... 156
8.4.2 Convergence ... 161

9 Digital Options ... 167
9.1 Introduction and Main Results ... 167
9.1.1 Digital Options ... 167
9.1.2 Main Results ... 168
9.2 Geometric Approach ... 173
9.2.1 Trivial Regions ... 173
9.2.2 Region of Interest ... 175
9.2.3 Synthesis ... 184
9.3 Viscosity Solution ... 186
9.3.1 The Discontinuity ... 186
9.3.2 Continuous Regions ... 189
9.4 Discrete Trading and Algorithms ... 192
9.4.1 Algorithms ... 192
9.4.2 Convergence ... 194

10 Validation ... 199
10.1 Numerical Results and Comparisons ... 199
10.1.1 Numerical Computations ... 199
10.1.2 Numerical Comparison with Black–Scholes ... 200
10.1.3 Explaining the Volatility Smile ... 202
10.2 Compared Strengths and Weaknesses ... 204
10.2.1 Strict Mathematical Properties ... 204
10.2.2 Robustness ... 205
10.3 Conclusion ... 211

Part IV Game-Theoretic Analysis of Rainbow Options in Incomplete Markets

11 Introduction 217
11.1 Introduction to Game-Theoretic Pricing 217
11.2 Related Works 218

12 Emergence of Risk-Neutral Probabilities from a Game-Theoretic Origin 221
12.1 Geometric Risk-Neutral Probabilities and Their Extreme Points 221
12.2 Game-Theoretic Origin of Risk-Neutral Laws: Preliminaries 226
12.3 Game-Theoretic Origin of Risk-Neutral Laws: Main Result 233
12.4 Nonlinear Extension 237
12.5 Infinite-Dimensional Setting and Finite-Dimensional Projections 240
12.6 Extension to a Random Geometry 243
12.7 Mixed Strategies with Linear Constraints 245

13 Rainbow Options in Discrete Time, I 249
13.1 Colored European Options as a Game Against Nature 249
13.2 Nonexpansion and Homogeneity of Solutions 253
13.3 Submodular Payoffs: Two Colors 254
13.4 Submodular Payoffs: Three or More Colors 256
13.5 Transaction Costs 258

14 Rainbow Options in Discrete Time, II 261
14.1 Rainbow American Options and Real Options 261
14.2 Path Dependence and Other Modifications 262
14.3 Upper and Lower Values for Intrinsic Risk 264
14.4 Cash-Back Methodology for Dealing with Intrinsic Risk 266
14.5 Degenerate or Random Geometry of Nonsimultaneous Jumps 267
14.6 Stochastic Interest Rates and Stochastic Volatility 269
14.7 Identification of Pre-Markov Chains 270

15 Continuous-Time Limits 273
15.1 Nonlinear Black–Scholes Equation 273
15.2 An Example with Two Colors 275
15.3 Transaction Costs in Continuous Time 278
15.4 Models with Stochastic Volatility 280
15.5 Fractional Dynamics 281

16 Credit Derivatives 285
16.1 Basic Model with No Simultaneous Jumps 285
16.2 Simultaneous Jumps: Completion by Tranching 288
16.3 Mean-Field Limit, Fluctuations, and Stochastic Law of Large Numbers 288

Part V Viability Approach to Complex Option Pricing and Portfolio Insurance

17 Computational Methods Based on the Guaranteed Capture Basin Algorithm ... 293
17.1 Guaranteed Capture Basin Method for Evaluating Portfolios ... 293
17.1.1 Classical Option Evaluation ... 293
17.1.2 Limits of Classic Evaluation Methods ... 294
17.2 The Dynamic System Underlying Financial Instruments ... 294
17.2.1 State and Control Variables ... 295
17.2.2 Viability Constraints and Target ... 295
17.2.3 Uncertainty of the Environment ... 297
17.2.4 Differential and Discrete Games Describing Portfolio Evolution ... 297
17.2.5 Guaranteed Capture Basin Algorithm ... 299
17.2.6 Approximation of Valuation Function ... 300
17.2.7 Implementing the Guaranteed Capture Basin Method to Evaluate a European Call ... 301
17.3 Extension of Capture Basin Methods to Evaluate Complex Instruments ... 302
17.3.1 Taking into Account Transaction Costs and Constraints ... 303
17.3.2 Approximation of Valuation Function in the Presence of Transaction Costs ... 305
17.3.3 Bermuda and Capped-Style Options: Two Examples of Constrained and Multitarget Problems ... 307
17.4 Evaluation of Complex Financial Instruments Using Impulse Systems ... 309
17.4.1 Hybrid Dynamic Systems and Viability Concepts ... 310
17.4.2 Guaranteed Hybrid Capture Basin Algorithm ... 311
17.4.3 Evaluation of Barrier Options ... 312
17.4.4 Evaluation of Options Using NGARCH Uncertainty Correction and in the Presence of Transaction Costs ... 314

18 Asset and Liability Insurance Management (ALIM) for Risk Eradication ... 319
18.1 Introduction ... 319
18.2 VPPI in Summer 2011 Crisis ... 324
18.2.1 Inputs ... 324
18.2.2 Outputs ... 326
18.2.3 Flow Chart of VPPI Software ... 329

18.3 Uncertainties 330
18.3.1 Stochastic Uncertainty 331
18.3.2 Tychastic Uncertainty 332
18.3.3 Historic Differential Inclusions 333
18.3.4 Nature and Measures of Insurance 334
18.4 Comparing VPPI and CPPI 335

References 337

Index 345

Part I
Revisiting Two Classic Results in Dynamic Portfolio Management

Author: Pierre Bernhard
INRIA Sophia Antipolis-Méditerranée,
France

The material presented in this part was developed for a course in portfolio management while the author was a professor at École Polytech'Nice, a department of the University of Nice Sophia Antipolis.

In this two-chapter part, we revisit the two most classic results in the theory of dynamic portfolio management: the Merton optimal portfolio in Chap. 1 and the famous Black and Scholes option pricing theory in Chap. 2.

In both cases we recall the classic result, to be used also as a reference for the remainder of the volume, but with nonclassic developments in the spirit of this volume attached to them. The "Merton" optimal portfolio problem is investigated with two different models: the classic one and a uniform-interval model. The Black and Scholes option pricing theory is dealt with in a robust control – or game theoretic – probability-free approach. A third very classic result is the discrete-time theory of Cox, Ross, and Rubinstein. It will be presented in the next part and fully revisited in Part IV.

Notation

- a^t: For any vector or matrix a, *a transposed*.

Universal constants

- $\mathbb{R}$: The real line.
- $\mathbb{N}$: The set of natural (positive) integers.
- $\mathbb{K} = \{0, 1, \ldots, K-1\}$.
- $\mathbb{1}$: A vector of any dimension with all entries equal to 1.

Main variables and parameters

- T: Horizon of finite horizon problems [Time]
- h: Time step of discrete trading theory [Time]
- $S_i(t)$, $i = 1,\dots,n$: Market price of risky asset i (without index if only one risky asset is present) [Currency]
- $S_0(t)$: Price of riskless asset [Currency]
- $R(t) = S_0(t)/S_0(T)$: End-time value coefficient [Dimensionless]
- $u_i(t) = S_i(t)/R(t)$: Normalized market price of asset i [Currency]
- μ_i: Drift coefficient in model for S_i [Time^{-1} (continuous), dimensionless (discrete)]
- μ_0: Expected return of riskless asset [Time^{-1} or dimensionless]
- $\lambda_i = \mu_i - \mu_0$ Excess expected return of asset i over riskless asset [Time^{-1} (continuous) or dimensionless (discrete)]
- σ_i: A line of coefficients defining the variability of S_i around its expected value [Time$^{-1/2}$ (continuous), dimensionless (discrete)]
- σ: Matrix whose lines are the σ_i [Time$^{-1/2}$ (continuous), dimensionless (discrete)]
- $\Sigma = \sigma\sigma^t$: Covariancelike matrix [Time^{-1} (continuous), dimensionless (discrete)]
- $X_i(t)$: Number of shares of asset i in portfolio [Dimensionless]
- $W(t)$: Portfolio worth [Currency]
- $w(t) = W(t)/R(t)$: Portfolio normalized worth. [Currency]
- $\varphi_i = X_i S_i/W = X_i u_i/w$: Fraction of portfolio invested in asset i [Dimensionless]
- $C(t)$: Rate of portfolio consumption [Currency$\times$Time^{-1}] (continuous) or stepwise consumption [Currency] (discrete)
- $c(t) = C(t)/R(t)$
- $\chi(t) = C(t)/W(t) = c(t)/w(t)$: Relative rate of withdrawal of funds from portfolio for consumption [Time^{-1} (continuous), dimensionless (discrete)]
- Π: Coefficient of bequest utility function [Currency]
- π: Coefficient of running utility function [Currency$\times$time^{-1} (continuous), currency (discrete)]
- P: Coefficient of Bellman function [Currency]
- $\gamma \in (0,1)$: Exponent of c and w in utility functions [Dimensionless]
- α ($\alpha^{1-\gamma}$ in discrete theory): Maximum (normalized in continuous theory) return rate of a portfolio [Dimensionless]
- δ: Discount rate for infinite horizon problem [Time^{-1} or dimensionless]
- $\beta = \delta - \mu_0$ (continuous) or $\delta^{h/(1-\gamma)}$ (discrete): Normalized discount rate [Time^{-1} or dimensionless]

Chapter 1
Merton's Optimal Dynamic Portfolio Revisited

1.1 Merton's Optimal Portfolio Problem

The problem considered here is that of managing a dynamic portfolio over a period of time $[0,T]$ – or over $[0,+\infty)$; we shall consider this infinite horizon case in a separate subsection – in a market where several assets are available, with differing and varying returns. The portfolio manager is allowed to sell parts of his portfolio to obtain an immediate utility. He is also interested in having sufficient wealth at the end of the period considered. In this section, we deal with the classic continuous trading formulation of [118].

1.1.1 Problem and Notation

1.1.1.1 The Market

We consider a market with n risky assets and one riskless asset. The riskless asset will always be referred to by the index 0, the risky assets by indices 1 to n. Let $S_i(t)$ be the market price of asset i, $i = 0, 1, \ldots, n$. We need a model of the market. We extend the model (0.1) to n assets in the following way: we assume that each risky asset obeys the Itô equation

$$\frac{\mathrm{d}S_i}{S_i} = \mu_i \mathrm{d}t + \sigma_i \mathrm{d}b. \tag{1.1}$$

Here, μ_i are known constants and σ_i are *lines* of coefficients, all of the same length $\ell \le n$. (But to make things simple, we may take $\ell = n$.) Accordingly, b is a standard Brownian motion of dimension ℓ (or n), i.e., a vector whose entries are independent standard Brownian motions. We let σ be a matrix whose line number i is σ_i, and

$$\Sigma := \sigma\sigma^t. \tag{1.2}$$

P. Bernhard et al., *The Interval Market Model in Mathematical Finance*, Static & Dynamic Game Theory: Foundations & Applications, DOI 10.1007/978-0-8176-8388-7_1,

The riskless asset satisfies

$$\frac{dS_0}{S_0} = \mu_0 dt.$$

In the finite horizon case, we shall rather use the dimensionless end-time value coefficient

$$R(t) = S_0(t)/S_0(T) = e^{\mu_0(t-T)}.$$

We shall assume that a reasonable portfolio does not contain assets whose expected return μ_i is less than the riskless return μ_0 since they would bring no value or risk alleviation. We shall call

$$\mu_i - \mu_0 = \lambda_i$$

the excess of expected return over the riskless rate. And we shall use the ratios

$$u_i(t) = \frac{S_i(t)}{R(t)},$$

which, in view of (1.1), satisfy the Itô equation

$$\frac{du_i}{u_i} = \lambda_i dt + \sigma_i db. \tag{1.3}$$

1.1.1.2 The Portfolio

A portfolio shall be defined by $n+1$ functions $X_i(t)$, $i = 0, 1, \dots, n$, giving the number of shares of each asset in the portfolio at time t. As a model simplification, we consider that the X_i are not restricted to being integers. They shall take their values in $\mathbb{R}$. Thus we also allow the portfolio to be "short" in some assets. Its worth is therefore

$$W(t) = \sum_{i=0}^{n} X_i(t) S_i(t).$$

We shall instead use

$$w(t) = \frac{W(t)}{R(t)} = X_0(t) S_0(T) + \sum_{i=1}^{n} X_i(t) u_i(t).$$

Transactions will be variations dX_i of the number of shares. Together with the variations in market prices, they produce variations in the worth of the portfolio:

$$dW = \sum_{i=0}^{n} (S_i dX_i + X_i dS_i) \qquad \text{or} \qquad dw = S_0(T) dX_0 + \sum_{i=1}^{n} (u_i dX_i + X_i du_i).$$

We allow such transactions to yield some excess cash, which the manager may want to use for immediate consumption. Let, therefore, $Cdt = cRdt$ be the cash taken from the portfolio by the transactions $\mathrm{d}X_i$. (If we wanted to allow discontinuous X_i, we could let $\mathrm{d}X_i$ be finite and $C(\cdot)$ contain a Dirac impulsion. But we will not need to do this because the solution we shall find has continuous X_i.) This is obtained through transactions satisfying $\mathrm{d}W + C\mathrm{d}t = 0$, or, dividing through by R,

$$S_0(T)\mathrm{d}X_0 + \sum_{i=1}^{n} u_i \mathrm{d}X_i + c\mathrm{d}t = 0.$$

Hence, we find that

$$\mathrm{d}w = \sum_{i=1}^{n} X_i \mathrm{d}u_i - c\mathrm{d}t,$$

or, using (1.3),

$$\mathrm{d}w = \sum_{i=1}^{n} X_i u_i (\lambda_i \mathrm{d}t + \sigma_i \mathrm{d}b) - c\mathrm{d}t.$$

It is customary to simplify this expression using the *fractions* φ_i and χ of the portfolio defined as

$$\varphi_i = \frac{X_i S_i}{W} = \frac{X_i u_i}{w}, \qquad \chi = \frac{C}{W} = \frac{c}{w}.$$

Since we allow short positions for the portfolio, φ is unconstrained in $\mathbb{R}^n$. Consumption, on the other hand, is assumed to be nonnegative; hence so must χ be.

We also use the vector notation $\lambda \in \mathbb{R}^n_+$ and $\varphi \in \Delta_n$ (the simplex of $\mathbb{R}^n$) for the n-vectors of the λ_i and φ_i, $i \geq 1$. We finally obtain

$$\mathrm{d}w = \left[(\varphi^t \lambda - \chi)\mathrm{d}t + \varphi^t \sigma \mathrm{d}b\right] w. \tag{1.4}$$

1.1.1.3 Utility

The utility derived by the manager, which he wants to maximize, is supposed to be the sum of a running utility (which we may, of course, write directly in terms of c instead of C; it is just a matter of lowering the discount rate by μ_0)

$$\int_0^T U(t, c(t))\,\mathrm{d}t$$

and a *bequest* utility $B(w(T))$. Hence, he seeks to maximize

$$J = \mathbb{E}\left[B(w(T)) + \int_0^T U(t, c(t))\,\mathrm{d}t\right]. \tag{1.5}$$

The utility functions U and B should be chosen to be increasing concave to model risk aversion and satiation effects. It turns out that a decision that will lead to a simple solution of the optimization problem can be made by choosing them to be the same fractional powers of c and w, respectively. Let $p(\cdot)$ be a given nonnegative function and Π a given nonnegative constant. Let us thus choose $\gamma \in (0,1)$ and

$$U(t,c) = p(t)^{1-\gamma} c^{\gamma} = p(t)^{1-\gamma} \chi^{\gamma} w^{\gamma}, \qquad B(w) = \Pi^{1-\gamma} w^{\gamma}. \tag{1.6}$$

We may, for instance, wish to have the future running utility of consumption in the form $\pi^{1-\gamma} C^{\gamma} \exp[\delta(T-t)]$, i.e., discounted by a factor $\delta = \mu_0 + \beta$. In that case, we just take

$$p(t) = \pi e^{\tilde{\beta}(T-t)}, \qquad \tilde{\beta} = \mu_0 + \frac{\beta}{1-\gamma}. \tag{1.7}$$

We are finally led to investigate a simple stochastic control problem of optimizing (1.5) (1.6) under the scalar dynamics (1.4).

1.1.2 Solution

We investigate first the finite horizon problem.

1.1.2.1 Finite Horizon

We apply a standard dynamic programming technique. The Bellman equation for the Value function $V(t,w)$ is, making use of the notation (1.2),

$$\begin{aligned}
&\forall (t,w) \in [0,T] \times \mathbb{R}, \\
&\frac{\partial V}{\partial t} + \max_{\varphi \in \mathbb{R}^n, \chi \in \mathbb{R}_+} \left[\frac{\partial V}{\partial w} (\varphi^t \lambda - \chi) w + \frac{w^2}{2} \varphi^t \Sigma \varphi \frac{\partial^2 V}{\partial w^2} + U(t, \chi w) \right] = 0, \\
&\forall w \in \mathbb{R}, \quad V(T,w) = B(w).
\end{aligned} \tag{1.8}$$

We replace U and B with (1.6) and look for a solution of the form

$$V(t,w) = P(t)^{1-\gamma} w^{\gamma}.$$

The simplifying fact is that now all individual terms in the equation have a coefficient w^{γ}, so that we may divide through by it, obtaining an ordinary differential equation for $P(t)$:

$$\begin{aligned}
\forall (t,w) \in [0,T] \times \mathbb{R}, \qquad & (1-\gamma) P(t)^{-\gamma} \dot{P}(t) + \max_{\varphi \in \mathbb{R}^n, \chi \in \mathbb{R}_+} \Big[\gamma P(t)^{1-\gamma} (\varphi^t \lambda - \chi) \\
& + \frac{1}{2} \varphi^t \Sigma \varphi \gamma (\gamma - 1) P(t)^{1-\gamma} + p(t)^{1-\gamma} \chi^{\gamma} \Big] = 0, \\
P(T) = \Pi.
\end{aligned}$$

Moreover, the maximizations in φ on the one hand and in χ on the other hand separate and yields further simplifications, giving with extremely simple calculations the optimal $\varphi^\star$ and $\chi^\star$ as

$$\varphi^\star(t) = \frac{1}{1-\gamma}\Sigma^{-1}\lambda, \qquad \chi^\star(t) = \frac{p(t)}{P(t)}. \tag{1.9}$$

And, finally, the equation for $P(t)$ is

$$\dot{P} + \frac{\gamma}{2(1-\gamma)^2}\lambda^t\Sigma^{-1}\lambda P + p = 0. \tag{1.10}$$

Let, therefore,

$$\alpha := \frac{\gamma}{2(1-\gamma)^2}\lambda^t\Sigma^{-1}\lambda, \tag{1.11}$$

to get

$$P(t) = \mathrm{e}^{\alpha(T-t)}\Pi + \int_t^T \mathrm{e}^{\alpha(s-t)}p(s)\,\mathrm{d}s. \tag{1.12}$$

Notice also that if $p(\cdot)$ is differentiable, then $\kappa = 1/\chi^\star$ can be directly obtained as the solution of a linear differential equation:

$$\dot{\kappa} + \left(\frac{\dot{p}}{p} + \alpha\right)\kappa + 1 = 0, \qquad \kappa(T) = \frac{\Pi}{\pi}. \tag{1.13}$$

Many comments are in order. More complete discussions of this classic result can be found in textbooks. We will make a minimum number of remarks.

Remark 1.1. 1. Formula (1.9) for $\varphi^\star$ yields a constant composition of the portfolio. Withdrawals for consumption should be made proportionally.
2. This formula is reminiscent of the corresponding formula in Markowitz's theory of static portfolio optimization (where S is a covariance matrix). See [110].
3. A "small" covariance matrix S tends to produce a "large" φ^*, leaving a smaller share $\varphi_0^\star = (1 - \sum_{i=1}^n \varphi_i^\star)$ for the riskless asset. Specifically, if $\langle \mathbb{1}, \Sigma^{-1}\lambda\rangle \geq 1-\gamma$, then the prescription is to borrow cash to invest in risky assets.
4. $\chi^\star$, though not constant, is also exceedingly simple, given by (1.9), (1.11), (1.12) or by $\chi^\star = 1/\kappa$ and (1.13) if p is differentiable.
5. It readily follows from (1.10) that $P(\cdot)$ is always decreasing, so that $P(t) \geq \Pi$ for all t. Hence if $\Pi > 0$, then $\chi^\star(t) \leq p(t)/\Pi$.
6. If $p(t)$ is chosen according to (1.7), then $\dot{p}/p = -\tilde{\beta}$ is constant, and $\chi^\star$ is obtained via (1.13) as a closed form. Let $\beta' = \tilde{\beta} - \alpha$. We shall see in the next subsection that it is desirable that it be positive. Then one obtains $\chi^\star = [(\Pi/\pi - 1/\beta')\exp[-\beta'(T-t)] + 1/\beta']^{-1}$. Noticeably, if $\beta' > \pi/\Pi$, then this ensures that $\pi/\Pi < \chi^\star < \beta'$.

7. A "large" Π and a small p make for a smaller $\chi^\star$. If one cares about the bequest to the next period, he should be parsimonious. In contrast, if $\Pi = 0$, then in the end, as $t \to T$, $\chi^\star(t) \to \infty$. The entire portfolio is sold for consumption.

1.1.2.2 Infinite Horizon

The concern for long-run wealth, represented by the bequest function, may be addressed by a utility performance index of the form

$$J = \mathbb{E}\pi^{1-\gamma}\int_0^\infty \mathrm{e}^{-(\beta+\mu_0)t} C^\gamma \,\mathrm{d}t.$$

(The coefficient $\pi^{1-\gamma}$ is there for the sake of preserving the dimension of J as a currency amount.)

To deal with that case, in the portfolio model of Sect. 1.1.1.2, we set $T = 0$. And we write the new criterion using $\tilde{\beta}$ as in (1.7):

$$J = \mathbb{E}\pi^{1-\gamma}\int_0^\infty \mathrm{e}^{-(1-\gamma)\tilde{\beta}t} c^\gamma \mathrm{d}t.$$

Equation (1.8) is now replaced by its stationary form:

$$(1-\gamma)\tilde{\beta}V = \max_{\varphi\in\mathbb{R}^n,\chi\in\mathbb{R}_+} \left[\frac{\partial V}{\partial w}(\varphi^t\lambda - \chi)w + \frac{w^2}{2}\varphi^t\Sigma\varphi\frac{\partial^2 V}{\partial w^2} + \pi^{1-\gamma}\chi^\gamma w^\gamma\right].$$

Calculations completely similar to those of the previous paragraph show that the optimum exists if and only if $\tilde{\beta} > \alpha$. (Otherwise, the portfolio may yield an infinite utility.) We find $P = \pi/(\tilde{\beta} - \alpha)$, so that we finally get

$$\varphi^\star = \frac{1}{1-\gamma}\Sigma^{-1}\lambda, \qquad \chi^\star = \tilde{\beta} - \alpha. \tag{1.14}$$

Similar remarks can be made as above. We leave them to the reader.

1.1.3 Logarithmic Utility Functions

Other forms of the utility functions lead to closed-form solutions. Such an instance is $U(C) = -\exp(-\gamma C)$, $\gamma > 0$. Yet it is considered less realistic in terms of representing the risk aversion of the portfolio manager. We refer to [118] for further discussion. We propose here a different extension.

Adding to the criterion to be maximized a number independent of the controls φ and χ clearly does not change the choice of optimal controls. The same holds if we multiply the criterion by a positive constant. Thus the criterion

$$\tilde{J}_\gamma = \mathbb{E}\left[\Pi^{1-\gamma}\frac{w^\gamma(T)-1}{\gamma} + \int_0^T p^{1-\gamma}(t)\frac{w^\gamma(t)-1}{\gamma}\,\mathrm{d}t\right]$$

leads to the same optimal controls as the original one. However, this new criterion presents the added feature that, as $\gamma \to 0$, it has a limit

$$\tilde{J}_0 = \mathbb{E}\left[\Pi \ln w(T) + \int_0^T p(t)\ln w(t)\,\mathrm{d}t\right].$$

We therefore expect that the same formulas for $\varphi^\star$ and $\chi^\star$, but with $\gamma = 0$, should hold for the criterion $\tilde{J}_0$ with logarithmic utility functions. This is indeed correct. However, the Value function is less simple. It is nevertheless a simple exercise to check that the following Value function $\widetilde{V}$, with

$$P(t) = \Pi + \int_t^T p(s)\,\mathrm{d}s,$$

satisfies the Bellman equation with the same formulas $\varphi^\star = \Sigma^{-1}\lambda$, $\chi^\star = p/P$. The Value function is

$$\widetilde{V}(t,w) = P(t)\ln w + \frac{\lambda^t\Sigma^{-1}\lambda}{2}\left[(T-t)\Pi + \int_t^T (s-t)p(s)\,\mathrm{d}s\right]$$
$$+ \int_t^T p(s)\left(\ln\frac{p(s)}{P(s)} - 1\right)\mathrm{d}s.$$

1.2 A Discrete-Time Model

We follow essentially the same path as in the continuous trading problem, but with a discrete-time model, generalizing somewhat Samuelson's solution [135]. But it will be convenient to postpone somewhat the description of the market model.

1.2.1 Problem and Notation

1.2.1.1 Dynamics and Portfolio Model

Market

We want to allow discrete transactions, with a fixed time step h between transactions, an integer submultiple of T. We set $T = Kh$ and $\mathbb{K} := \{0,1,\ldots,K-1\}$. Let, therefore, $t_k = kh$, $k \in \mathbb{K}$ be the trading instants.

As previously, the index 0 denotes a riskless asset for which

$$S_0(t_k) = \exp(\mu_0(k-K)h)S_0(T) = R(t_k)S_0(T).$$

For $i = 1, \dots, n$, let, as previously, $u_i(t_k) = S_i(t_k)/R(t_k)$, and let

$$\tau_i(t) = \frac{u_i(t+h) - u_i(t)}{u_i(t)}$$

be the relative price increment of asset i in one time step,[1] so that we have

$$u_i(t_{k+1}) = (1 + \tau_i(t_k))u_i(t_k).$$

The n-vector of τ_i is as usual denoted by τ.

Portfolio

Our portfolio is, as in the continuous trading theory, composed of X_i shares of asset number i, $i = 0, 1, \dots, n$. Its worth is again

$$W(t_k) = \sum_{i=0}^{n} X_i(t_k)S_i(t_k).$$

We prefer to use

$$w(t_k) = \frac{W(t_k)}{R(t_k)} = \sum_{i=0}^{n} X_i(t_k)u_i(t_k).$$

We allow the portfolio manager to change the X_i at each time t_k. Hence, we must distinguish values before and after the transactions. We let $X_i(t_k)$, $W(t_k)$, and $w(t_k)$ denote the values *before the transactions* of time t_k and, when needed, $X_i(t_k^+)$, $W(t_k^+)$, and $w(t_k^+)$ be their values *after the transactions* of time t_k [with $X_i(t_k^+) = X_i(t_{k+1})$]. One exception to this rule is that $\varphi_i(t_k)$ will denote the fractions *after the transactions*. The transactions of time t_k may decrease the worth of the portfolio by an amount $C(t_k) = R(t_k)c(t_k)$ available for immediate consumption. Hence

$$\sum_{i=0}^{n} (X_i(t_k^+) - X_i(t_k))u_i(t_k) + c(t_k) = 0$$

and

$$W(t_k^+) = W(t_k) - C(t_k), \qquad w(t_k^+) = w(t_k) - c(t_k).$$

Let

---

[1] We choose to consider τ_i as dimensionless, but this is an increment *per time step*, so that it might be considered the inverse of a time. Avoiding that ambiguity complicates the notation.

$$\varphi_i(t_k) = \frac{X_i(t_k^+)S_i(t_k)}{W(t_k^+)} = \frac{X_i(t_k^+)u_i(t_k)}{w(t_k^+)} \quad \text{and} \quad \chi(t_k) = \frac{C(t_k)}{W(t_k)} = \frac{c(t_k)}{w(t_k)}$$

be the decision variables of the manager. As previously, φ may lie anywhere in $\mathbb{R}^n$, while now, χ is constrained to be nonnegative and no more than one.

One easily obtains the dynamics of the portfolio as

$$w(t_{k+1}) = [1+\varphi^t(t_k)\tau(t_k)][1-\chi(t_k)]w(t_k). \tag{1.15}$$

1.2.1.2 Utility

We assume that the portfolio manager wants to maximize a weighted sum of the expected utility of future consumption and of the expected utility of the portfolio worth at final time $T = t_K = Kh$; hence a performance index of the form

$$J = \mathbb{E}\left[B(w(T)) + \sum_{k=0}^{K-1} U(t_k, c(t_k))\right]. \tag{1.16}$$

And as in the continuous trading theory, we shall specialize the analysis to fractional power utility functions. Let Π be a given nonnegative constant and $\{p_k\}_{k\in\mathbb{K}}$ a given sequence of nonnegative numbers. We set

$$U(t_k, c) = p_k^{1-\gamma}c^\gamma = p_k^{1-\gamma}\chi^\gamma w^\gamma, \qquad B(w) = \Pi^{1-\gamma}w^\gamma. \tag{1.17}$$

In formula (1.16), p_K is not used. It will be convenient to define it as

$$p_K = \Pi. \tag{1.18}$$

We shall consider the logarithmic utility in Sect. 1.2.2.3.

1.2.2 *Solution*

1.2.2.1 Finite Horizon

We have to optimize criterion (1.16), (1.17) with the dynamics (1.15). We do this via dynamic programming. Bellman's equation reads

$$\forall (k,w) \in \mathbb{K}\times\mathbb{R}_+, V(t_k, w)$$
$$= \max_{\varphi\in\mathbb{R}^n, \chi\in[0,1]} \left\{\mathbb{E}V\Big(t_{k+1}, (1+\varphi^t\tau(t_k))(1-\chi)w\Big) + p_k^{1-\gamma}\chi^\gamma w^\gamma\right\}, \tag{1.19}$$

with the terminal condition

$$\forall w \in \mathbb{R}_+, \qquad V(T,w) = \Pi^{1-\gamma} w^{\gamma}. \tag{1.20}$$

Assume that, for some number P_{k+1},

$$V(t_{k+1},w) = P_{k+1}^{1-\gamma} w^{\gamma},$$

which is true for $k = K-1$ with $P_K = \Pi$. Then (1.19) yields

$$V(t_k,w) = \max_{\varphi\in\mathbb{R}^n,\chi\in[0,1]} \left\{ \mathbb{E}P_{k+1}^{1-\gamma}\Big([1+\varphi^t\tau(t_k)](1-\chi)\Big)^{\gamma} + p_k^{1-\gamma}\chi^{\gamma} \right\} w^{\gamma},$$

hence $V(t_k,w) = P_k^{1-\gamma} w^{\gamma}$, with

$$P_k^{1-\gamma} = \max_{\varphi\in\mathbb{R}^n,\chi\in[0,1]} \left\{ \mathbb{E}P_{k+1}^{1-\gamma}\Big([1+\varphi^t\tau(t_k)](1-\chi)\Big)^{\gamma} + p_k^{1-\gamma}\chi^{\gamma} \right\}.$$

This recurrence formula for P_k may be simplified as follows. Notice first that it can be written as

$$P_k^{1-\gamma} = \max_{\chi\in[0,1]} \left\{ P_{k+1}^{1-\gamma} \max_{\varphi\in\mathbb{R}^n} \mathbb{E}[1+\varphi^t\tau(t_k)]^{\gamma}(1-\chi)^{\gamma} + p_k^{1-\gamma}\chi^{\gamma} \right\}.$$

In the preceding equation, on the right-hand side, the market parameters enter only the term

$$L(\varphi) := \mathbb{E}[1+\varphi^t\tau(t_k)]^{\gamma}. \tag{1.21}$$

Let α be defined as

$$\alpha^{1-\gamma} := \max_{\varphi\in\mathbb{R}^n} L(\varphi). \tag{1.22}$$

This is a characteristic of the market. With this notation, the recursion for P_k becomes

$$P_k^{1-\gamma} = \max_{\chi\in[0,1]} [P_{k+1}^{1-\gamma}\alpha^{1-\gamma}(1-\chi)^{\gamma} + p_k^{1-\gamma}\chi^{\gamma}].$$

We now use the following "little lemma."

Lemma 1.2. *Let p, q, and r be positive numbers, and $\gamma \in (0,1)$. Then*

$$\max_{x\in[0,r]} \{p^{1-\gamma}x^{\gamma} + q^{1-\gamma}(r-x)^{\gamma}\} = (p+q)^{1-\gamma}r^{\gamma},$$

it is obtained for

$$x = \frac{p}{p+q}r, \qquad 1-x = \frac{q}{p+q}r.$$

Proof. It suffices to equate the derivative with respect to x to zero,

$$\gamma p^{1-\gamma} x^{\gamma-1} - \gamma q^{1-\gamma}(r-x)^{\gamma-1} = 0,$$

to get

$$\frac{r-x}{x} = \frac{q}{p},$$

hence $x = rp/(p+q)$, which lies in $(0,r)$, and place this back in the quantity to maximize. We check the second derivative:

$$\gamma(\gamma-1)[p^{1-\gamma}x^{\gamma-2} + q^{1-\gamma}(r-x)^{\gamma-2}]$$

is negative for all $x \in [0,1]$ since $\gamma - 1 < 0$. □

As a consequence, we find that

$$P_k = \alpha P_{k+1} + p_k, \qquad P_K = \Pi = p_K;$$

hence, recalling (1.18) and (1.20),

$$P_k = \sum_{\ell=k}^{K} \alpha^{(\ell-k)} p_\ell.$$

We also obtain that the optimal consumption ratio $\chi^\star$ is

$$\chi^\star(t_k) = \frac{p_k}{P_k}.$$

The optimal $\varphi^\star$, as well as the precise value of α, depends on the probability law we adopt in the market model. We shall consider that question hereafter, but we may nevertheless make some remarks similar to those for the continuous-time theory.

Remark 1.3. 1. $\varphi^\star$, maximizing $L(\varphi)$, is constant, depending only on the market model.
2. α is a measure of the efficiency of the market. The P_k are increasing in α. Hence $\chi^\star$ is decreasing in α. As α goes to 0, χ^* goes to 1.
3. If Π is large and the p_k, $k < K$, are small, then χ^* is small. In contrast, if $\Pi = 0$, then $\chi^\star(t_{K-1}) = 1$. The entire portfolio is sold for consumption in the last step.

1.2.2.2 Infinite Horizon

We investigate now the formulation in an infinite horizon, which is another way of addressing the long-run worth of the portfolio. Therefore, let a discount constant δ be given, and let the performance index be

$$J = \mathbb{E}\sum_{k=0}^{\infty} \delta^{-kh} c(t_k)^\gamma = \mathbb{E}\sum_{k=0}^{\infty} \delta^{-kh} \chi(t_k)^\gamma w(t_k)^\gamma.$$

(Again, a coefficient $p^{1-\gamma}$, to parallel the finite horizon case, is useless since it would only multiply the performance index by a positive constant.) We use a normalized discount factor β defined by

$$\delta^h = \beta^{1-\gamma}.$$

The dynamic programming equation now reads

$$\beta^{1-\gamma} V(w) = \max_{\varphi \in \mathbb{R}^n \chi \in [0,1]} \mathbb{E}\left\{V\Big((1+\varphi^t \tau)(1-\chi)w\Big) + \chi^\gamma w^\gamma\right\}.$$

Again, we look for a solution of the form $V(w) = P^{1-\gamma} w^\gamma$. Hence

$$\beta^{1-\gamma} P^{1-\gamma} = \max_{\chi \in [0,1]} \left\{P^{1-\gamma}\left[\max_{\varphi \in \mathbb{R}^n} \mathbb{E}(1+\varphi^t \tau)\right]^\gamma (1-\chi)^\gamma + \chi^\gamma\right\},$$

or, using the same notation (1.22) as previously,

$$\beta^{1-\gamma} P^{1-\gamma} = \max_{\chi \in [0,1]} \{P^{1-\gamma} \alpha^{1-\gamma} (1-\chi)^\gamma + \chi^\gamma\}.$$

We use the same "little lemma" to conclude that a solution exists provided that $\beta > \alpha$ as

$$P = \frac{1}{\beta - \alpha}, \qquad \chi^\star = \frac{\beta - \alpha}{\beta}.$$

The determination of φ^* and α presents the same difficulties as in the finite horizon case.

1.2.2.3 Logarithmic Utility Function

As in the continuous-time case, we expect to be able to replace the utility functions in powers $\gamma < 1$ by logarithms and obtain the limit as $\gamma \to 0$ of the preceding formulas as the solution.

Indeed, we leave it to the reader to check that if we set $U(t_k, c) = p_k \ln C$ and $B(w) = p_K \ln w$, then the value function takes the form $V(t_k, w) = P_k \ln w + q_k$, where

$$P_k = \sum_{i=k}^{K} p_k.$$

The optimal portfolio composition $\varphi^\star$ is obtained by solving $\max_{\varphi \in \mathbb{R}^n} \mathbb{E} \ln(\varphi^t \tau)$, but it is not necessary to actually compute this maximum to obtain the optimal consumption ratio $\chi^\star$, which is given by the same formula $\chi^\star = p_k / P_k$ as previously.

1.2.3 Market Models

The critical step in calculating the optimal portfolio composition is to carry out $\max_\varphi L(\varphi)$, where, for the fractional power utilities, $L(\varphi) = \mathbb{E}[(1+\varphi^t\tau)^\gamma]$. The expectation, relative to the random variable τ, depends on the market model.

1.2.3.1 Uniform Interval Market Model

We introduce now a typical interval market model, called *uniform* because we use a uniform distribution for ω. Notice that in the following parts, there will be no a priori probability associated to similar models.

We assume that we have an empirical measure of $\mathbb{E}\tau = \lambda$ and of the covariance $\mathbb{E}(\tau-\lambda)(\tau-\lambda)^t$. For convenience, we call this covariance $(1/3)\Sigma$, where Σ is necessarily nonnegative definite. Let σ be a square root of Σ. The choice of square root is left to the modeler, but we furthermore request that, for all i,

$$\sum_{j=1}^{n} |\sigma_{ij}| \leq 1 + \lambda_i.$$

And we use as our *uniform interval model*

$$\tau = \lambda + \sigma\omega,$$

where the coordinates of ω are independent random variables uniformly distributed over $[-1,1]$. In such a model, the vector τ lies in the parallelotope $\lambda + \sigma\mathscr{C}$, where $\mathscr{C}$ is the hypercube $[-1,1]^n$.

In that model, it is possible to give a closed-form formula for L (1.21):

$$L(\varphi) = \frac{1}{2^n}\int_{\mathscr{C}} (1+\varphi^t(\lambda+\sigma\omega))^\gamma \mathrm{d}\omega.$$

We need the following notation: $\sigma^t\varphi = \psi$. Let $\widehat{\mathscr{C}}$ be the set of vertices of $\mathscr{C}$, i.e., all vectors (denoted by $\hat{\omega}$) whose coordinates are either -1 or 1. To make the formula readable, we let

$$\varsigma(\hat{\omega}) = \prod_{i=0}^{n} \hat{\omega}_i,$$

where $\varsigma(\hat{\omega})$ is $+1$ if the number of positive coordinates of $\hat{\omega}$ is even, and -1 if it is odd. Repeated use of Fubini's theorem lets one integrate explicitly to obtain

$$L(\varphi) = \frac{1}{2^n\prod_{i=1}^{n}(\gamma+i)\psi_i} \sum_{\hat{\omega}\in\widehat{\mathscr{C}}} \varsigma(\hat{\omega})[1+\varphi^t(\lambda+\sigma\hat{\omega})]^{\gamma+n}.$$

Of course, this is not simple enough to let one calculate explicitly the optimizing $\varphi^\star$. But one can compute the gradient of L with respect to φ for use in a numerical procedure. Denote by $(\frac{1}{\psi})$ a vector whose coordinates are $1/\psi_i$. One finds

$$\nabla_\varphi L(\varphi) = \frac{1}{2^n \prod_{i=1}^n (\gamma+i)\psi_i} \sum_{\hat{\omega}\in\widehat{\mathscr{C}}} (\lambda+\sigma\hat{\omega})\varsigma(\hat{\omega})\left(1+\varphi^t(\lambda+\sigma\hat{\omega})\right)^{\gamma+n-1}$$
$$-\sigma\left(\frac{1}{\psi}\right)L(\varphi).$$

This expression is not very appealing. One can, however, check somewhat its consistency with the continuous-time case or Markovitz's mean-variance analysis in the case where there is only one risky asset, an academic example and a bridge to subsequent parts where the portfolio will often consist of one risky asset and a riskless bond. See in particular Chap. 5. Note that in any meaningful application, λ^2 will be very small as compared to Σ, a fact we use to get approximations of the exact formulas. We leave as an exercise to the reader to check that if $\gamma = 1/2$, one finds

$$\varphi^\star = 2\frac{\lambda}{\lambda^2+\Sigma} \simeq \frac{1}{1-\gamma}\Sigma^{-1}\lambda,$$

the same formula as (1.14) obtained in the continuous-time theory. Moreover, if $\gamma = 1/3$, then one finds

$$\varphi^\star = \frac{\sigma^2+3\lambda^2-\sigma\sqrt{\sigma^2-3\lambda^2}}{\lambda(\lambda^2+\sigma^2)} \simeq \frac{(3/2)\lambda}{\sigma^2} = \frac{1}{1-\gamma}\Sigma^{-1}\lambda,$$

i.e., the same formula as for the approximation.

1.2.3.2 An Empirical Market Model

To stress the large choice open to the modelization, and in keeping with the philosophy of avoiding to infer a probability law reflecting information beyond that actually available, we propose here a market model that seems to add as little information as conceivably possible to the available information on past price histories.

Assume that at time t_k, a sequence of past prices of length ℓ is known, from which we derive the sequence $\{\tau(t_{k-j})\}$, $j = 1,\dots,\ell$. Choose a single parameter: a *forget factor* a, such that a^ℓ is small. Use as probability law for τ a law whose (finite) support is just that set of past $\tau(k-j)$, each with a probability decreasing by a factor a as we go one step backward, i.e., proportional to a^j:

$$\mathscr{P}(\tau_k = \tau_{k-j}) = \frac{1-a}{1-a^\ell}a^{j-1}.$$

This provides a very frugal probability law. The drawback is that no analytical help is available to perform the computation of $L(\varphi)$ and to minimize it, and computations must be carried out entirely numerically at each time step.

Chapter 2
Option Pricing: Classic Results

2.1 Introduction

Several authors have proposed a nonstochastic version of the famous Black–Scholes theory.

McEneaney [112] may have been the first to replace the stochastic framework with a robust control approach. He derives the so-called stop-loss strategy for bounded variation trajectories, as we do here. He also recovers the Black–Scholes theory, but this is done at the price of artificially modifying the portfolio model with no other justification than recovering the Itô calculus and the Black–Scholes partial differential equation (PDE).

Cox, Ross, and Rubinstein [57] introduced a nonstochastic approach to the theory of option pricing in a discrete-time setting. We will discuss their approach as compared to the "interval market model" in the next chapter. Their discrete model clearly involves no claim of being realistic for any finite time step. Its only objective is to converge, as the time step vanishes, to a continuous random walk, to recover either the Black–Scholes theory or another one with possible price jumps, depending on how the market model behaves in that limiting process. This approach has been generalized and extended by Kolokoltsov [95,97], as will be explained in Part IV of the book.

The crucial point, usually attributed to Robert Merton, in the Black–Scholes theory of option pricing [46] is that of finding a portfolio together with a self-financed trading strategy that "replicates" (ensures the same return as) the option to be priced. Hence, if no "arbitrage" (riskless profitable trading) is to exist in that market, the price of the option should be equal to that of the replicating portfolio.

What is requested is that the portfolio and strategy constructed replicate the option, i.e., yield the same payment to the owner *for all possible outcomes of the underlying stock's value*. As has been stressed by several authors, this statement is not in terms of probabilities, and therefore the precise (probabilistic) model adopted

P. Bernhard et al., *The Interval Market Model in Mathematical Finance*, Static & Dynamic Game Theory: Foundations & Applications, DOI 10.1007/978-0-8176-8388-7_2,

for a stock's price should be irrelevant. As a matter of fact, it is known that if one adopts the classic Samuelson model,

$$\frac{dS}{S} = \mu dt + \sigma dB, \tag{2.1}$$

with $B(t)$ a Wiener process, then the famous Black–Scholes equation and formula do not contain μ. Explaining this fact has been a concern of many an article or textbook. In our formulation, μ just does not appear in the problem statement. We will further argue that the volatility σ appears only as a characteristic of the set of allowable histories $S(\cdot)$, not as a probabilistic entity.

We first show an elementary theory that emphasizes this point and let us discuss a zero-volatility, yet stochastic – in the sense that $S(\cdot)$ is a priori unknown and can thus be thought of as stochastic or, in Aubin's words, *tychastic*: see part V – model. This model leads to the naive "stop-loss" strategy, which lacks robustness against transaction costs. This naive theory is useful to set the stage for further discussion.

As in subsequent chapters, we adopt a control theoretic viewpoint where the value of a portfolio is seen as a dynamic system, influenced by two exogenous inputs: the underlying stock's price and the trading strategy of the owner. From this viewpoint, it is only natural to reinterpret the goal of replication as one of controllability in the presence of disturbances. Now, a recent trend in control theory is to deal with uncertain disturbances with little modeling but to try to ensure a desired outcome against all possible disturbance histories within a prescribed set of possible such histories. Often called the "robust control" approach, it leads to a problem in dynamic games. This is the path we will follow here and, more prominently, in the rest of the book.

In this chapter, we aim to recover the Black–Scholes formula. Hence our model, although not stochastic, cannot be very different from Samuelson's model. Indeed, in contrast to a stochastic model, it formulates fewer hypotheses on the price process since it implies nothing about repeated experiments – no such thing as a law of large numbers. But this is of little help since we are in a domain where experiments cannot be repeated. (This in itself can be seen as an epistemological obstruction to any valid probabilistic model.)

The reader should be warned that, to keep this short chapter as simple as possible, several choices of notation are inconsistent with those in subsequent chapters using the "interval model," in contrast to this chapter.

2.2 Problem Formulation

The time variable, always denoted t, ranges over a continuous time interval $[0,T]$ of the real line. A given stock is assumed to have a time-dependent, unpredictable, market price $S(t)$ at time t. What we will consider the model of this process, or *market model*, will be the set Ω of assumed possible time functions $S(\cdot)$.

There also exists in that economy a riskless bond, the value of one unit of which at time t is $R(t)$, characterized by $R(T) = 1$ and its rate μ_0. Thus

$$R(t) = \mathrm{e}^{\mu_0(t-T)}$$

is a current value factor.

We are interested in replicating a security whose value at time T is a given function M of $S(T)$. As is well known, in the case of a classic "vanilla call" of exercise price (or "strike") $\mathcal{K}$,

$$M(s) = \max\{s - \mathcal{K}, 0\} =: [s - \mathcal{K}]_+,$$

and in the case of a put,

$$M(s) = [\mathcal{K} - s]_+.$$

In these cases, M is convex and has a finite growth rate at infinity, properties discussed subsequently in more detail.

We will consider a portfolio composed of X shares of the stock and Y riskless bonds. Its value at time t is thus

$$w(t) = X(t)S(t) + Y(t)R(t).$$

We will consider trading strategies of the form

$$X(t) = \varphi(t, S(t)) \tag{2.2}$$

and discuss the choice of the function, or strategy, φ. Whether it is feasible to instantly and continuously implement such a strategy is debatable, but it is a standard assumption of the classic Black–Scholes theory. On the other hand, we stress that we will always restrict the functions $S(\cdot)$ to a set Ω of continuous functions so that it is mathematically unambiguous.

Trading in the riskless bonds will always be decided, beyond the initial time, by the requirement that the portfolio must be self-financed. In the absence of transaction costs, as assumed in this chapter as in the Black–Scholes theory, this means that the amount of shares bought at time t, $\mathrm{d}X(t)$, at price $S(t)$ should exactly balance the amount of trading in the bond, $\mathrm{d}Y(t)$, at price $R(t)$. Therefore, we should have the following *portfolio model*:

$$S(t)\mathrm{d}X(t) + R(t)\mathrm{d}Y(t) = 0. \tag{2.3}$$

As a result, inherent in our calculations will be that

$$\mathrm{d}w(t) = X(t)\mathrm{d}S(t) + Y(t)\mathrm{d}R(t),$$

or, using the fact that $\mathrm{d}R(t) = \mu_0 R(t)\mathrm{d}t$ and $Y(t)R(t) = w(t) - X(t)S(t)$,

$$\mathrm{d}w(t) = X(t)\mathrm{d}S(t) + \mu_0(w(t) - X(t)S(t))\mathrm{d}t. \tag{2.4}$$

The use we will make of that differential form will be made clear later.

2.3 Stop-Loss Strategy

In this section, we assume that the set Ω of possible market price histories $S(\cdot)$ is that of continuous bounded-variation positive functions. One possible instance would be a stochastic process driven by (2.1), with a *stochastic* drift μ and zero volatility σ. In spite of its zero volatility, this can be a very unpredictable stochastic process, of very high frequency, depending upon the stochastic process μ. But we will not need that interpretation.

We want to find a function $W(t,s)$ and a trading strategy $\varphi(t,s)$ such that the use of (2.2) will lead to $w(t) = W(t,S(t))$ for all $t \in [0,T]$, and this for all $S(\cdot) \in \Omega$. If this is possible, and if W is of the class C^1, then we must have

$$\mathrm{d}W(t) = \frac{\partial W}{\partial s}(t,S(t))\mathrm{d}S(t) + \frac{\partial W}{\partial t}(t,S(t))\mathrm{d}t = \mathrm{d}w(t), \tag{2.5}$$

where $\mathrm{d}w(t)$ is to be taken in (2.4) and the differential calculus is to be taken in the sense of Stieltjes. We have a way to make this hold for every $S(\cdot) \in \Omega$ by equating the terms in $\mathrm{d}S$ through the choice

$$X(t) = \frac{\partial W}{\partial s}(t,S(t)), \tag{2.6}$$

and further equating the remaining terms in $\mathrm{d}t$ yields

$$\mu_0(w(t) - X(t)S(t)) = \frac{\partial W}{\partial t}(t,S(t)).$$

Using the previous equality again, we see that this will be satisfied if, $\forall (t,s)$,

$$\frac{\partial W}{\partial t}(t,s) - \mu_0 W(t,s) + \mu_0 s \frac{\partial W}{\partial s}(t,s) = 0. \tag{2.7}$$

If this partial differential equation has a solution that furthermore satisfies

$$\forall s \in \mathbb{R}^+, \quad W(T,s) = M(s), \tag{2.8}$$

then a portfolio of total value $W(0,S(0))$ at time 0, driven by the strategy thus computed, indeed replicates the security considered. An equilibrium price for the option in that model is thus $W(0,S(0))$.

The unique solution of the PDE (2.7), (2.8) is the (discounted) *parity value*:

$$W(t,s) = \mathrm{e}^{\mu_0(t-T)} M(\mathrm{e}^{\mu_0(T-t)} s). \tag{2.9}$$

In the case of a call, say, the associated naive hedging strategy is just $X = 0$ if the call is "out of the money": $S(t) < \exp(\mu_0(t-T)\mathscr{K})$, and $X = 1$ otherwise. The solution is symmetrical for a put. This stop-loss strategy is easily seen to be indeed self-financed and replicating the option.

Its undesirable feature shows up if the price of the underlying stock oscillates close to the discounted value of the exercise price. Then the owner is perpetually in doubt as to whether the price will rise, in which case he must buy a share, or fall, in which case he must not. Any friction, such as transaction costs, ruins that strategy. We argue that the weakness of that model, which would yield an essentially free insurance mechanism, is in the fact that we ignored transaction costs in the portfolio model, *not* in the choice of Ω, which may be more realistic than the more classic next choice.

2.4 Black–Scholes Theory

2.4.1 Black–Scholes Equation

Assume now that the set Ω of allowable price processes still contains continuous positive functions, but now of unbounded total variation, and all of a given quadratic relative variation. That is, a positive number σ is given, and for any $t \in [0,T]$ and for any infinite family of divisions indexed by $N \in \mathbb{N}$, $0 = t_0 < t_1 < \cdots < t_N = t$, with a diameter – the largest interval $t_{k+1} - t_k$ – that goes to zero as N goes to infinity, we require that

$$\lim_{N\to\infty} \sum_{k=1}^{N} \left(\frac{S(t_{k+1}) - S(t_k)}{S(t_k)} \right)^2 = \sigma^2 t.$$

Almost all trajectories generated by (2.1) have that property. But the drift has no effect on the set of possible trajectories and, hence, does not appear here.

For this class of functions, we have the following lemma (see the next section).

Lemma 2.1. *Let $V(t,s) : \mathbb{R}\times\mathbb{R} \to \mathbb{R}$ be twice continuously differentiable. Let $S(\cdot) \in \Omega$, and assume that $(\partial V/\partial s)(t,S(t)) = 0$ for all $t \in [0,T]$. Then $\forall t \in [0,T]$,*

$$V(t,S(t)) = V(0,S(0)) + \int_0^t \left(\frac{\partial V}{\partial t}(\tau,S(\tau)) + \frac{\sigma^2}{2} S(\tau)^2 \frac{\partial^2 V}{\partial s^2}(\tau,S(\tau)) \right) \mathrm{d}\tau.$$

We apply the lemma to $V(t,s) = W(t,s) - X(t)s - Y(t)R(t)$, trying to keep $V(t,S(t))$ equal to zero. First, ensure that $(\partial V/\partial s)(t,S(t)) = 0$ through the choice (2.6). Using our portfolio model (2.3) and again $Y\dot{R} = \mu_0(w - XS)$, we ensure that $V(t,S(t))$ remains constant along any trajectory if,

$$\forall (t,s) \in [0,T] \times \mathbb{R}^+, \quad \frac{\partial W}{\partial t} - \mu_0 W + \mu_0 s \frac{\partial W}{\partial s} + \frac{\sigma^2}{2} s^2 \frac{\partial^2 W}{\partial s^2} = 0,$$

and that this constant is zero through the boundary condition (2.8).

Not surprisingly, this is exactly the Black–Scholes equation. We recall, for the sake of completeness, its famous semiexplicit solution:

$$W(t,s) = s\mathcal{N}(f(T-t,s)) - \mathcal{K}\mathrm{e}^{\mu_0(t-T)}\mathcal{N}(g(T-t,s)), \tag{2.10}$$

where

$$\mathcal{N}(x) = \frac{1}{\sqrt{2\pi}} \int_{-\infty}^{x} \mathrm{e}^{-\frac{\xi^2}{2}} \, \mathrm{d}\xi$$

is the normal Gaussian distribution and

$$f(\tau,s) = \frac{1}{\sigma\sqrt{\tau}} \left(\ln \frac{s}{\mathcal{K}} + \left(\mu_0 + \frac{\sigma^2}{2} \right) \tau \right), \tag{2.11}$$

$$g(\tau,s) = \frac{1}{\sigma\sqrt{\tau}} \left(\ln \frac{s}{\mathcal{K}} + \left(\mu_0 - \frac{\sigma^2}{2} \right) \tau \right). \tag{2.12}$$

It is a simple matter to check that it indeed converges to (2.9) as $\sigma \to 0$ and that the corresponding hedging strategy also converges to the stop-loss strategy.

Let us also note that formula (2.6) now yields

$$X(t) = \mathcal{N}\Big(f(T-t,S(t))\Big).$$

This is so because, as shown by direct calculation,

$$s\mathcal{N}'(f(T-t,s)) = \mathcal{K}\mathrm{e}^{-\mu_0(T-t)}\mathcal{N}'(g(T-t,s)). \tag{2.13}$$

2.4.2 *Proof of Lemma*

We prove here a lemma of deterministic Itô calculus, which yields Lemma 2.1 upon placing $b(t,S) = S^2(\partial^2 V/\partial s^2)(t,S)$. This is a direct proof of a particular case of Föllmer's lemma [73].

Lemma 2.2. *Let $\sigma(\cdot)$ be a measurable real function and $z(\cdot)$ a continuous real function, both defined over the interval $[0,T]$, such that for any $t \in [0,T]$ and any* *sequence of divisions indexed by N,* $0 = t_0 < t_1 < t_2 < \cdots < t_N = t$ *with a diameter h going to* 0 *as* $N \to \infty$, *it holds that*

$$\lim_{N\to\infty} \sum_{k=0}^{N-1} (z(t_{k+1}) - z(t_k))^2 = \int_0^t \sigma^2(\tau)\,\mathrm{d}\tau. \tag{2.14}$$

Let $f(t,x)$ *be a function from* $[0,T] \times \mathbb{R}$ *to* $\mathbb{R}$ *twice continuously differentiable. And assume that, for all* $t \in [0,T]$,

$$\frac{\partial f}{\partial x}(t,z(t)) = 0.$$

Then, for all $t \in [0,T]$,

$$f(t,z(t)) = f(0,z(0)) + \int_0^t \left(\frac{\partial f}{\partial t}(\tau,z(\tau)) + \frac{\sigma(\tau)^2}{2}\frac{\partial^2 f}{\partial x^2}(\tau,z(\tau)) \right) \mathrm{d}\tau.$$

Proof. Consider a division $0 = t_0 < t_1 < t_2 < \cdots < t_N = t$. Using a Taylor expansion to second order with exact rest, we have

$$f(t_{k+1},z(t_{k+1})) - f(t_k,z(t_k)) = \left(\frac{\partial f}{\partial t}(t_k,z(t_k)) \right)(t_{k+1} - t_k)$$

$$+ \left(\frac{\partial f}{\partial x}(t_k,z(t_k)) \right)(z(t_{k+1}) - z(t_k)) + \frac{1}{2}\frac{\partial^2 f}{\partial t^2}(t_{k+1} - t_k)^2$$

$$+ \frac{\partial^2 f}{\partial t \partial x}(t_{k+1} - t_k)(z(t_{k+1}) - z(t_k)) + \frac{1}{2}\frac{\partial^2 f}{\partial x^2}(z(t_{k+1}) - z(t_k))^2,$$

where all second partial derivatives are evaluated at a point (t_k', z_k') on the line segment $[(t_k,z(t_k)) \quad (t_{k+1},z(t_{k+1}))]$.

By assumption, the second term on the right-hand side is equal to zero. We sum these expressions for $k = 0$ to $N-1$. The left-hand side is just $f(t,z(t)) - f(0,z(0))$. We want to investigate the limit of the four remaining sums on the right-hand side.

The first sum is elementary: since $t \mapsto (\partial f/\partial t)(t,z(t))$ is continuous, we obtain the following proposition.

Proposition 2.3.

$$\lim_{N\to\infty} \sum_{k=0}^{N-1} \left(\frac{\partial f}{\partial t}(t_k,z(t_k)) \right)(t_{k+1} - t_k) = \int_0^t \frac{\partial f}{\partial t}(\tau,z(\tau))\,\mathrm{d}\tau.$$

Let us examine the quadratic terms. Again, it is a trivial matter to check that

Proposition 2.4.

$$\lim_{N\to\infty}\sum_{k=0}^{N-1}\left(\frac{\partial^2 f}{\partial t^2}(t'_k,z'_k)\right)(t_{k+1}-t_k)^2=0.$$

Let us show

Proposition 2.5.

$$\lim_{N\to\infty}\sum_{k=0}^{N-1}\left(\frac{\partial^2 f}{\partial t\partial x}(t'_k,z'_k)\right)(t_{k+1}-t_k)(z(t_{k+1})-z(t_k))=0.$$

Proof. This is hardly more complicated than the previous fact, but we will deal carefully with terms involving $z(\cdot)$. We know that that function is continuous over $[0,t]$ and, hence, uniformly so. Therefore, for any positive ε there exists a sufficiently small h such that if the diameter of the division is less than h, then $|z(t_{k+1})-z(t_k)|\le\varepsilon$ for all k. Moreover, $z(t)$ remains within a compact, and thus so do all z'_k. Hence, f being of class C^2, its second derivative evaluated in (t'_k,z'_k) is bounded by a number C. Therefore, for a small enough diameter, the absolute value of the preceding sum is less than

$$\sum_{k=0}^{N-1}C\varepsilon(t_{k+1}-t_k)=C\varepsilon t,$$

and hence we obtain the result claimed. □

We now want to prove the following proposition.

Proposition 2.6.

$$\lim\sum_{k=0}^{N-1}\left(\frac{\partial^2 f}{\partial x^2}(t'_k,z'_k)\right)(z(t_{k+1})-z(t_k))^2=\int_0^t\sigma(\tau)^2\frac{\partial^2 f}{\partial x^2}(\tau,z(\tau))\mathrm{d}\tau.$$

To that end, we show two intermediary facts.

Fact 1

For any continuous real function $a(\cdot)$, we have that, for $t_k\le t'_k\le t_{k+1}$,

$$\lim\sum_{k=0}^{N-1}a(t'_k)(z(t_{k+1})-z(t_k))^2=\int_0^t a(\tau)\sigma(\tau)^2\,\mathrm{d}\tau.$$

Proof. Notice that the result trivially follows from assumption (2.14) whenever the function a is piecewise constant. As a matter of fact, in the limit as $h\to 0$, only a finite number of intervals $[z(t_k), z(t_k+1)]$ contain a discontinuity of a, and their weight in the sum vanishes. For all the other ones, just piece together the intervals where a is constant. There the differences are multiplied by a constant number.

Now, both the finite sums of the left-hand side and the integral are continuous with respect to $a(\cdot)$ for uniform convergence. [The finite sums are linear in $a(\cdot)$. Check the continuity at zero.] And since a is continuous over $[0,T]$, it is uniformly continuous and can be approximated arbitrarily well, in the distance of the uniform convergence, by a piecewise continuous function. The result follows.

Fact 2

For any continuous function $b(t,z)$ we have for $t_k \le t'_k \le t'_{k+1}$ and $z(t_k) \le z'_k \le z(t_{k+1})$

$$\lim \sum_{k=0}^{N-1} b(t'_k, z'_k)(z(t_{k+1}) - z(t_k))^2 = \int_0^t \sigma(\tau)^2 b(\tau, z(\tau)) \mathrm{d}\tau.$$

Proof. First, replace z'_k by $z(t'_k)$ as the second argument of b on the left-hand side. Then just let $a(t) = b(t, z(t))$ in Fact 1 above, and the limit follows. Now, as $h \to 0$, and because b and z are continuous, $b(t'_k, z'_k) - b(t'_k, z(t'_k))$ converges to zero uniformly in k. [They both approach $b(t_k, z(t_k))$ uniformly.] The result follows.

Set $b = \partial^2 f / \partial x^2$ in Fact 4.2 above to obtain Proposition 2.6.

The four preceding propositions together yield the lemma. □

2.5 Digital Options

We consider now a "cash or nothing" digital option, i.e., one whose terminal payment is $M(S(T))$ with

$$M(s) = \begin{cases} 0 \text{ if } s < \mathscr{K}, \\ D \text{ if } s \ge \mathscr{K}, \end{cases} \tag{2.15}$$

and D is a given amount, part of the contract.[1] Since the theory for vanilla options yields a perfect replication, this payment may be approximated with that obtained by holding n call options at a strike $\mathscr{K} - D/n$ and short selling n such options at a strike $\mathscr{K}$. In that case, the terminal payment function is as shown in the graph of Fig. 2.1, approximating the step function (2.15). Let us write Eq. (2.10) with (2.11),

[1] We could have, without loss of generality, taken $D = 1$ and then considered D such options. We resisted this simplification to avoid losing the dimensionality: like $\mathscr{K}$, D is an amount in some currency.

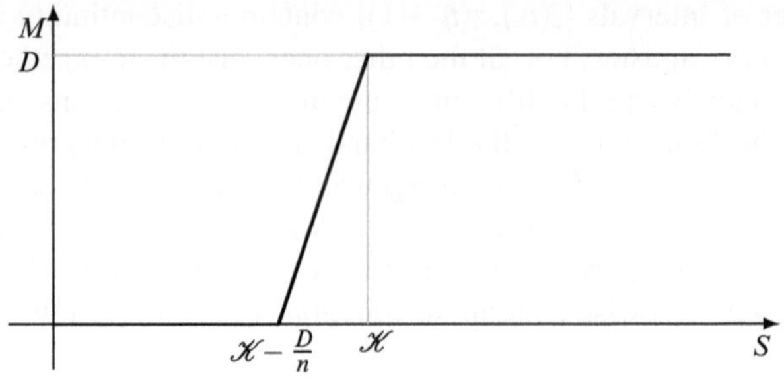

Fig. 2.1 Approximating a step payment with n long calls and n short calls

(2.12) as $W(\mathscr{K};t,s)$. Accordingly, the above combined option leads to a premium of $W_n^d(\mathscr{K};0,S(0))$, with

$$W_n^d(\mathscr{K};s,t) = nW(\mathscr{K} - D/n;t,s) - nW(\mathscr{K};t,s).$$

Its limit, as $n \to \infty$, is clearly

$$W^d(\mathscr{K};t,s) = -D\frac{\partial W(\mathscr{K};t,s)}{\partial \mathscr{K}}.$$

Hence, we derive from the Black–Scholes theory for vanilla options the following formula for a digital call [using again relation (2.13)]:

$$W^d(\mathscr{K};t,s) = De^{-\mu_0(T-t)}\mathscr{N}(g(T-t,s)). \tag{2.16}$$

Part II
Hedging in Interval Models

Authors:
Jacob Engwerda
Department of Econometrics and Operations Research,
Tilburg University,
The Netherlands

Berend Roorda
School of Management and Governance,
University of Twente,
The Netherlands

Hans Schumacher
Department of Econometrics and Operations Research,
Tilburg University,
The Netherlands

Trading strategies designed to reduce risk (i.e., hedging strategies) are a widely studied topic of research in finance. Usually this design is based on stochastic models for the underlying assets. In this part, we introduce in a discrete-time setting the deterministic modeling framework that will be used, in various forms including continuous time versions, in this volume.

After a short introduction on hedging we introduce the basic underlying concept of this framework, namely, the interval model. The interval model assumes that prices at the next time instant can fluctuate between an upper and lower bound, which are given.

We discuss the pricing of derivatives in interval models and optimal hedging under robust-control constraints. Numerical algorithms are provided to calculate the corresponding hedging strategies.

The first chapter of this part, Chap. 3, is introductory and contains well-known material that can be found, for instance, in Hull [88] and Neftci [121]. The second chapter, Chap. 4, discusses pricing in interval models, and the third chapter, Chap. 5, deals with optimal hedging under robust-control constraints. Most of the material presented in Chap. 4 has appeared in a paper published in 2005 in Kybernetika [132]; the material presented in Chap. 5 has not been published before. The work of Berend Roorda on the topics discussed here was mainly done while he was at Tilburg University, supported by a grant from the Netherlands Organization for Scientific Research (NOW) through MaGW/ESR Project 510-01-0025.

Notation

Universal constants

- $\mathbb{R}$: Set of real numbers
- $\mathbb{R}^+$: Set of nonnegative real numbers

Main variables and parameters

- $\mathbb{B}^{u,d}$: Binomial tree model with proportional jump factors u and d
- BC^g: Best-case costs under strategy g
- $\mathrm{BC}^*(S_0,V)$: Best-case costs under RCC limit V on worst-case cost
- $\mathrm{co}I$: Smallest convex subset containing subset I
- $E_j[S]$: Expectation of S conditional on the information available at time t_j
- $F(.)$: Payoff function of option
- $f_j(S_j)$: Option price at time t_j if the price of the asset at time t_j is S_j
- FPI($\mathbb{M}$, F, S): Fair price interval
- $\mathbb{G}$: Set of admissible hedging strategies
- $\mathbb{G}^V$: Set of admissible hedging strategies under RCC limit V
- $g_i(S_0,\ldots,S_i)$: Amount of underlying asset held at time t_j
- H_j: Realized hedge costs
- $\mathbb{I}^{u,d}$: Interval model with maximal and minimal growth factor over each time step u and d, respectively
- $I^g(\mathbb{M},F,S)$: Cost range of strategy g
- $\mathrm{LPR}(f,V)$: Loss-profit ratio with option premium f and RCC limit V
- $\mathbb{M}$: Model, i.e., sequence of $N+1$ numbers in $\mathbb{R}^+$
- $N(\mu,\sigma^2)$: Normal distribution with expectation μ and variance σ^2
- $\Phi(d)$: Cumulative standard normal distribution evaluated at d
- $Q^g(F,\mathscr{S})$: Total cost of hedging and closure
- RCC: Robust-cost constraint
- S: Asset price path $\{S_0,\ldots,S_N\}$
- S_j: Price of asset at time t_j
- σ: Volatility (standard deviation) of stock price
- T: Expiration time
- VaR: Value-at-Risk condition

- WC^g: Worst-case costs under strategy g
- X: Strike price
- x^T: Transpose of vector x
- $[Z]^+$: Maximum of values Z and 0

Chapter 3
Introduction

3.1 Why Hedge?

In many markets, companies face risks that are imposed from outside. For instance, a company producing toys and selling them abroad is faced with a currency risk. To protect the company from bankruptcy caused by this kind of risk, the company might look for trading strategies that reduce this risk. A trading strategy that is designed to reduce risk is called a hedging strategy. To reduce risk, hedgers can trade futures, forward, and option contracts. Both *futures* and *forward contracts* are agreements to buy or sell an asset at a future time T for a certain price (the so-called *strike price*). Thus both parties commit themselves to some action at time T. The difference between both contracts is that forward contracts are agreements between private institutions/persons, whereas futures contracts are contracts that are traded on an exchange. An *option contract* gives the holder the right to buy/sell an asset by a certain date T for a certain price. An option that gives the holder the right to buy an asset is called a *call option*, and one that gives the holder the right to sell an asset is called a *put option*. Unlike with futures and forward contracts, holders of an option are not obligated to exercise their right. For instance, with a call option, say the right to buy some raw material at time T for a price of 2, if it turns out that at time T the actual price of the material is 1, then a company holding this option will not exercise its right to buy the material for a price of 2.

Forward contracts are designed to neutralize risk by fixing the price that the hedger will pay or receive for the underlying asset. Option contracts provide insurance. With an option a company can protect itself against, for example, unfavorable price swings while benefiting from favorable ones. As in the preceding example, the company holding the call option insures itself that it will not have to pay more than 2 for its raw material at time T, and it can buy the raw material for the actual price at time T if it is smaller than 2.

Another distinction between futures/forward contracts and option contracts is that it costs nothing to enter into a futures contract, whereas the holder of an option contract has to pay a price for it up front.

P. Bernhard et al., *The Interval Market Model in Mathematical Finance*, Static & Dynamic Game Theory: Foundations & Applications, DOI 10.1007/978-0-8176-8388-7_3,

Notice that a contract always involves two parties – the one writing the contract and the one buying the contract. An important point to make about the smooth functioning of the futures, forwards, and options markets is that there is a mechanism to guarantee that both parties of a contract will honor the contract. That is, there are mechanisms (like daily settlements) in place so that if one of the parties does not live up to the agreement, the other party will not have to resort to costly lawsuits. Furthermore, the markets should be such that for each side of a contract there is someone that is prepared to take the opposite position in the contract. Usually this means that in futures markets two other types of traders take positions too, i.e., speculators and arbitrageurs. *Speculators* are willing to take on the risk of a contract. *Arbitrageurs* take offsetting positions in different markets to lock in a profit without taking any risk.

Hedging is used to avoid unpleasant surprises in price movements. This can be appropriate if one owns an asset and expects to sell it at some future time T (like a farmer who grows grain) or if one has to buy a certain asset at time T and wants to lock in a price now (like the company who needs raw material at time T). Another reason for hedging can be that one is planning to hold a portfolio for a long period of time and would like to protect oneself against short-term market uncertainties. High transaction costs of selling and buying the portfolio back later might be a reason to use this strategy. In that case one can use stock index futures to hedge market risk.

However, in practice many risks are not hedged. One reason is that risk hedging usually costs money. Another reason is that one should look at all the implications of price changes for a company's profitability. It may happen that different effects of a price change on the profitability of a firm will offset each other. That is, the company is already hedged internally for this price change.

Problems that may arise in hedging include the hedger's not knowing the exact date the asset will be bought or sold, a mismatch between the expiration date of the contract and the date required by the hedger, a hedger's ability to hedge only a proxy of the asset on the market.

Also, situations exist where one would like to mitigate a risk that will arise far into the future at time T but there exist no futures contracts to hedge this risk (like a pension fund that makes commitments to pay pensions in the distant future). A usual approach to tackling such a situation is to roll the hedge forward by closing out one futures contract and taking the same position in a futures contract with a later delivery date and repeating this procedure until one arrives at time T.

As indicated previously, the main reason that hedging was introduced was to reduce trading risk, that is, to shift (a part of) the risk to another trader who either has greater expertise in dealing with that risk or who has the capability to shoulder the risk. An important issue in the context of the latter case is that for large traders it is in practice not always clear what the exact risk position is they have taken. Clearly one should try to improve on this situation. One should avoid situations where large traders cannot meet their commitments. How to improve on this is an ongoing discussion. One line of thinking is to formulate more explicit rules traders must follow. Within this context one should keep in mind that optimal trading strategies often occur at the boundaries of what is allowed. So these rules should anticipate such behavior.

3.2 A Simplistic Hedging Scheme: The Stop-Loss Strategy

A well-known simple hedging strategy is the so-called *stop-loss strategy*. To illustrate the basic idea, consider a hedger who has written a call option with a strike price of X to buy one unit of a stock. To hedge his position, the simplest procedure the hedger could follow is to buy one unit of the stock when its price rises above X and to sell this unit again when its price drops below X. In this way the hedger makes sure that at the *expiration time* T of the option he will be in a position where he owns the stock if the stock price is greater than X. Figure 3.1 illustrates the selling and buying procedure.

Note that basically four different situations can occur. Denoting the stock price at time t by $S(t)$, (1) $S(0)$ and $S(T)$ are less than X; (2) $S(0)$ and $S(T)$ are greater than X; (3) $S(0) > X$ and $S(T) < X$; or (4) $S(0) < X$ and $S(T) > X$. Denoting $[K]^+ = \max\{K, 0\}$, it follows directly that the total revenues from hedging and closure under these four different scenarios are as follows:

$$\begin{aligned}
&(1)\ -[S(T)-X]^+ = 0,\\
&(2)\ -S(0)+S(T)-[S(T)-X]^+ = X-S(0),\\
&(3)\ -S(0)+X-[S(T)-X]^+ = X-S(0),\\
&(4)\ -X+S(T)-[S(T)-X]^+ = 0,
\end{aligned}$$

respectively. Notice that in cases (1) and (4), $S(0) < X$. Therefore, we can rewrite the total revenues from hedging and closure in compact form as $-[S(0)-X]^+$. We state this result formally in a theorem.

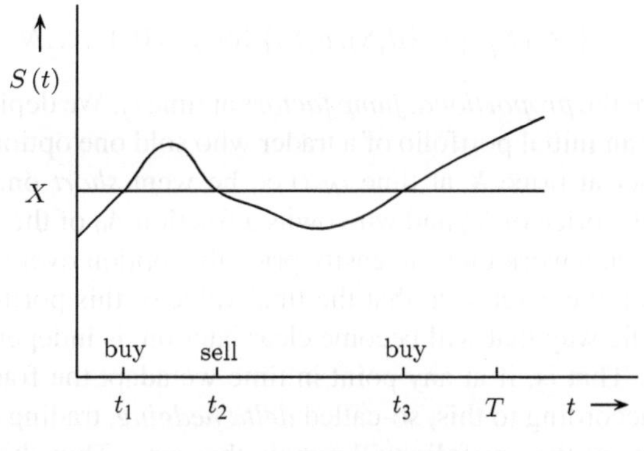

Fig. 3.1 Stop-loss strategy

Theorem 3.1. *The total costs of hedging and closure for a call option using a stop-loss strategy is* $Q^{\text{stop-loss}}(S_0) = [S(0) - X]^+$. □

Or, stated differently, the total cost of hedging and closure equals the intrinsic value of the option.

However, notice that we ignored transaction costs associated with buying and selling the stock under this strategy. Furthermore, if we assume that trading takes place continuously in time, then an important issue is that the hedger cannot know whether, when the stock price equals X, it will then rise above or fall below X. These issues imply that in practice this hedging scheme usually does not work as well as one might have hoped. For a further discussion on this issue we refer the reader to, for example, [88].

3.3 Risk-Free Hedging in the Binomial Tree Model

In this section we recall the well-known binomial tree model that was analyzed by Cox et al. [57] to price options under the assumption that there exist *no arbitrage opportunities*.[1] For a more extensive treatment of this subject, we refer the reader to, for example, Hull [88, Chap. 12].

Consider a market with a single underlying asset. Assume a discrete-time setting where time points are indicated by $t_j,\ j = 0, 1, 2, \ldots$. The price of the asset at time t_j will be denoted by S_j. An *asset price path* is a sequence

$$S = \{S_0, \ldots, S_N\},$$

where the initial price S_0 is fixed throughout and t_N represents the time horizon, which is also assumed to be fixed. The *binomial tree* model $\mathbb{B}^{u,d}$ consists of all price paths that just allow one specific upward and downward price movement at any point in time:

$$\mathbb{B}^{u,d} := \{\mathscr{S} \mid S_{j+1} \in \{d_j S_j, u_j S_j\} \text{ for } j = 0, 1, \ldots, N-1\}.$$

Here u_j and d_j are the *proportional jump factors* at time t_j. We depict this in Fig. 3.2.

Now consider an initial portfolio of a trader who sold one option contract at time t_0 to buy the asset at price X at time t_N (i.e., he went *short* one *European*[2] *call option* with a strike price of X) and who owns a fraction Δ_0 of the asset. Within this binomial model framework one can easily price this option over time and design a trading strategy on the asset such that the final value of this portfolio, where Δ_0 is chosen in a specific way that will become clear later on, is independent of the price path of the asset. That is, if at any point in time we adapt the fraction of the asset in our portfolio according to this, so-called *delta hedging*, trading strategy, then the (net present) value of the portfolio will remain the same. Thus this trading strategy

[1] That is, it is not possible to earn a profit on securities that are mispriced relative to each other.

[2] A European style option contract can be exercised only at the option's expiration date.

tells us at any point in time how many units of the stock we should hold for each option contract in order to create a portfolio whose value does not change over time. Such a *risk-free portfolio* can be set up because the price of the asset and option contract have the same underlying source of uncertainty: the change in asset prices.

To determine this option contract's price and a strategy to trade it over time, we proceed as follows. Let $f_j(S_j)$ denote the value (price) of the option contract at time t_j if the price of the asset at time t_j is S_j. Assume that our portfolio consists at time t_j of Δ_j shares of the asset and the option contract. Then, since the trader has the obligation to pay the buyer of the option contract the value of the contract at t_N, the value of his portfolio at time t_{j+1} is

$$\Delta_j u_j S_j - f_{j+1}(u_j S_j) \text{ if } S_{j+1} = u_j S_j \text{ and } \Delta_j d_j S_j - f_{j+1}(d_j S_j) \text{ if } S_{j+1} = d_j S_j,$$

if the stock price moves up/down, respectively.

Thus the portfolio has the same value in both scenarios if $\Delta_j u_j S_j - f_{j+1}(u_j S_j) = \Delta_j d_j S_j - f_{j+1}(d_j S_j)$, that is, if we choose Δ_j as follows:

$$\Delta_j = \frac{1}{S_j} \frac{f_{j+1}(u_j S_j) - f_{j+1}(d_j S_j)}{u_j - d_j}. \tag{3.1}$$

Stated differently, if we choose Δ_j as the ratio of the change in the price of the stock option contract to the change in the price of the underlying stock [cf. (3.1)], then the portfolio is risk free and must therefore earn the *risk-free*[3] *interest rate* r_j. Thus, denoting the time elapsed between t_{j+1} and t_j by Δt_j, we obtain the present value of the portfolio at time t_j as

$$(\Delta_j u_j S_j - f_{j+1}(u_j S_j))\mathrm{e}^{-r_j \Delta t_j}.$$

On the other hand, we know that this value equals $\Delta_j S_j - f_j(S_j)$. So we get

$$(\Delta_j u S_j - f_{j+1}(u_j S_j))\mathrm{e}^{-r_j \Delta t_j} = \Delta_j S_j - f_j(S_j).$$

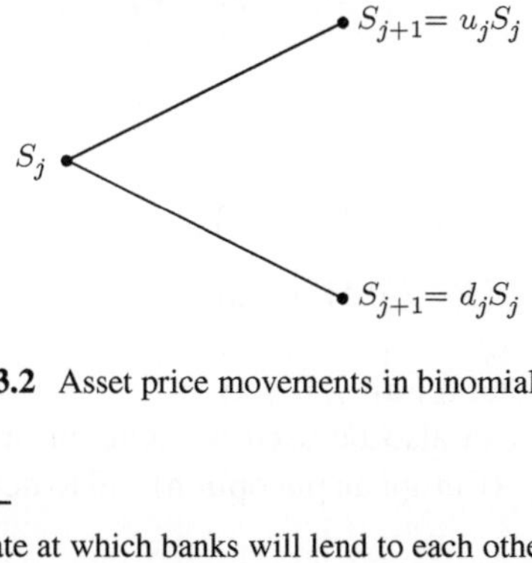

Fig. 3.2 Asset price movements in binomial tree

[3]Usually this is the interest rate at which banks will lend to each other.

Substitution of Δ_j from (3.1) then yields the following backward recursion formula for the option price f_j:

$$f_j(S_j) = \frac{f_{j+1}(u_jS_j)(1-d_je^{-r_j\Delta t_j}) + f_{j+1}(d_jS_j)(u_je^{-r_j\Delta t_j}-1)}{u_j-d_j}$$

$$= q_jf_{j+1}(u_jS_j) + (e^{-r_j\Delta t_j} - q_j)f_{j+1}(d_jS_j), \text{ with} \tag{3.2}$$

$$f_N(S_N) = [S_N - X]^+. \tag{3.3}$$

Here $q_j := \frac{1-d_je^{-r_j\Delta t_j}}{u_j-d_j}$.

From this recursion formula for the price (3.2) we can now also derive directly the following recursion formula for the corresponding *delta-hedging* trading strategy:

$$\Delta_j(S_j) = \lambda_j\Delta_{j+1}(u_jS_j) + (1-\lambda_j)\Delta_{j+1}(d_jS_j), \text{ with} \tag{3.4}$$

$$\Delta_{N-1}(S_{N-1}) = \frac{[u_{N-1}S_{N-1}-X]^+ - [d_{N-1}S_{N-1}-X]^+}{(u_{N-1}-d_{N-1})S_{N-1}}. \tag{3.5}$$

Here $\lambda_j = u_jq_j$.

We will just show the correctness of (3.4). That (3.5) is correct is easily verified. Substitution of (3.2) into (3.1) gives

$$\Delta_j(S_j) = \frac{1}{S_j}\frac{f_{j+1}(u_jS_j) - f_{j+1}(d_jS_j)}{u_j-d_j}$$

$$= \frac{1}{S_j(u_j-d_j)}\Big\{q_jf_{j+2}(u_j^2S_j) + (e^{-r_j\Delta t_j}-q_j)f_{j+2}(u_jd_jS_j)$$

$$-q_jf_{j+2}(u_jd_jS_j) - (e^{-r_j\Delta t_j}-q_j)f_{j+2}(d_j^2S_j)\Big\}$$

$$= \frac{q_j}{S_j(u_j-d_j)}\{f_{j+2}(u_j^2S_j) - f_{j+2}(u_jd_jS_j)\}$$

$$+\frac{e^{-r_j\Delta t_j}-q_j}{S_j(u_j-d_j)}\{f_{j+2}(u_jd_jS_j) - f_{j+2}(d_j^2S_j)\}$$

$$= \frac{u_jq_j}{(u_jS_j)(u_j-d_j)}\{f_{j+2}(u_j(u_jS_j)) - f_{j+2}(d_j(u_jS_j))\} + \frac{d_j(e^{-r_j\Delta t_j}-q_j)}{(d_jS_j)(u_j-d_j)}$$

$$\{f_{j+2}(u_j(d_jS_j)) - f_{j+2}(d_j(d_jS_j))\}$$

$$= \lambda_j\Delta_{j+1}(u_jS_j) + (1-\lambda_j)\Delta_{j+1}(d_jS_j).$$

Remark 3.2. (1) Notice that for all j, $0 \le \Delta_j \le 1$.

(2) The same procedure can also be used to value an option to sell an asset at a certain price at time t_N (European put option) and to determine a trading strategy

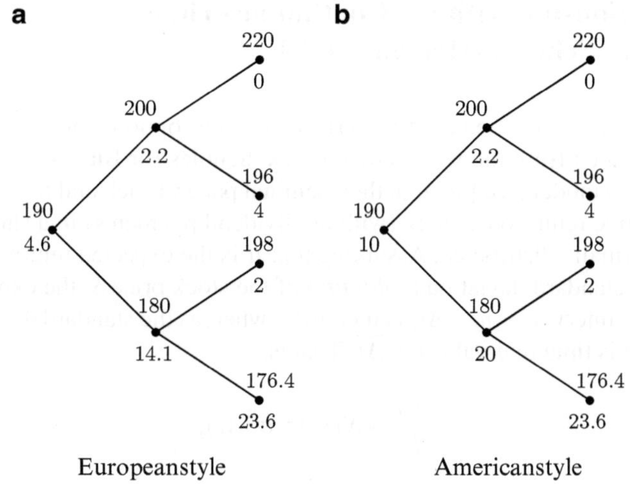

Fig. 3.3 Put option valuation in binomial tree

such that the net present value of a portfolio consisting of the option and a number of shares does not change over time.

(3) The presented formulas can be used also to value so-called *American style option contracts*, i.e., option contracts that can be exercised at any point t_j in time. Working backward in time, the value of such an option at time t_j is the maximum of the value given by (3.2) at t_j and the payoff from exercise at t_j. We illustrate this in Example 3.3 below. □

Example 3.3. Since European and American call options (with no dividend payments for the stock) yield the same price, we will consider in this example the valuation of a put option contract in a two-step binomial model. The initial price of the corresponding stock is 190 euros and the strike price is 200 euros. We assume that each time step is 3 months long and the risk-free annual interest rate is 12%. In the first time step the price may go up by a factor $u_0 = \frac{20}{19}$ and down by a factor $d_0 = \frac{18}{19}$. In the second time step the potential growth factor is $u_1 = 1.1$ and the potential decline factor is $d_1 = 0.98$. This leads to the stock prices illustrated in Fig. 3.3. The upper number at each node indicates the stock price.

The payoff from the European put option is at time t_N given by $[X - S_N]^+$. At time $t_j,\ j < N$ its value is determined by the backward recursion (3.2), where $q_1 = 0.408, q_0 = 0.766$, and $r_j \Delta_j = 0.03$. At each node of the tree the lower number indicates the option price. In Fig. 3.3a the price of the European style option is indicated. Figure 3.3b shows how prices are affected if early exercise of the option is allowed. □

3.4 Relationship with the Continuous-Time Black–Scholes–Merton Model

The binomial model is often used numerically to value options and other derivatives. This is motivated from the well-known Black–Scholes (or Black–Scholes–Merton or Samuelson) model (see [46]). In their seminal paper, Black and Scholes assumed that the relative return on a stock (with no dividend payments) in a short period of time was normally distributed. Assuming that μ is the expected return on the stock and σ is the standard deviation (*volatility*) of the stock price S, the expected return over the time interval $[t_0, t_0 + \Delta t]$ is then $\mu \Delta t$, whereas the standard deviation of the return over this time interval is $\sigma \sqrt{\Delta t}$. That is,

$$\frac{\Delta S}{S} \sim N(\mu \Delta t, \sigma^2 \Delta t), \tag{3.6}$$

where ΔS is the change in the stock price S from $t = t_0$ to $t = t_0 + \Delta t$, μ is the expected return on the stock, and σ is the standard deviation of the stock price.

Following Merton's approach (e.g., [118]) this can be motivated as follows. Assuming that the expectation mentioned below exists, consider the random variable

$$\Delta W_j = (S_j - S_{j-1}) - E_{j-1}[S_j - S_{j-1}].$$

Here, $E_{j-1}[S]$ is the expectation of S conditional on the information that is available at time t_{j-1}.

Thus ΔW_j is the part in $S_j - S_{j-1}$ that cannot be predicted given the available information at time t_{j-1}. Moreover, we assume that ΔW_j can be observed at time t_j, that is, $E_j[\Delta W_j] = \Delta W_j$, and that the ΔW_j are uncorrelated across time. ΔW_j is called the *innovation term* of the stock price because

$$S_j = S_{j-1} + E_{j-1}[S_j - S_{j-1}] + \Delta W_j.$$

Now let $V_j = E_0[(\Delta W_j)^2]$ denote the variance of ΔW_j and $V = E_0[(\sum_{j=1}^{N} \Delta W_j)^2]$ the variance of the cumulative errors. Since the ΔW_j are uncorrelated across time, it follows that

$$V = \sum_{j=1}^{N} V_j.$$

In finance the next three assumptions on V_k and V are widely accepted.

Assumption 3.4. Consider a fixed time interval $[t_0, t_0 + T]$, where stock prices are observed at N equidistant points in time t_j, $j = 0, \ldots, N$. Then there exist three positive constants $c_i > 0$, $i = 1, 2, 3$, that are independent of the number of points N such that:

1. $V \geq c_1 > 0$, that is, increasing the number of observations of stock prices will not completely eliminate risk. There always remains uncertainty about stock prices.

2. $V \le c_2 < \infty$, that is, if more observations of stock prices, and therefore more trading, occurs, then the pricing system will not become unstable.
3. $\frac{V_j}{\max\{V_j,\ j=1,\ldots,N\}} \ge c_3,\ j = 1,\ldots,N$, that is, market uncertainty is not concentrated in some special periods. Whenever markets are open, there is at least some volatility. □

Merton [118] used these three assumptions to prove that the innovation term ΔW_j has a variance that is proportional to the length of the time interval Δt that has elapsed between t_j and t_{j-1} (see also [121] for a proof of the next result).

Theorem 3.5. *Under Assumption 3.4 there exist finite constants σ_j that are independent of Δt such that $V_j = \sigma_j^2 \Delta t$. The σ_j depend on the available information at time t_{j-1}.* □

The next step to motivate (3.6) is to give an approximation for the conditional expectation of the change in stock prices $E_{j-1}[S_j - S_{j-1}]$. Notice that this expectation depends both on the available information at time t_{j-1}, which we will denote by I_j, and on the length of the time interval Δt. Assuming that this is a smooth function $f(I_{j-1}, \Delta t)$ we can use Taylor's theorem to approximate this expectation as follows:

$$E_{j-1}[S_j - S_{j-1}] = f(I_{j-1}, 0) + \frac{\partial f(I_{j-1}, \Delta t)}{\partial \Delta t} \Delta t + h((\Delta t)^2),$$

where $h(.)$ contains the higher-order terms in Δt. Now, if $\Delta t = 0$, then time will not pass and the predicted change in stock prices will be zero, i.e., $f(I_{j-1}, 0) = 0$. Therefore, neglecting the higher-order terms in Δt we have that

$$E_{j-1}[S_j - S_{j-1}] \approx \frac{\partial f(I_{j-1}, \Delta t)}{\partial \Delta t} \Delta t. \tag{3.7}$$

Therefore, assuming additionally that the increments have a normal distribution,[4] we arrive at (3.6).

Assumption (3.6) implies that the stock price $S(t)$ has a *lognormal distribution*. That is, given the price of the stock at time $t = 0$ is S_0, the distribution of the natural logarithm of the stock at time t is

$$\ln(S(t)) \sim N(\ln(S_0) + \left(\mu - \frac{\sigma^2}{2}\right) t, \sigma^2 t).$$

Thus the expectation and variance of $S(t)$ are

$$E[S(t)] = S_0 e^{\mu t} \text{ and } \sigma^2[S(t)] = S_0^2 e^{2\mu t (e^{\sigma^2 t} - 1)}, \text{ respectively.}$$

[4]Together with the previous assumptions made on W_j this implies that W_j is a *Brownian motion*.

Furthermore, we conclude with 95% confidence that

$$\ln(S_0)+\left(\mu-\frac{\sigma^2}{2}\right)t-z_{\frac{\alpha}{2}}\sigma\sqrt{t}<\ln(S(t))<\ln(S_0)+\left(\mu-\frac{\sigma^2}{2}\right)t+z_{\frac{\alpha}{2}}\sigma\sqrt{t},$$

where $z_{\frac{\alpha}{2}} \approx 1.96$ is the number that satisfies $\Phi(z_{\frac{\alpha}{2}}) = \frac{\alpha}{2} = 0.975$. Here $\Phi(d)$ is the *cumulative standard normal distribution* evaluated at d.[5] This implies that

$$S_0 * d := \mathrm{e}^{\ln(S_0)+(\mu-\frac{\sigma^2}{2})t-z_{\frac{\alpha}{2}}\sigma\sqrt{t}} < S(t) < \mathrm{e}^{\ln(S_0)+(\mu-\frac{\sigma^2}{2})t+z_{\frac{\alpha}{2}}\sigma\sqrt{t}} =: S_0 * u.$$

Thus, there is a 95% probability that the stock price will lie between S_0d and S_0u. These numbers u and d give, then, some educated guesses for the corresponding numbers in the interval model we will discuss in Sect. 3.5.2.

In practice when the binomial model is used to value derivatives, and consequently Δt is small, one often uses $u = \frac{1}{d} = \mathrm{e}^{\sigma\sqrt{\Delta t}}$. This choice has the advantage that the tree recombines at the nodes, that is, an up movement followed by a down movement leads to the same stock prices as a down movement followed by an up movement. Furthermore, since $ud = 1$, one can easily calculate the price at any node. Notice that within the foregoing context with $z_{\frac{\alpha}{2}} = 1$, this choice implies that there is a 16% probability that the stock price will be lower, a 16% probability that it will be higher, and a 68% probability that it will be between these upper and lower bounds.

Black, Scholes, and Merton also derived pricing formulas for European calls and puts under the assumption that stock prices change continuously under the assumption of (3.6). They showed that the corresponding unique arbitrage-free prices for call and put options are

$$\begin{aligned} f_0(S_0) &= \Phi(d_1)S_0 - \mathrm{e}^{-r(t_N-t_0)}\Phi(d_2)X \text{ and} \\ f_0(S_0) &= \mathrm{e}^{-r(t_N-t_0)}\Phi(-d_2)X - S_0\Phi(-d_1), \end{aligned} \tag{3.8}$$

respectively, where $d_1 = \frac{\ln(\frac{S_0}{X})+(r+\frac{\sigma^2}{2})(t_N-t_0)}{\sigma\sqrt{t_N-t_0}}$, $d_2 = d_1 - \sigma\sqrt{t_N-t_0}$.

Example 3.6. Consider the pricing of a European call option when both the stock and strike prices are 50 euros, the risk-free interest rate is 10% per year, the volatility is 40% per year, and the contract ends in 3 months. Then, with $r = 0.1$, $\sigma = 0.4$, $t_N - t_0 = 3/12$, and $S_0 = X = 50$, the price of this call option is, according to (3.8), $f_0 = 4.58$. In Fig. 3.4 we illustrate the pricing of this option using (3.2), (3.3) in a corresponding binomial tree with $N = 3$, which implies $\Delta t = 1/12$ and $u = \frac{1}{d} = \mathrm{e}^{\sigma\sqrt{\Delta t}} = 1.1224$. The price that results in this case is $f_0 = 4.77$. If we take a smaller grid $N = 6$, implying $\Delta = 1/24$ and $u = \frac{1}{d} = \mathrm{e}^{\sigma\sqrt{\Delta t}} = 1.1224$, a price of 4.42 results.

---

[5] Or, the probability that a variable with a standard normal distribution will be less than d.

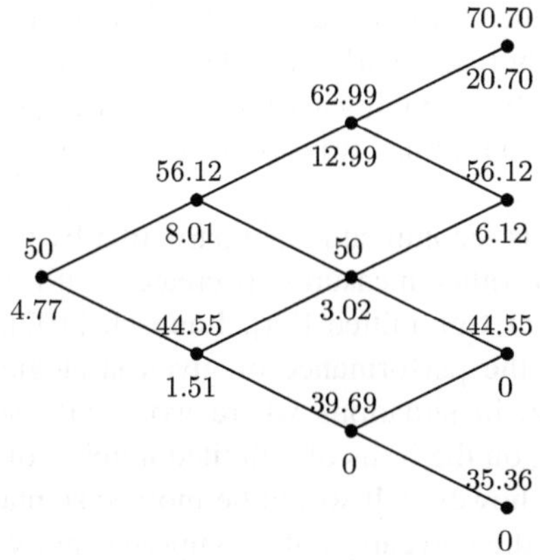

Fig. 3.4 Call option valuation in binomial tree

It can be shown in general that by increasing the number of grid points N, the price in the binomial model will converge to the continuous-time-model price (3.8). □

3.5 Risk Assessment Models

3.5.1 Current Models

Since the publication of the Black–Scholes formula [46], the theory of option pricing has gone through extensive developments in both theory and applications. Today it is the basis of a multibillion-dollar industry that covers not only stock options but also contracts written on interest rates, exchange rates, and so on. The theory has implications not only for the pricing of derivatives, but also for the way in which the risks associated with these contracts can be hedged by taking market positions in related assets. In fact the two sides of the theory are linked together inextricably since the theoretical price of an option is usually based on model assumptions that imply that all risk can be eliminated by suitable hedging. In daily financial practice, hedging is a theme that is at least as important as pricing; indeed, probably greater losses have been caused by misconstrued hedging schemes than by incorrect pricing.

Given the size of the derivatives markets, it is imperative that the risks associated with derivative contracts be properly quantified. The idealized model assumptions that usually form the basis of hedging constructions are clearly not enough to create a reliable assessment of risk. *Value-at-Risk* (VaR) was introduced by Morgan [119] as a way of measuring the sensitivity of the value of a portfolio to typical changes in asset prices. Although the VaR concept has been criticized on theoretical grounds (see, for instance, Artzner et al. [2]), it has become a standard that is used by regulatory authorities worldwide. For portfolios with a strong emphasis on derivative contracts, the normality assumptions underlying the VaR methodology may not be suitable, and additional ways of measuring risk are called for to generate a more complete picture.

Often, *stress testing* is recommended, in particular by practitioners, as a method that should supplement other measures to create a full picture of portfolio risk (see, for instance, Basel Committee [24], Laubsch [107], and Greenspan [78]). The method evaluates the performance of given strategies under fairly extreme scenarios. In particular, in situations where worst-case scenarios are not easily identified, stress testing on the basis of a limited number of selected scenarios may be somewhat arbitrary, however. It would be more systematic, although also more computationally demanding, to carry out a comprehensive worst-case search among all scenarios that satisfy certain limits.

Major concerns associated with worst-case analysis are firstly, as already mentioned, the computational cost and, secondly, the dependence of the results on the restrictions placed on scenarios. The latter problem cannot be avoided in any worst-case setting; in the absence of restrictions on scenarios, the analysis would not lead to meaningful results. To some extent, the second problem may be obviated (at the cost of increased computational complexity) by looking at the results as a function of the imposed constraints. Among an array of risk management tools that are likely to be used jointly in practice, worst-case analysis may be valued as a method that is easily understood also by nonexperts.

In the standard Black–Scholes model, there is one parameter that is not directly observable, *volatility*. When the value of this parameter is inferred from actual option prices, quite a bit of variation is seen both over time and across various option types. It is therefore natural that uncertainty modeling in the context of option pricing and hedging has concentrated on the volatility parameter. In particular, the so-called *uncertain volatility model* has been considered by a number of authors [18, 108, 146]. In this model, volatility is assumed to range between certain given bounds, and prices and hedges are computed corresponding to a worst-case scenario.

The uncertain volatility model as proposed in the cited references assumes continuous trading, which is of course an idealization. In the following sections, we consider a discrete-time version that we call the *interval model*. In this model, the relative price changes of basic assets from one point in time to the next are bounded below and above, but no further assumptions concerning price movements are made.

3.5.2 Interval Model

Interval models naturally arise in the context of markets where uncertainty instead of risk plays a dominant role, that is, if the uncertainty cannot be quantified in, e.g., a probability distribution. For instance if one would like to launch a completely new product for which there as yet no market, it is almost impossible to assess the involved risk of price changes. Also, in actuarial science well-known variables that are uncertain are, for example, life expectancy, evolution of wages, and interest rates. Further, in general, model risk cannot always be quantified in a stochastic framework. Therefore, we will approach price uncertainty here differently. We will assume that tomorrow's prices can fluctuate between some upper and lower bounds, which are given. For the rest we do not have a clue as to which price in this interval will be realized.

Formally, an *interval model* is a model of the form

$$\mathbb{I}^{u,d} := \{\mathscr{S} \mid S_{j+1} \in [dS_j, uS_j] \text{ for } j = 0, 1, 2, \dots\}, \tag{3.9}$$

where u and d are given parameters satisfying $d < 1 < u$. The following figure illustrates a typical step in the price path of an interval model.

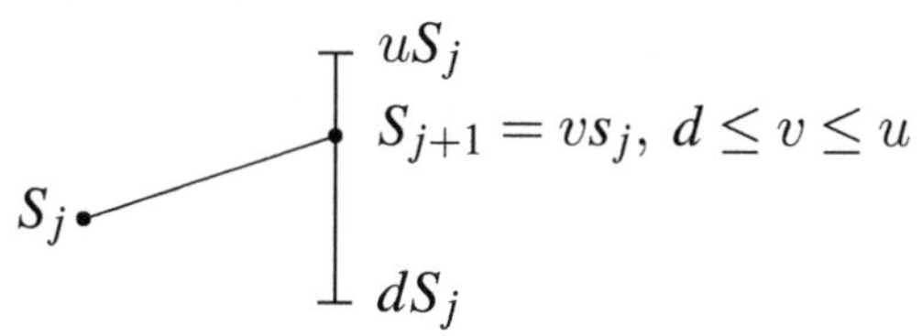

The model parameters u and d denote respectively the maximal and minimal growth factor over each time step.

An important issue is how these models relate to the binomial tree model and the continuous-time Black–Scholes–Merton model considered in Sect. 3.3.

The interval model may be compared to the standard binomial tree model with parameters u and d [57]:

$$\mathbb{B}^{u,d} := \{\mathscr{S} \mid S_{j+1} \in \{dS_j, uS_j\} \text{ for } j = 0, 1, 2, \dots\}.$$

The binomial tree model just allows one specific upward and downward price movement. It provides boundary paths for the interval model $\mathbb{I}^{u,d}$. As already mentioned in Sect. 3.3, binomial models are motivated mainly by the fact that they can be used to approximate continuous-time models by letting the time step tend to zero. In contrast, the interval model may be taken seriously on its own, even for time steps that are not small.

Compared to the continuous-time modeling framework of Black, Scholes, and Merton, we recall from Sect. 3.4 that the continuous time models postulate a lognormal distribution for future prices. That is, with $t_0 = 0$,

$$\ln\left(\frac{S(t)}{S_0}\right) \sim N\left(t\left(\mu - \frac{\sigma^2}{2}\right), \sigma\sqrt{t}\right).$$

The stepwise comparison with interval models is straightforward. For a given price S_0 at time $t_0 = 0$, the statement on the next price S_1 at time t_1 is

$$\ln\left(\frac{S_1}{S_0}\right) \in [\ln(d), \ln(u)] \tag{3.10}$$

according to the interval model, while the continuous-time model postulates

$$\ln\left(\frac{S_1}{S_0}\right) \sim N\left(\mu - \frac{\sigma^2}{2}, \sigma\right). \tag{3.11}$$

The first statement is nondeterministic, though it may be interpreted in a stochastic sense, with σ-field $\{\emptyset, [\ln(d), \ln(u)]\}$ and their complements in $\mathbb{R}$, and probability one assigned to the interval. Under Assumption (3.11), the statement (3.10) is true with probability $\Phi(\frac{\ln(u)-\mu}{\sigma} - \frac{\sigma^2}{2}) - \Phi(\frac{\ln(d)-\mu}{\sigma} - \frac{\sigma^2}{2})$. In particular, under the extra symmetry condition $ud = 1$, u and d are fixed by specifying a confidence level for that probability. It is harder to compare the models globally over several time steps.

Chapter 4
Fair Price Intervals

4.1 Fair Price Interval of an Option: The General Discrete-Time Case

Consider again the discrete-time setting where time points are indicated by t_j, $j = 0,1,2,\ldots$. We consider in this section a market with a single underlying asset. There are no conceptual difficulties, however, in extending the analysis to a situation with multiple assets. To simplify formulas, we assume zero interest rates; this assumption is not essential.

Our basic framework is nonprobabilistic. Let $\mathscr{S}$ denote the asset price path

$$\mathscr{S} = \{S_0, S_1, S_2, \ldots, S_N\}, \tag{4.1}$$

where t_N represents the time horizon, which will be fixed in the subsequent discussion. A *model* $\mathbb{M}$ is a collection of such sequences of real numbers,

$$\mathbb{M} \subset (\mathbb{R}^+)^{N+1}; \tag{4.2}$$

no probability structure is imposed at the outset. A *European derivative* maturing at time t_N is specified by a *payoff function* $F(\cdot)$; the value of the derivative at time t_N for a path $\{S_0, \ldots, S_N\}$ is $F(S_N)$. We will consider models in which asset prices are always positive, and so we can look at the payoff function as a function from $(0,\infty)$ to $\mathbb{R}$. We note that if such a function is convex, it is also continuous.

We consider a portfolio consisting of one option owed (short position) and a quantity γ of the underlying asset held (*long position*). Positions are closed at the expiry of the derivative. A *strategy* is a collection of *strategy functions*

$$\{g_0(S_0), g_1(S_0,S_1), \ldots, g_{N-1}(S_0, \ldots, S_{N-1})\}$$

that at each time t_j determine the quantity of the underlying asset to be held.

P. Bernhard et al., *The Interval Market Model in Mathematical Finance*, Static & Dynamic Game Theory: Foundations & Applications, DOI 10.1007/978-0-8176-8388-7_4,

Note that strategies do not require knowledge of the future, that is, they are nonanticipating by definition. A specific subclass of strategies we will consider are the so-called *path-independent* strategies. Path-independent strategies take only the current price of the underlying into account and can therefore be characterized by strategy functions $g_j(S_j)$, or, stated differently, the trader's *information structure* is of the *closed-loop perfect state* pattern when strategies in general are used, whereas in the case of path-independent strategies a trader just has *feedback perfect state* information on the price process.
As an example we recall from previous sections the following two strategies:

- The (left-continuous) *stop-loss* strategy: $g_j(S_j) = 0$ if $S_j \leq X$, and $g_j(S_j) = 1$ if $S_j > X$, where X is the strike price, which may be used to hedge, e.g., a short European call option;
- The *delta* strategy [see (3.4), (3.5), where $F(.)$ is the payoff function for a European call option] with parameters $F(\cdot)$, u, and d, which is given by strategy functions Δ_j that are defined recursively by

$$\Delta_{N-1}(S_{N-1}) = \frac{F(uS_{N-1}) - F(dS_{N-1})}{(u-d)S_{N-1}}, \tag{4.3}$$

$$\Delta_j(S_j) = \lambda \Delta_{j+1}(uS_j) + (1-\lambda)\Delta_{j+1}(dS_j), \tag{4.4}$$

where $\lambda := \frac{u(1-d)}{u-d}$, which may be used to construct a risk-free portfolio consisting of a short European call option and a fraction Δ of the asset.

A strategy g is said to be *continuous* if the strategy functions are continuous functions of their arguments. The delta strategy is continuous; the stop-loss strategy is not.

To a given hedging strategy $g := \{g_0(S_0), \ldots, g_{N-1}(S_0, \ldots, S_{N-1})\}$ and a given price path $\mathscr{S} := \{S_0, \ldots, S_N\}$ we associate the *total cost of hedging and closure* defined by

$$Q^g(F, \mathscr{S}) := F(S_N) - \Sigma_{j=0}^{N-1} g_j(S_0, \ldots, S_j)(S_{j+1} - S_j). \tag{4.5}$$

The first term represents the cost of closure of a short position in the derivative at the time of expiry, and the second term (appearing with a minus sign) represents the gains from trading in the underlying according to the hedging strategy. For a given model $\mathbb{M}$ and a given initial price S of the underlying asset, the *cost range* of a strategy g is defined as the set of all possible total costs for paths in the model that start at the given initial price:

$$I^g(\mathbb{M}, F, S) := \{Q^g(F, \mathscr{S}) \mid \mathscr{S} = (S_0, \ldots, S_N) \in \mathbb{M},\ S_0 = S\}. \tag{4.6}$$

Given some initial value S for the underlying asset, a price f for a European derivative with payoff function F is said to be a *fair price within the model* $\mathbb{M}$ if for all strategies g there are paths $\mathscr{S}_1$ and $\mathscr{S}_2$ in $\mathbb{M}$ such that

$$Q^g(F, \mathscr{S}_1) \leq f \leq Q^g(F, \mathscr{S}_2). \tag{4.7}$$

For any given subset I of $\mathbb{R}$, let co$I$ denote the smallest convex subset of $\mathbb{R}$ containing I. Then the preceding definition of a fair price may also be expressed as

$$f \in \cap_g \mathrm{co} I^g(\mathbb{M}, F, S), \tag{4.8}$$

where the intersection takes place over all strategies. The right-hand side in (4.8) is an interval, which could reduce to a single point. Since this intersection of the cost intervals associated to all strategies can also be interpreted as the set of option premiums that are consistent with arbitrage pricing, we refer to it as the *fair price interval* FPI($\mathbb{M}$, F, S) corresponding to the model $\mathbb{M}$, the payoff function $F(\cdot)$, and the initial price S. From the definition it follows that

$$\text{if } \mathbb{M}_1 \subset \mathbb{M}_2, \text{ then } \mathrm{FPI}(\mathbb{M}_1, F, S) \subset \mathrm{FPI}(\mathbb{M}_2, F, S). \tag{4.9}$$

Intervals of fair prices are discussed by Pliska [127, Sect. 1.5] in a single-period setting and also appear in a stochastic continuous-time context; see, for instance, El Karoui and Quenez [71].

Remark 4.1. In the preceding definition, a price f can be fair even if there exists a strategy that generates costs that are equal to f along some (but not all) paths and that are less than f along all other paths. It should be noted, though, that in our nonprobabilistic setting, no positive statement is made concerning the probability that a path with costs less than f will occur. We believe that, among the various possible definitions of the notion of a "fair price," the one proposed above must be chosen if one wants to capture both the usual Cox–Ross–Rubinstein price in the binomial model and the monotonicity property (4.9). □

4.2 Fair Price Intervals in Interval Models

In this section we first show, in Sect. 4.2.1, that the fair price interval in interval models is given by an interval. In Sect. 4.2.2 we show what strategies yield the upper and lower bounds in this interval. This is illustrated with an example. Our last subsection connects these results with the stochastic literature and provides a characterization of this interval in terms of martingale measures.

4.2.1 Fair Price Interval

Our first result of this section states that if asset prices behave according to an interval model, then the cost range of any strategy is an *interval* I. That is, if $x, y \in I$ and $x \le z \le y$, then $z \in I$. So if two numbers belong to I then so does every number between them. Stated differently, an interval consists of a convex set in $\mathbb{R}$ that may be closed, open, or half-open or may consist of a single point.

Assuming just the continuity of the payoff function of the European derivative, Proposition 4.2 below shows that for any hedging strategy used in an interval model the set of all possible total costs is an interval. A proof can be found in the appendix.

Proposition 4.2. *Consider an interval model $\mathbb{I}^{u,d}$. For any strategy g with respect to a European derivative with continuous payoff function $F(\cdot)$ and for any initial price S_0, the cost range $I^g(\mathbb{I}^{u,d}, F, S_0)$ is an interval. If, moreover, the strategy g is continuous, then this cost range interval is closed.* □

Example 4.3. An example of a cost interval that is not closed is provided by the stop-loss strategy (Sect. 4.1) in the case of a two-period model with $u = 1.1, d = 0.8$, and $S_0 = 100$, applied to a short position in a European call option with exercise price $X = 80$.

Clearly, $\gamma_0 = 1$. Furthermore, $\gamma_1 = 1$ for all prices except $S_1 = 80$, where $\gamma_1 = 0$. In the latter case where $\gamma_0 = 1$ and $\gamma_1 = 0$, the total *costs* of hedging and closure are

$$Q^g(F, 100, 80, S_2) := [S_2 - 80]^+ - (80 - 100) \in [28, 20].$$

If $\gamma_0 = 1$ and $\gamma_1 = 1$, then the total costs of hedging and closure are

$$Q^g(F, \mathscr{S}) := [S_2 - 80]^+ - (S_1 - 100) - (S_2 - S_1) = [S_2 - 80]^+ + 100 - S_2.$$

Therefore, in the case $S_1 \in (80, 110]$, $Q^g(F, 100, S_1, S_2) = 100 - S_2$, if $S_2 < 80$, and $Q^g(F, 100, S_1, S_2) = 20$, if $S_2 \geq 80$. Notice that the interval of all obtainable prices S_2 smaller than 80 that are compatible with $\gamma_i = 1$ is the open interval $(64, 80)$. Consequently, $Q^g(F, 100, S_1, S_2) \in (20, 36)$ in this case. Combining these results we then get that the cost range $I^{\text{stop-loss}}(\mathbb{I}^{1.1,0.8}, [S_2 - 80]^+, 100)$ is the half-open interval $[20, 36)$. □

A direct consequence of Proposition 4.2 is that the set of fair prices is also an interval. We state this result formally in the next corollary.

Corollary 4.4. *Consider a European derivative with continuous payoff function F. Then the fair price interval FPI($\mathbb{I}^{u,d}$, F, S_0) is an interval.* □

In the next subsection we show that the fair price interval is in fact a compact interval if one additionally assumes that the payoff function F is convex.

4.2.2 Characterization of the Fair Price Interval in Terms of Strategies

Below, we will be interested in two particular strategies. The first is the standard binomial delta strategy for the binomial tree with the same parameters as the given interval model. We call this the *extreme delta strategy* because it corresponds to paths that at each time step exhibit the largest possible jump that is allowed by the interval model in either the upward or the downward direction. Before introducing

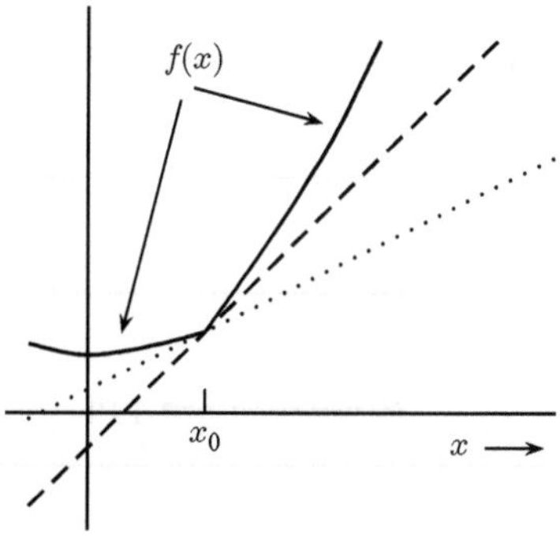

Fig. 4.1 Tangent lines corresponding to subderivatives at x_0

the second strategy, we first recall the notion of *subderivative*. Consider the convex function f at x_0 in Fig. 4.1. Clearly, f is not differentiable at x_0. However, many lines go through $(x_0, f(x_0))$ and are in an open neighborhood of x_0 everywhere either touching or below the graph of f. The slope of such a line is called a subderivative. The set of all subderivatives is called the *subdifferential* $\partial f(x_0)$ of f at x_0. This concept can be generalized for functions from $\mathbb{R}^n$ to $\mathbb{R}$. That is, for a given convex function $F : \mathbb{R}^n \mapsto \mathbb{R}$, the *subdifferential* $\partial F(x)$ of F at $x \in \mathbb{R}^n$ is defined as the set of all vectors $\gamma \in \mathbb{R}^n$ such that $F(y) \geq F(x) + \gamma^{\mathrm{T}}(y - x)$ for all y. The subdifferential is always a nonempty convex compact set. The elements of $\partial F(x)$ are called in this case the *subgradients* of F at x. We call a *subgradient strategy* for a European derivative with convex payoff F any strategy g such that $g(S_j) \in \partial F(S_j)$. The stop-loss strategy, for instance, is a subgradient strategy for the European call option.

The special role played by the extreme delta and the subgradient strategies is indicated in the theorem below. In the theorem, we place ourselves in the position of an institution that holds a short position in a certain derivative and that is looking for a hedging strategy. We will identify strategies that minimize worst-case costs and strategies that maximize best-case costs. The first are, of course, simple to interpret; the latter strategies are more easily viewed as the opposites of strategies that maximize worst-case gain for a party holding a long position in the derivative. The theorem states that, in a situation described by an interval model, an institution holding a short position in a European option with a convex payoff can minimize its downward risk by hedging as if maximal volatility were going to occur. On the other hand, an institution holding a long position will minimize its downward risk by hedging as if minimal (actually zero) volatility were going to occur. Part I of the theorem below can also be found in Kolokoltsov [95]. The theorem is visualized in Fig. 4.2.

Theorem 4.5. *Consider a frictionless market in which the price paths of an underlying asset follow an interval model with parameters u and d, where $d < 1 < u$; the initial value S_0 of the underlying is given. Let $F(\cdot)$ be the payoff function of a European derivative, and assume that F is convex. We consider portfolios that*

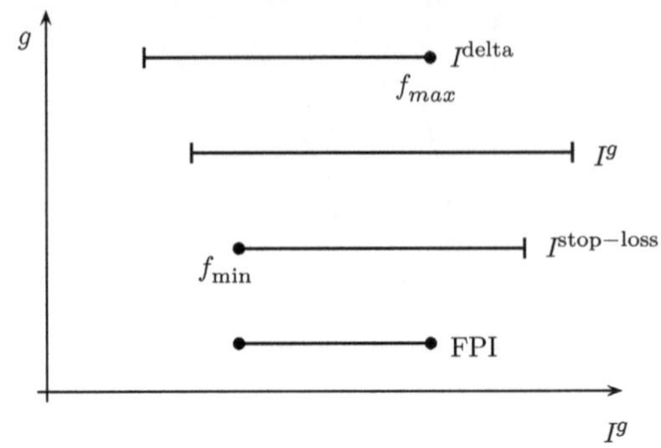

Fig. 4.2 Fair price interval. This figure illustrates the cost range of three strategies: the stop-loss, the delta, and another arbitrary strategy g. The fair price interval (FPI) is obtained as the intersection of the cost range intervals of the stop-loss and delta strategy. The cost range of strategy g always includes the FPI

consist of (1) a given short position in the option and (2) a position in the underlying asset that is determined at each time point by a trading strategy.

1. *Lowest worst-case costs are generated by the extreme delta-hedging strategy. The corresponding costs, which we denote by $f_{\max}$, are given by the Cox–Ross–Rubinstein price of the derivative in the binomial tree model with the same parameters as the interval model. Worst-case costs are achieved for paths in this tree model.*
2. *Highest best-case costs are generated by any subgradient strategy. The corresponding costs are equal to $f_{\min} := F(S_0)$ and are realized along the constant path.*
3. *The FPI for the derivative is $[f_{\min}, f_{\max}]$.* □

In the case of a call option, the stop-loss strategy is best in the worst-case sense for a party holding a long position, and the corresponding worst-case paths are those in which the strike level is not crossed. More generally, it can be easily verified that if we have a piecewise linear payoff function, then the worst-case paths for a party holding a long position in the derivative and following a subgradient hedge strategy are those in which the successive values of the underlying are confined to one of the regions where the payoff function behaves linearly.

Remark 4.6. Notice that on the trivial range of prices that can no longer cross the exercise level,

$$S_{N-j} \leq \frac{X}{u^j} \text{ or } S_{N-j} \geq \frac{X}{d^j},$$

delta hedging coincides with the stop-loss strategy. Consequently, if this occurs, the FPI for the derivative reduces to one point. Or, stated differently, the price of the derivative is under those conditions uniquely determined. □

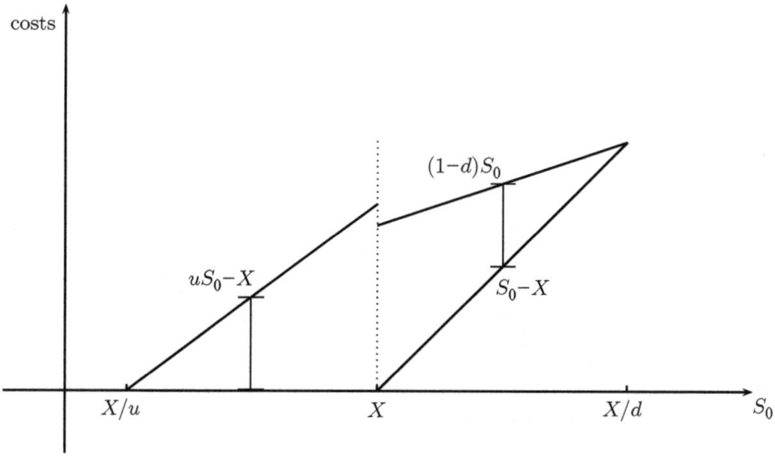

Fig. 4.3 Cost range stop-loss strategy in one-step interval model

4.2.3 Example

For a simple illustration of the foregoing results, consider a call option in a one-step interval model. In such a model the choice of a strategy comes down to the choice of a real number that indicates the position to be taken in the underlying at time 0. The total cost of hedging and closure is in this case

$$Q^g = [S_1 - X]^+ - g_0(S_0)(S_1 - S_0). \tag{4.10}$$

For the stop-loss trading strategy this yields

$$Q^{\text{stop-loss}} = \begin{cases} [S_1 - X]^+ & \text{if } X \geq S_0 \\ [S_1 - X]^+ - (S_1 - S_0) & \text{if } X < S_0. \end{cases}$$

Thus the cost range of the stop-loss strategy for the one-step interval model is

$$I^{\text{stop-loss}} = \begin{cases} [uS_0 - X, 0] & \text{if } X \geq S_0 \\ [S_0 - X, (1-d)S_0] & \text{if } X < S_0. \end{cases}$$

In Fig. 4.3 we illustrate the cost range for the stop-loss strategy as a function of the initial value S_0. For two specific choices of S_0 we mark this interval "I." Notice the discontinuity of the cost range at $S_0 = X$.

Similarly, we obtain from (4.10) that the total cost of hedging and closing using the extreme delta trading strategy is

$$Q^{\text{delta}} = [S_1 - X]^+ - \frac{[uS_0 - X]^+ - [dS_0 - X]^+}{(u-d)S_0}(S_1 - S_0).$$

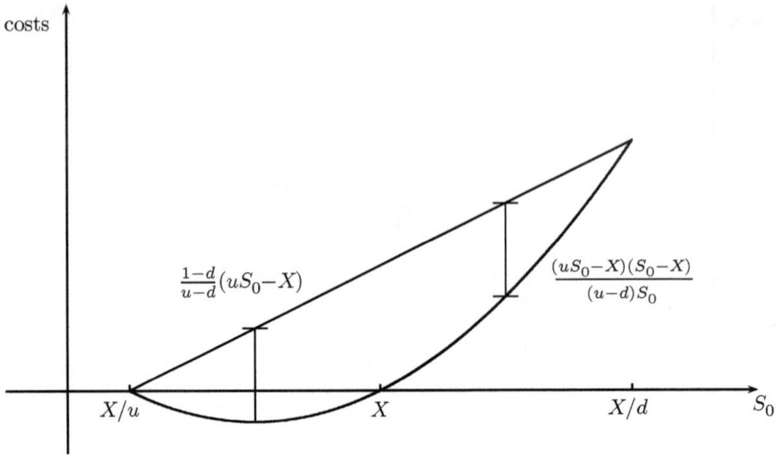

Fig. 4.4 Cost range delta strategy in one-step interval model

Under the assumption that the strike price X will also be situated somewhere in the interval $[dS_0, uS_0]$, we have that $[uS_0 - X]^+ = uS_0 - X$ and $[dS_0 - X]^+ = 0$. Thus $Q^{\text{delta}} = [S_1 - X]^+ - \frac{(uS_0-X)}{(u-d)S_0}(S_1 - S_0)$. With $S_1 \in [dS_0, uS_0]$, we have that for a fixed X, Q^{delta} attains its maximum value for $S_1 = uS_0$ and its minimum value for $S_1 = X$. Consequently, the cost range of the extreme delta strategy for the one-step interval model is

$$I^{\text{delta}} = \left[\frac{(uS_0 - X)(S_0 - X)}{(u-d)S_0}, \frac{1-d}{u-d}(uS_0 - X) \right].$$

In Fig. 4.4 we illustrate the cost range for the extreme delta strategy as a function of the initial value S_0. Again, for two specific choices of S_0, this interval is marked "I" in the graph.

Finally, both graphs are merged in Fig. 4.5. Recall from Theorem 4.5 that for the call option the boundary points of the FPI are given by the intersection of both cost ranges. So for any S_0 the upper bound $f_{\max}$ of the FPI is given by $\frac{(uS_0-X)(S_0-X)}{(u-d)S_0}$ and the lower bound of this interval $f_{\min}$ equals $[S_0 - X]^+$. From Fig. 4.5 this FPI can be read off for each value of S_0 as the intersection of the cost intervals of the two strategies. We specify for two values of S_0 this FPI as "I" in the graph.

4.2.4 Characterization of Fair Price Interval in Terms of Martingale Measures

We now introduce martingale measures. We consider price paths of a fixed length $N+1$ with a given initial value S_0, and so the measures that we will consider can be thought of as probability measures on the vector space $\mathbb{R}^N$. Any such measure $\mathbf{Q}$ will

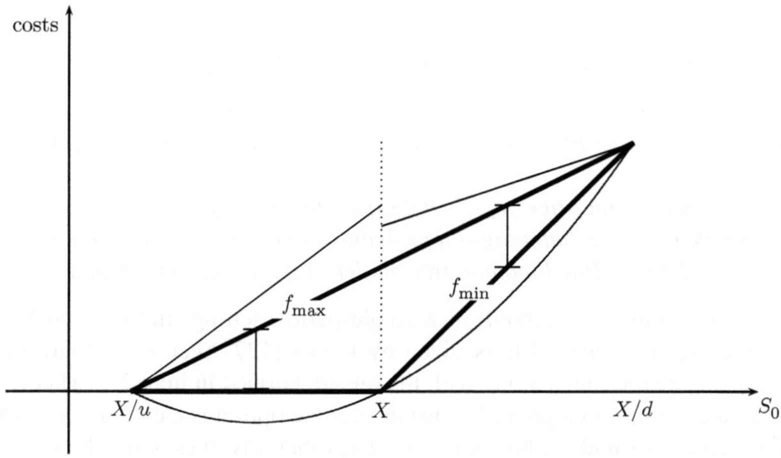

Fig. 4.5 Fair price intervals in one-step models

be called a *martingale measure* for the model $\mathbb{M}$ with initial value S_0 if it assigns probability 1 to the paths in the model $\mathbb{M}$ with initial value S_0 and if the martingale property holds, that is, $E_{\mathbf{Q}}(S_{j+k} \mid S_j, S_{j-1}, \ldots, S_0) = S_j$ for all j and $k \geq 0$. The set of all martingale measures for a model $\mathbb{M}$ with initial condition S will be denoted by $\mathscr{Q}(\mathbb{M}, S)$. The most important property of martingale measures that we will need is the fact that the expected gain from any trading strategy under a martingale measure is zero. From this it follows immediately [see (4.5) and (4.6)] that, for any hedging strategy g applied to a European derivative with payoff function F, we have

$$E_{\mathbf{Q}}F \in \operatorname{co} I^g(\mathbb{M}, F, S)$$

for any martingale measure $\mathbf{Q} \in \mathscr{Q}(\mathbb{M}, S)$. Consequently, we can write

$$\{E_{\mathbf{Q}}F \mid \mathbf{Q} \in \mathscr{Q}(\mathbb{M}, S)\} \subset \cap_g \operatorname{co} I^g(\mathbb{M}, F, S),$$

where the intersection is taken over all strategies.

It is clear that interval models allow many martingale measures. For instance, for an interval model with parameters u and d, all martingale measures associated to binomial tree models with parameters u' and d' satisfying $d \leq d' < 1 < u' \leq u$ are also martingale measures for the interval model. We have already shown that if the payoff function F is convex, then the FPI is closed in interval models. In the following theorem, we show that all fair prices in that case are generated by martingale measures, and we indicate the measures that generate the extreme points of the FPI.

Theorem 4.7. *Let an interval model $\mathbb{I}^{u,d}$ and an initial asset value S_0 be given, and let $\mathscr{Q}$ denote the set of all martingale measures that can be placed on the collection of paths in $\mathbb{I}^{u,d}$ that start at S_0. Consider a European derivative with convex payoff $F(\cdot)$, and denote the fair price interval for the derivative by $[f_{\min}, f_{\max}]$.*

1. *We have*

$$\{E_{\mathbf{Q}}[F(S_N)] \mid \mathbf{Q} \in \mathscr{Q}\} = [f_{\min}, f_{\max}].$$

2. *The minimal option price $f_{\min}$ is the expected value of the derivative under the martingale measure that assigns a probability of one to the constant path $S_j = S_0$ for all j.*
3. *The maximal option price $f_{\max}$ is the expected value of the derivative under the martingale measure that assigns a probability of one to the collection of paths in the submodel $\mathbb{B}^{u,d}$ (the binomial tree model with parameters u and d).* □

For general incomplete markets in a single-period setting, the relation between martingale measures and FPIs is given by Pliska [127, Sect. 1.5]. There may be many martingale measures along with the one mentioned in item 2 of Theorem 4.7 that generate the minimal price; for instance, if the option is a call option, then any martingale measure under which there is zero probability of crossing the strike level will generate this price. On the other hand, the maximal price is generated uniquely by the measure indicated in item 3, except in the (trivial) case in which the payoff function $F(\cdot)$ is linear; for instance, if the option is a call option, the measure is unique until the asset price in a path becomes too high or low for crossing the exercise level.

All intermediate prices are generated by many different martingale measures, and, unlike the extreme prices, they obviously allow for "equivalent martingale measures," in the sense that every set of paths in the interval model with positive (Lebesgue) measure is assigned a positive probability.

4.3 Computation of the Fair Price Interval for Path-Independent Strategies

Assuming that we just consider path-independent strategies, the computation of the FPI for a European derivative with a continuous payoff function amounts to determining the best- and worst-case costs over all price paths in a given interval model, and algorithms can be designed according to the principles of dynamic programming. We briefly sketch the standard idea.

Let θ_j denote a state variable at time t_j that summarizes all information over the strict past $t_0, \ldots, t_{j-1}$ that is relevant to a given strategy g. Replacing past prices by θ_j in the argument of the strategy functions g_j we obtain the state space system

$$\begin{aligned} \theta_{j+1} &= f_j(\theta_j, S_j), \ \theta_0 \text{ fixed}, \\ \gamma_j &= g_j(\theta_j, S_j), \end{aligned} \tag{4.11}$$

where f_j is a state evolution function and γ_j is the hedge position at t_j according to strategy g. Now determine, for every time instant, the value functions $V^{\max}$ and $V^{\min}$ that assign to a state (θ_j, S_j) the worst-case and best-case costs respectively over all paths starting in S_j at t_j that satisfy the restrictions of the given interval model. Starting at expiry with boundary conditions

$$V^{\min}(N,S_N,\theta_N) = V^{\max}(N,S_N,\theta_N) := F(S_N)$$

we are led to a backward recursive optimization

$$V^{\min}(j,S_j,\theta_j) := \min_{v\in[d,u]} V^{\min}(j+1,vS_j,f_j(S_j,\theta_j)) - g_j(S_j,\theta_j)(v-1)S_j$$

and

$$V^{\max}(j,S_j,\theta_j) := \max_{v\in[d,u]} V^{\max}(j+1,vS_j,f_j(S_j,\theta_j)) - g_j(S_j,\theta_j)(v-1)S_j,$$

respectively. For discontinuous strategies the preceding minima and maxima need not exist; taking infima and suprema instead, we actually compute the closure of the cost interval.

The complexity of the algorithm depends on the number of state variables in the hedge strategy and the number of underlyings. The number of required operations is quadratic in NK, where N is the number of time steps and K is the number of grid points in the state space of the θ and S variables. For regular grids, K depends exponentially on the dimension of θ and S. Variations of the preceding algorithm such as using a forward rather than a backward recursion do not fundamentally affect this complexity. We will restrict our discussion here to path-independent hedging strategies for options on a single underlying.

Algorithm 4.8 (Cost Intervals).

Data: Initial asset price S_0, an interval model for assets $\mathbb{I}^{u,d}$, a continuous payoff function $F(S_N)$, and a strategy g in state space form (4.11).

Step 1: Initialization

Define for $S_N \in [d^N S_0, u^N S_0]$, and arbitrary values of θ_N,

$$V^{\min}(N,S_N,\theta_N) = V^{\max}(N,S_N,\theta_N) := F(S_N).$$

Step 2: Backward recursion

Determine for $j = N-1,\ldots,0$, $S_j \in [d^j S_0, u^j S_0]$, and a suitable domain for θ_j,

$$V^{\min}(j,S_j,\theta_j) := \min_{v\in[d,u]} V^{\min}(j+1,vS_j,f(S_j,\theta_j)) - g_j(S_j,\theta_j)(v-1)S_j \quad (4.12)$$

and

$$V^{\max}(j,S_j,\theta_j) := \max_{v\in[d,u]} V^{\max}(j+1,vS_j,f(S_j,\theta_j)) - g_j(S_j,\theta_j)(v-1)S_j, \quad (4.13)$$

respectively. Define $v_j^{\min}(S_j,\theta_j)$ and $v_j^{\max}(S_j,\theta_j)$ as respectively the minimum and maximum location for v.

Step 3: Cost interval and extreme cost paths

Define

$$I := [V^{\min}(0,S_0,\theta_0), V^{\max}(0,S_0,\theta_0)], \tag{4.14}$$

and for $j = 0,\dots,N-1$,

$$S_{j+1}^{\min} = v_j^{\min}(S_j,\theta_j)S_j;\ S_{j+1}^{\max} = v_j^{\max}(S_j,\theta_j)S_j;\ S_0^{\min} = S_0^{\max} = S_0. \tag{4.15}$$

Result: I is the cost interval $I^g(\mathbb{I}^{u,d},F,S)$; best-case costs are attained for $S^{\min}$ and worst-case costs for $S^{\max}$. □

Alternatively, the cost intervals may be computed by the following *path filter* that selects best- and worst-case paths *forward* in time.

Algorithm 4.9 (Path Filter).

Data: Initial asset price S_0, an interval model for assets $\mathbb{I}^{u,d}$, a path-independent continuous strategy $g_j(S_j)$, and a continuous payoff function $F(S_N)$.

Step 1: Initialization

Define

$$H^{\min}(0,S_0) = H^{\max}(0,S_0) := 0,$$
$$H^{\min}(1,S_1) = H^{\max}(1,S_1) := -g_0(S_0)(S_1 - S_0).$$

Step 2: Forward recursion

Determine for $j = 1,\dots,N-1$,

$$H^{\min}(j+1,S_{j+1}) := \min_{S_j \in I_j} H^{\min}(j,S_j) - g_j(S_j)(S_{j+1} - S_j), \tag{4.16}$$

with $S_{j+1} \in [d^{j+1}S_0, u^{j+1}S_0]$, $I_j := [d^jS_0, u^jS_0] \cap [\frac{S_{j+1}}{u}, \frac{S_{j+1}}{d}]$. Define $H^{\max}$ similarly, with min replaced by max.

Step 3: Best- and worst-case price paths:

Determine

$$S_N^{\min} := \operatorname{argmin}_{S_N \in [d^N S_0, u^N S_0]} H^{\min}(N,S_N) + F(S_N), \tag{4.17}$$

$$S_j^{\min} := \operatorname{argmin}_{S_j \in I_j} H^{\min}(j,S_j^{\min}) - g_j(S_j)(S_{j+1}^{\min} - S_j)., \tag{4.18}$$

and $S^{\max}$ similarly, with min replaced by max.

Step 4: Cost interval

Define

$$I := [H^{\min}(N, S_N^{\min}) + F(S_N^{\min}), H^{\max}(N, S_N^{\max}) + F(S_N^{\max})]. \tag{4.19}$$

Result: I is the cost interval $I^g(\mathbb{I}^{u,d}, F, S)$; best-case costs are attained for $S^{\min}$ and worst-case costs for $S^{\max}$. □

The main difference between both procedures is in the intermediate results. In the forward algorithm (with empty state variable θ_j removed), $[V^{\min}(j, S_j), V^{\max}(j, S_j)]$ is the "updated" cost interval, with S_j considered as "initial" price at t_j, while in the backward algorithm $[H^{\min}(j, S_j), H^{\max}(j, S_j)]$ denotes the interval of realized hedge costs with (j, S_j) considered as end condition. Notice that for all time instants t_j, it holds that $I = [H^{\min}(j, S_j) + V^{\min}(j, S_j), H^{\max}(j, S_j) + V^{\max}(j, S_j)]$, so cost intervals can also be computed by a mixture of both algorithms.

The difference in time direction may also become relevant in extending the procedures to early-exercise possibilities and path dependency (in strategies or options), although we do not consider these types of strategies and options here.

A general advantage of working backward in time is that early-exercise possibilities in options can be accounted for in a straightforward way. In particular, in complete markets the unique arbitrage-free option price is available at each time t_j, and it is obvious when early exercise is profitable. Interval models, however, are not complete, and at each time t_j only an interval of costs is available, and rules for early exercise should be based on additional considerations. On the other hand, forward procedures are more easily adapted for handling path dependency, which can then be represented by adding extra state variables.

4.4 Worst-Case Analysis

In this section we compare the results obtained from an interval model with those obtained from a simpler model (the standard binomial model) in a number of test cases.

4.4.1 Introduction

The derivative that we consider is a European call option. One may, of course, in principle envisage many hedging strategies, but we will restrict ourselves to delta strategies derived from binomial tree models. Specifically, we denote by Δ^σ the standard hedge for the binomial tree model with parameters u_σ and d_σ, where for each given number $\sigma > 0$ the parameters u_σ and d_σ are chosen such that $d_\sigma = 1/u_\sigma$ and the price of the option in the tree model with parameters u_σ and d_σ is equal to the price in the continuous-time Black–Scholes model with volatility parameter σ. In this way we have a one-parameter family of strategies that we will test.

The tests will be carried out in an interval model. As always when carrying out a worst-case analysis, one must specify the range of situations that will be considered; for an interval model this comes down to the choice of the parameters u and d. The results of the test will depend on this choice; the choice is, however, to some extent arbitrary. One way out is to carry out tests for a range of parameter values. In view of the moderate computational demands associated to path-independent hedging of derivatives on single assets, we will in fact proceed in this manner. We will consider interval models with parameters u_τ and d_τ that are determined by the single parameter τ in the same way as previously. These interval models will be denoted by $\mathbb{I}^\tau$, and the tree models with the same parameters will be denoted by $\mathbb{B}^\tau$.

As the nominal situation we consider a binary tree $\mathbb{B}^\sigma$, a corresponding delta-hedging strategy Δ^σ, and the resulting option price f_σ.

Now suppose the volatility may drop below σ and need not be constant over time. This is accounted for by considering, in addition to the binary tree paths in $\mathbb{B}^\sigma$, also interior paths in the interval model $\mathbb{I}^\sigma$, which may have smaller jumps at any moment. The outcome of costs for these interior paths need not be equal to f_σ, and the question arises as to how large this difference can be.

First, reconsider the one-step model. Recall from Sect. 4.2.3 that with $u = u_\sigma$ and $d = \frac{1}{u_\sigma}$ the cost range of the delta strategy in a one-step interval model is given by

$$I^{\Delta^\sigma} = \left[\frac{(u_\sigma S_0 - X)(S_0 - X)}{(u_\sigma - \frac{1}{u_\sigma})S_0}, \frac{1 - \frac{1}{u_\sigma}}{u_\sigma - \frac{1}{u_\sigma}}(u_\sigma S_0 - X)\right].$$

Costs may fall to this lower bound in "quiet" interior paths, with not all jumps at the limits. This fall is zero for $S_0 \le \frac{X}{u_\sigma}$ and $S_0 \ge \frac{X}{d} = u_\sigma X$ and has a maximum value of $\frac{(u-1)(1-d)}{u-d}$ for $S_0 = X$ (Fig. 4.4). The best-case costs are even smaller than the minimum of the corresponding FPI, and the difference is greatest for $S_0 = \frac{X}{\sqrt{u}}$ for out-of-the-money options and $S_0 = \frac{X}{\sqrt{d}}$ for in-the-money options (Fig. 4.5).

A second analytic result, valid for any number of steps, concerns the worst-case costs: they remain equal to f_σ as a consequence of Theorem 4.5. Thus a (temporary) decrease in volatility leads to a decrease in costs.

Next consider the case that *overhedging* occurs, that is, actual volatility is below the volatility for which the delta strategy is designed. Or, more formally, actual prices are in $\mathbb{I}^\tau$ with $\tau < \sigma$. Let the corresponding cost range interval of the delta strategy be

$$I^{\Delta^\sigma} =: [f_{\text{low}}, f_{\text{high}}]. \tag{4.20}$$

Then we notice that

$$f_\sigma \le f_{\text{high}} < f_\tau,$$

as $\mathbb{I}^\tau \subset \mathbb{I}^\sigma$, while f_τ is the minimum worst-case costs in $\mathbb{I}^\tau$.
In fact, the worst-case path in $\mathbb{I}^\tau$ is the one that has constant maximum volatility τ. This is shown in the next proposition, whose proof can be found in the appendix.

Proposition 4.10. *The worst-case price path in $\mathbb{I}^\tau$ under overhedging Δ^σ, with $\sigma > \tau$, is in $\mathbb{B}^\tau$.* □

This implies that, in the case of overhedging, there is no extra loss in interval models as compared to the binary trees. In fact, the analysis could take place entirely on the level of binary trees by considering worst-case costs of delta hedging based on a too high volatility.

Next consider the case of *underhedging*, that is, if the hedge strategy underestimates the volatility of assets. Thus we analyze the performance of the delta-hedging strategy Δ^σ, assuming that the actual price paths are in $\mathbb{I}^\tau$, with $\tau > \sigma$. Denoting the cost range interval I^{Δ^σ} again by (4.20) we have

$$f_\sigma < f_\tau < f_{\text{high}},$$

as Δ^τ is the unique strategy with minimal worst-case costs in $\mathbb{I}^\tau$.

In contrast to overhedged options, worst-case paths in $\mathbb{I}^\tau$ under Δ^σ need not have the maximum constant volatility. Stated differently, it turns out that the paths with the largest possible jumps are not always the ones that generate the worst costs. Costs are maximal for paths that cross the exercise level as often as possible with extreme jumps. In general, to cross the exercise level as often as possible, price paths should not use the largest amplitude at all times. Therefore, the cost of hedging predicted by interval models may be considerably higher than the cost derived from using a binomial tree model. This is demonstrated in the following simple example.

4.4.2 A Nonextremal Path with Worst-Case Cost

Consider an at-the-money European call option with exercise price $X = S_0 = 100$ in a two-period model. Let $u_\sigma = 1.20$ and $u_\tau = 1.25$. So we consider a situation where the hedge strategy is based on $\sigma = 0.16$, whereas the actual volatility in the model is $\tau = 0.19$.

From (4.3), (4.4) it follows that the price of the option in the tree model $\mathbb{B}^\sigma$ is $f_{\max} = 9.09$, whereas in the tree model $\mathbb{B}^\tau$ the price is $f_\tau = 11.11$. Recall from Theorem 4.5 that the latter quantity also represents the maximal worst-case costs in $\mathbb{I}^\tau$, which are achieved by the extreme delta hedge Δ^τ. If, however, the strategy Δ^σ is applied in the model $\mathbb{I}^\tau$, then the worst-case costs are found to be $f_{\max} = 13.26$. The corresponding worst-case path is $\{S_0, S_1, S_2\} = \{100, 83.3, 104.2\}$. Clearly this is not an extreme path. If we limit paths to the tree $\mathbb{B}^\tau$ and compute the costs for the strategy Δ^σ in this model, then we find the value $f_{\text{bin}} = [S_2 - X]^+ - \Delta_0(S_0)(S_1 - S_0) - \Delta_1(S_1)(S_2 - S_1) = 56.25 - \Delta_0(100) * 25 - \Delta_1(125) * 31.25$ along the path

$\{100, 125, 156.25\}$. Using (4.3), (4.4) again, with $u = 1.2$ and $d = \frac{5}{6}$, one can easily compute that $\Delta_0(100) = \lambda = \frac{6}{11}$ and $\Delta_0(125) = 1$. This yields, then, $f_{\text{bin}} = 11.36$. Similarly, one can show that also along the path $\{100, 125, 100\}$, f_{bin} attains this value, whereas along the other paths a smaller value is attained. So the worst-case costs in this model are $f_{\text{bin}} = 11.36$, and there are two corresponding worst-case paths, namely, $\{100, 125, 100\}$ and $\{100, 125, 156.25\}$.

The conclusions from this example may be summarized as follows. The worst-case costs for the strategy Δ^σ in the interval model $\mathbb{I}^\sigma$ are equal to 9.09. A worst-case analysis in the tree model $\mathbb{B}^\tau$ suggests that this figure may increase to 11.36 if the actual volatility turns out to be $\tau = 0.19$ rather than $\sigma = 0.16$. However, if the analysis is carried out in the interval model $\mathbb{I}^\tau$ rather than in the tree model $\mathbb{B}^\tau$, then it turns out that costs may actually increase up to 13.26. So if the option is sold for 9.09 corresponding to the implied volatility $\sigma = 0.16$ but the actual implied volatility parameter turns out to be $\tau = 0.19$, then the potential loss in an interval model with volatility parameter $\tau = 0.19$ is almost twice as large as the loss suggested by the corresponding binomial tree model. This illustrates that replacing the assumption of constant volatility by limited volatility could considerably increase the sensitivity of costs to underhedging.

Notice that the key value in the worst-case path is $S_1 = 83.3$, corresponding to a nonextreme first jump in $\mathbb{I}^\tau$. This is the highest asset price that maneuvers the optimistic hedge Δ^σ into an uncovered position, thereby preparing for large costs in the second step.

4.4.3 Worst Cases in Interval Models Versus Tree Models

In a more extensive experiment, we consider the hedging of a European call option in a ten-period model for several combinations of hedging strategies and interval models. The following parameter values are used:

Initial price:	$S_0 = 100$,
Exercise price:	$X = 100$,
Exercise time:	$T = 1$,
Interest rate:	$r = 0$,
Time step:	$h = 0.1$, so $N = 10$.

Think of an option with exercise date 1 year away and adaptation of the hedge portfolio every 5 weeks. As our main reference point we take $\sigma^* = \tau^* = 0.2$, which means an annual variance of asset prices of 20%. We compute the worst-case costs of hedging strategies Δ^σ in the actual models $\mathbb{I}^\tau$, with σ and τ ranging from 0.1 to 0.3 in steps of 0.05. Worst cases are determined as indicated in Sect. 4.3, where the one-dimensional optimizations are implemented on a grid for the logarithms of prices. The results are shown in Fig. 4.6.

Because all paths of an interval model with a given volatility parameter are also contained in interval models with a larger volatility parameter, the worst-case costs

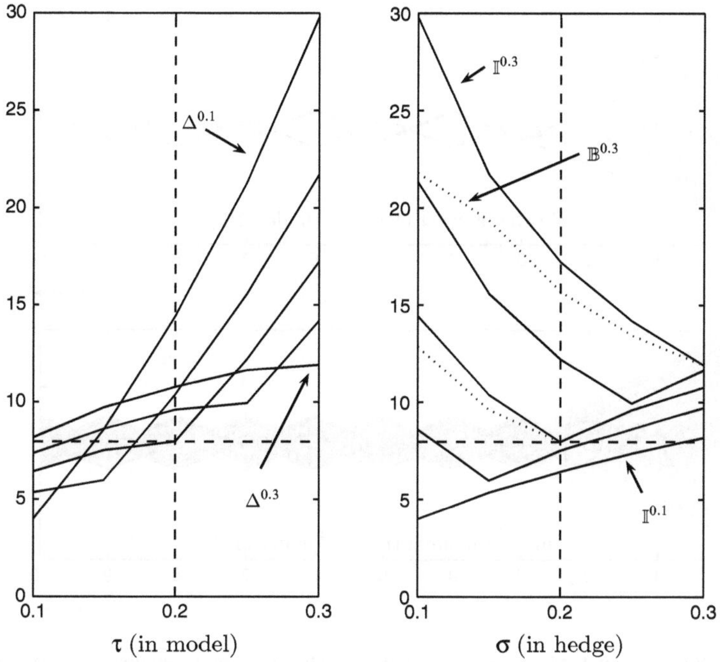

Fig. 4.6 Worst-case costs for Δ^σ in $\mathbb{I}^\tau$. In the *left* plot each line corresponds to worst-case costs under a fixed strategy Δ^σ for a range of interval models. In the *right* plot every line denotes the worst-case costs in a fixed interval model $\mathbb{I}^\tau$ for a range of hedging strategies Δ^σ. The *dotted lines* denote worst-case costs if Δ^σ is used in the binomial trees $\mathbb{B}^{0.2}$ and $\mathbb{B}^{0.3}$; on the left these are not shown

corresponding to a fixed strategy must be nondecreasing as a function of τ; this is seen in the left-hand plot. Both plots also show the optimality in a worst-case sense of Δ^σ within the model $\mathbb{I}^\sigma$; for $\sigma^* = 0.2$ this is indicated by dashed lines. There is a striking asymmetry between overhedging and underhedging: the loss due to underhedging according to $\Delta^{0.1}$ in the interval model $\mathbb{I}^{0.3}$ is much larger than the loss due to overhedging according to $\Delta^{0.3}$ in $\mathbb{I}^{0.1}$. The dotted lines in the right-hand plot again show that worst-case analysis in a binomial tree setting may produce results that are a bit more optimistic than the results obtained from an interval model, especially when the hedge strategy is based on a value of the volatility that is considerably too low. If we compare, for instance, in the right-hand plot the worst-case costs of $\Delta^{0.1}$ in the interval model $\mathbb{I}^{0.3}$ with that of the binomial tree model $\mathbb{B}^{0.3}$, then we see a difference of approximately eight units, whereas this difference in worst-case costs using $\Delta^{0.1}$ is less than two units. So the risk associated with a too low specification of volatility can be considerably higher when in the actual model volatility is nonconstant than the risk implied by the delta strategy used in a higher, but constant, volatility model. Further, this discrepancy seems to grow exponentially the larger the misspecification of volatility is.

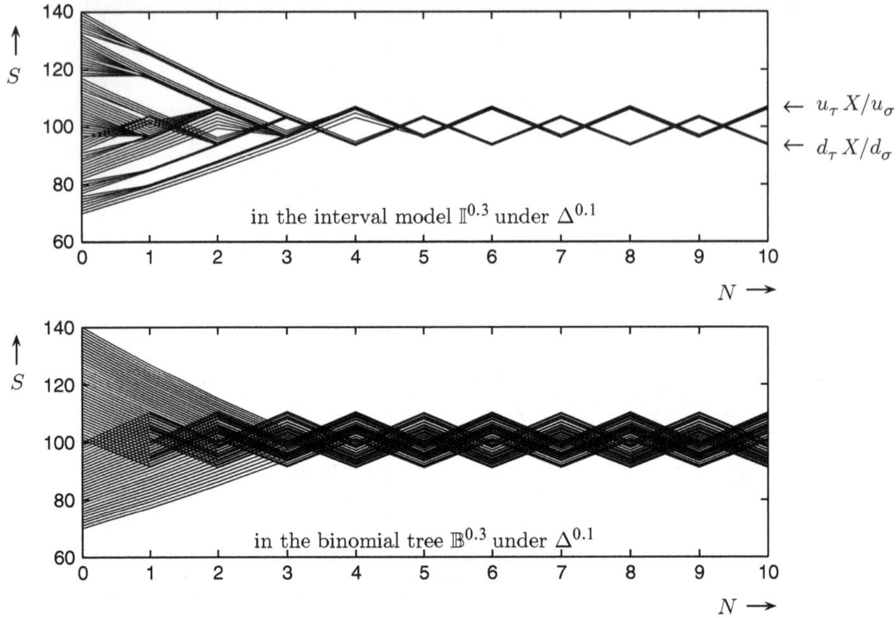

Fig. 4.7 Worst-case paths for various initial prices. The *upper plot* contains worst-case paths in $\mathbb{I}^\tau$ under Δ^σ for $\tau = 0.3$ and $\sigma = 0.1$, with initial price ranging from 70 to 140. The *lower plot* shows the worst-case paths in $\mathbb{B}^\tau$

The question may arise as to how nonextreme price fluctuations contribute to extreme costs. To discover a pattern, we consider several worst-case paths for a range of initial prices S_0 and all other parameters kept constant. These are compared with worst-case paths in the binomial tree model in Fig. 4.7. The graphs indicate that in both models, costs are maximal for paths that cross the exercise level as often as possible with extreme jumps. They differ, however, in the levels of the peaks in the end regime. In the binomial model (with $u = 1/d$) all prices are of the form $u^j S_0$, where j may be positive or negative, and hence the peak levels are at $u^j S_0$, where j is the smallest integer such that $u^j S_0 > X$. Nonextreme jumps allow a change in the level of peaks, and this extra freedom in interval models may increase the cost substantially. The simulations suggest that worst-case costs are achieved for upward peaks at X/u_σ or downward peaks at X/d_σ; a formal statement in this direction remains to be proven, however. The graphs clearly suggest that most of the freedom allowed by interval models is used in the first few time steps. This in turn suggests that a reduction in computational load of a worst-case search may be achieved by using an interval model for the first few time steps (or even just for the first one) and a binomial model thereafter.

A similar effect is apparent when the exercise level X is varied, with initial prices kept fixed. This is illustrated in Fig. 4.8, in which worst-case costs in interval models and binomial trees are compared for various exercise prices X. There is considerable

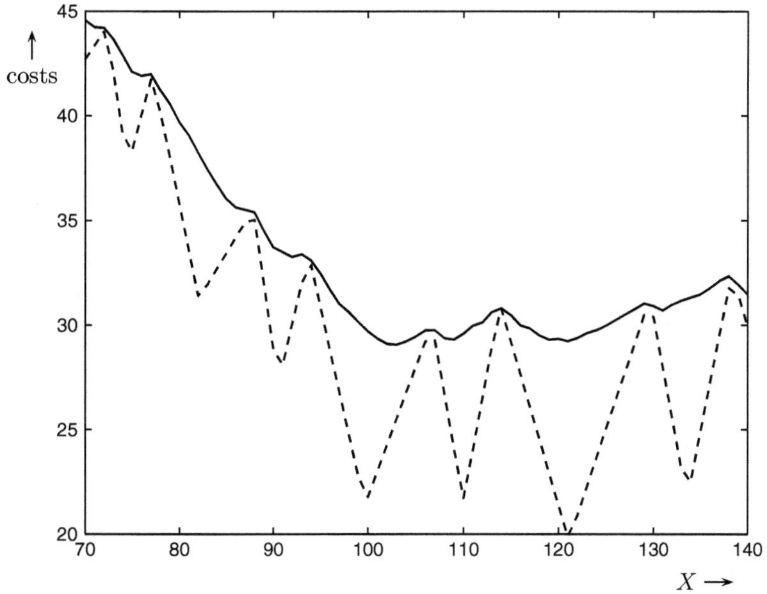

Fig. 4.8 Worst-case costs for various exercise levels. The *solid line* corresponds to worst-case costs in $\mathbb{I}^{0.3}$ under $\Delta^{0.1}$, with X ranging from 70 to 140 (by unit steps). The *dashed line* represents worst-case costs in $\mathbb{B}^{0.3}$

variation in the size of the underestimation of worst-case costs by binomial models as compared to interval models. Again the irregular pattern for the binomial tree is explained by the fact that worst-case paths are restrained to fixed grid points $u^j S_0$. In particular, it seems that for strike prices X above the initial price S_0 the underestimation by binomial models can be more profound.

Chapter 5
Optimal Hedging Under Robust-Cost Constraints

5.1 Introduction

In this chapter we analyze hedging of a short position in a European call option by an optimal strategy in the underlying asset under a *robust cost constraint* (RCC), that is, under the restriction that the worst-case costs do not exceed a certain a priori given upper bound. This relates to a Value-at-Risk (VaR) condition, which is usually defined for stochastic models as the maximum costs for a specified confidence level. As compared to VaR, an RCC denotes a level of worst-case costs that *cannot* be exceeded within a given interval model.

More specifically, the asset is modeled by an interval model $\mathbb{I}^{u,d}$ in N equal time steps from current time to expiry, cf. (3.9). Recall from Proposition 4.2 that a short position in the option, kept under a hedge strategy g, yields an outcome of costs in an interval I^g. For discontinuous strategies this interval I^g is not necessarily closed, and therefore *best-* and *worst-case costs* are defined as the infimum and supremum of costs:

$$\mathrm{BC}^g := \inf I^g = \inf_{S\in\mathbb{I}^{u,d}} Q^g(S).$$

$$\mathrm{WC}^g := \sup I^g = \sup_{S\in\mathbb{I}^{u,d}} Q^g(S).$$

We refer to $-\mathrm{BC}^g$ also as the *maximum profit* under g.

The RCC condition simply limits the worst-case costs WC^g. In this section we analyze the impact of such a restriction for the set of admissible hedging strategies and provide an algorithm to solve this constrained optimization problem. To that end we first introduce some notation.

The set of all strategies with price paths S in some interval model is denoted by $\mathbb{G}$. Thus

$$\mathbb{G} := \{g = (g_0,\dots,g_{N-1}) \mid g_j : (S_0,\dots,S_j) \to \gamma_j \in \mathbb{R}\}.$$

P. Bernhard et al., *The Interval Market Model in Mathematical Finance*, Static & Dynamic Game Theory: Foundations & Applications, DOI 10.1007/978-0-8176-8388-7_5,

Let V denote the RCC limit; then the set of all admissible strategies under this RCC limit is defined by

$$\mathbb{G}^V := \{g \in \mathbb{G} | \, \mathrm{WC}^g \le V\}.$$

Furthermore, by Δ_j we will denote the delta-hedging strategy [see (3.4), (3.5)]

$$\Delta_j(S_j) = \lambda \Delta_{j+1}(uS_j) + (1-\lambda)\Delta_{j+1}(dS_j), with$$
$$\Delta_{N-1}(S_{N-1}) = \frac{[uS_{N-1} - X]^+ - [dS_{N-1} - X]^+}{(u-d)S_{N-1}},$$

where $\lambda = \frac{u(1-d)}{u-d}$, and by $f_j(S_j)$ we will denote the corresponding Cox–Ross–Rubinstein option premium [see (3.2), (3.3)]

$$f_N(S_N) = [S_N - X]^+,$$
$$f_j(S_j) = q f_{j+1}(uS_j) + (1-q) f_{j+1}(d_j S_j), \tag{5.1}$$

where $q := \frac{1-d}{u-d}$.

5.2 Effect of Cost Constraints on Admissible Strategies

Since delta hedging yields the lowest upper bound of costs among all strategies in $\mathbb{G}$ (Theorem 4.5), we have the next result.

Proposition 5.1. *If $V < f_0(S_0)$, then $\mathbb{G}^V$ is empty. If $V \ge f_0(S_0)$, then the delta-hedging strategy belongs to $\mathbb{G}^V$.* □

So the arbitrage-free Cox–Ross–Rubinstein price of the option C in the binomial tree model $\mathbb{B}^{u,d}$ is the smallest RCC limit that is achievable for a hedged short position in the call option C with underlying asset $S \in \mathbb{I}^{u,d}$.

As may be expected, for RCC beyond this minimal level, the space of admissible strategies is centered around the delta-hedging strategy. To formulate the precise result, we introduce the following concepts. For a given strategy, H_j denotes the *realized hedge costs* at t_j:

$$H_j := -\Sigma_{k=0}^{j-1} \gamma_k (S_{k+1} - S_k). \tag{5.2}$$

In view of the previous result we also define the *current latitude*

$$\bar{V}_j(S_j, H_j) := V - H_j - f_j(S_j), \tag{5.3}$$

which is the excess of the RCC limit V over the past hedge costs H_j plus the minimal future worst-case costs (given by the Cox–Ross–Rubinstein price) $f_j(S_j)$, or, equivalently, the current wealth offset to the total minimal cost-to-go $f_j(S_j)$.

The next theorem shows that $\bar{V}$ indeed determines the extent to which admissible strategies may differ from delta hedging.

Theorem 5.2. *The set $\mathbb{G}^V$ of strategies admissible under the RCC level $V \geq f_0(S_0)$ for $S \in \mathbb{I}^{u,d}$ is given by*

$$\{g \in \mathbb{G} | \gamma_j^{\min}(S_j, H_j) \leq g_j(S_0, \ldots, S_j) \leq \gamma_j^{\max}(S_j, H_j)\}, \tag{5.4}$$

with

$$\gamma_j^{\min}(S_j, H_j) := \Delta_j(S_j) - \frac{\bar{V}_j(S_j, H_j)}{(u-1)S_j},$$

$$\gamma_j^{\max}(S_j, H_j) := \Delta_j(S_j) + \frac{\bar{V}_j(S_j, H_j)}{(1-d)S_j},$$

with $\bar{V}$ defined by (5.3) and H_j the realized hedge costs (5.2). □

See the appendix for a proof.

Summarizing, the consequence of including a restriction on the worst-case costs is that the set of admissible strategies is restricted to an interval around delta hedging, with fixed proportional centering determined by u and d and time-varying interval length determined by realized hedge costs.

5.3 Calculating Maximum Profit Under a Cost Constraint

In this section we present a numerical algorithm to maximize profits under a limit for worst-case costs in a given interval model for the asset. Or, stated differently, with an RCC limit V on worst-case costs, we look for a strategy g that provides the highest lower bound for best-case costs. We will denote this lower bound for best-case costs by

$$\text{BC}^*(S_0, V) := \inf_{\{g \in \mathbb{G}^V, S \in \mathbb{I}^{u,d}\}} Q^g(S), \tag{5.5}$$

with $Q^g(S)$ as defined in (4.5).

Thus $\text{BC}^*(S_0, V)$ is a lower bound for best-case costs (and hence an upper bound on maximum profit) under the RCC level V and with asset prices in $\mathbb{I}^{u,d}$ with initial price S_0. Notice that by choosing g equal to the stop-loss strategy, we in fact obtain just the opposite of what we want, i.e., a maximum profit that is outperformed by any other admissible strategy under optimal conditions.

To determine solutions, we first analyze the recursive structure of this minimization (5.5). It amounts to the dynamic programming problem

$$\begin{aligned} \text{Minimize } J &:= \Sigma_{j=0}^{N-1} F(j,x_j,u_j,v_j) + G(x_N) \\ \text{with } x_{j+1} &= h(j,x_j,u_j,v_j) \\ \text{for } (u_j,v_j) &\in D(j,x_j), \end{aligned}$$

with the following definitions of the variables:

$$\begin{aligned} x_j &:= \begin{bmatrix} S_j \\ H_j \end{bmatrix}, \\ u_j &:= \gamma_j, \\ v_j &:= S_{j+1}/S_j, \end{aligned}$$

with the domain $D(j,x_j)$ the rectangle specified by the conditions

$$\begin{aligned} u_j &\in \Gamma_j(S_j,H_j) := [\gamma_j^{\min}(S_j,H_j), \gamma_j^{\max}(S_j,H_j)], \\ v_j &\in [d,u], \end{aligned}$$

with $\gamma_j^{\min}$ and $\gamma_j^{\max}$ defined as in Theorem 5.2, and with the state recursion and cost function given by

$$\begin{aligned} \begin{bmatrix} S_{j+1} \\ H_{j+1} \end{bmatrix} &= \begin{bmatrix} v_j S_j \\ H_j - u_j(v_j-1)S_j \end{bmatrix}; \begin{bmatrix} S_0 \\ H_0 \end{bmatrix} = \begin{bmatrix} S_0 \\ 0 \end{bmatrix}; \\ J &= -\Sigma_{j=0}^{N-1} u_j(v_j-1)S_j + [S_N - X]^+. \end{aligned}$$

Notice the path dependence of the criterion, which becomes apparent in the occurrence of the realized hedge costs in the state. Equivalently, the criterion is path dependent through "current wealth" $V - H_j$, where the RCC limit V is interpreted as the initial wealth.

The corresponding *value function* consists of the best-case costs conditioned on the current asset price and past hedge costs:

$$\begin{aligned} \mathrm{BC}_N(S_N,H_N) &:= [S_N - X]^+ + H_N, \\ \mathrm{BC}_j(S_j,H_j) &:= \min_{\{\gamma_j \in \Gamma_j, S_{j+1} \in [dS_j, uS_j]\}} \left(\mathrm{BC}_{j+1}(S_{j+1}, H_j - \gamma_j(S_{j+1} - S_j) \right), \end{aligned} \tag{5.6}$$

with the domain for BC_j taken as

$$\{(S_j,H_j) | S_j > 0, H_j \geq V - f_j(S_j)\} \tag{5.7}$$

to avoid minimization over an empty domain. Not coincidentally, this definition of domain is consistent with the recursion in (5.6) because Γ_j is determined just to guarantee that $\bar{V}_j = V - H_j - f_j(S_j)$ remains nonnegative.

Thus BC_j denotes the best-case costs, given that at t_j the asset price is S_j and that past hedge costs accumulated to H_j or, equivalently, that $x_j = (S_j, H_j)$.

Before we go into computations, we show that the optimization problem has a solution. A proof of the next result can be found again in the appendix.

Proposition 5.3. *There exists an optimal strategy g^* and a best-case price path S^* such that $Q^{g^*}(S^*) = \mathrm{BC}^*(S_0, V)$.* □

As it seems too complicated to obtain closed-form solutions, we develop a numerical procedure that exploits some specific features of the dynamic programming problem, enabling a relatively simple *forward* recursion for a "frontier function" in one variable, with known initial conditions. Note that this approach differs from the standard numerical solution of the dynamic programming problem, which would amount to a backward recursion for a function in two variables, conditioned on *unknown* final values of asset prices S_N and realized hedge costs H_N. The following method hence avoids the use of a rather large grid matrix of sample points.

First a frontier function of minimal realized hedge costs is determined as a function of asset prices, then the best-case asset price path is determined by backward recursion, and finally the optimal strategy is reconstructed.

Algorithm 5.4 (Maximum profit under RCC).

Data: Initial asset price S_0, an interval model for assets $\mathbb{I}^{u,d}$, excercise price X and time T of a European call option, and an RCC limit on worst-case total costs at expiry, $\mathrm{WC}^g \leq V$.

Step 1: Determine the "frontier function" of minimal realized hedge costs $H^*(j, S_j)$ by

$$H^*(0, S_0) := 0, \tag{5.8}$$

$$H^*(1, S_1) := -\gamma_0^\sharp (S_1 - S_0) \text{ for } S_1 \in [dS_0, uS_0], \tag{5.9}$$

$$H^*(j+1, S_{j+1}) := \min_{S_j \in I_j} [H^*(j, S_j) - (S_{j+1} - S_j)\gamma_j^\sharp(S_j, H^*(j, S_j))], \tag{5.10}$$

with $S_j \in [d^j S_0, u^j S_0]$, $I_j := [d^j S_0, u^j S_0] \cap [\frac{S_{j+1}}{u}, \frac{S_{j+1}}{d}]$ and $\gamma_j^\sharp$ defined by

$$\gamma_j^\sharp := \begin{cases} \gamma_j^{\max}(S_j, H^*(j, S_j)) & \text{for } S_{j+1} > S_j, \\ \gamma_j^{\min}(S_j, H^*(j, S_j)) & \text{for } S_{j+1} \leq S_j. \end{cases} \tag{5.11}$$

Step 2: Determine the optimal price path S^* recursively by

$$S_N^* := \operatorname{argmin}_{S_N \in [d^N S_0, u^N S_0]} [H^*(N, S_N) + [S_N - X]^+], \tag{5.12}$$

$$S_j^* := \operatorname{argmin}_{S_j \in I_j} [H^*(j, S_j^*) - \gamma_j^\sharp (S_{j+1}^* - S_j)], \tag{5.13}$$

with $\gamma_j^\sharp$ defined as in (5.11) with S_{j+1} replaced by S_{j+1}^*.

Step 3: Determine the optimal strategy g^* by

$$g_j^*(S_0,\ldots,S_j) = \begin{cases} \gamma_j^{\max}(S_j,H_j) & \text{if } S_{j+1}^* > S_j^*, \\ \gamma_j^{\min}(S_j,H_j) & \text{if } S_{j+1}^* \leq S_j^*. \end{cases} \tag{5.14}$$

Result: The strategy g^* yields the maximum profit $-\mathrm{BC}^g$ under the restriction that worst-case costs WC^g are at most V and asset prices are in accordance with the interval model $\mathbb{I}^{u,d}$. These best-case costs are achieved for the price path S^* under strategy g^*. □

Remark 5.5. From (5.14) we see that the optimal hedge depends on the realized hedge costs in the past and that at each time step all gained reserves beyond the RCC limit are put at risk. This is a typical feature of the modeling we have used so far. If one is unhappy with these kinds of strategies because they are too risky, one should take this into account explicitly in the modeling. At this moment there is no incentive in the modeling to avoid this kind of behavior. We will return to this issue later on. □

That Algorithm 5.4 indeed achieves the advertised result is shown in the appendix. Before illustrating the algorithm with an example, we give a brief explanation. Initial hedge costs are set to zero in (5.8), and (5.9) simply denotes the realized hedge costs under optimal hedging $\gamma_0^\#$ as a function of the current price S_1 after the first step. For the second time step, $H^*(2,S_2)$ denotes optimal realized hedge costs, which now not only involve optimization over hedge position γ_1 but also over all paths in $\mathbb{I}^{u,d}$ starting at S_0 and ending at a fixed value S_2. The interval I_1 specifies all possible values for the asset price at t_1 for such paths. It is important to note that the algorithm postpones optimization over current prices, so hedge costs $H^*(j,S_j)$ are *conditionally* optimal, assuming an arbitrary fixed price level S_j at time t_j. Optimal price paths are then determined by a backward recursion (5.13) starting at an easy-to-evaluate final condition (5.12).

Example 5.6. We consider an at-the-money European call option with exercise price $X = 1$. We assume that the underlying asset follows a price path in the interval model $\mathbb{I}^{u,d}$, with $u = 5/4$ and $d = 4/5$, $N = 4$ time steps, and initial asset price $S_0 = 1$. Sampling of asset prices is done with a logarithmically regular grid, with 51 points on $[d,u]$. A further decrease of this mesh hardly affects the outcome of the algorithm.

The unique Cox–Ross–Rubinstein arbitrage-free option price in the corresponding binomial tree $\mathbb{B}^{u,d}$ is given by $f_0(S_0) = 0.1660$, cf. (5.1). With the aid of the algorithm we computed that maximum profit, under strategies that guarantee this limit, are given by $1/9$ and are achieved for the price path

$$S^* = (S_0^*, S_1^*, S_2^*, S_3^*, S_4^*) = (1,1,1,5/4,25/16). \tag{5.15}$$

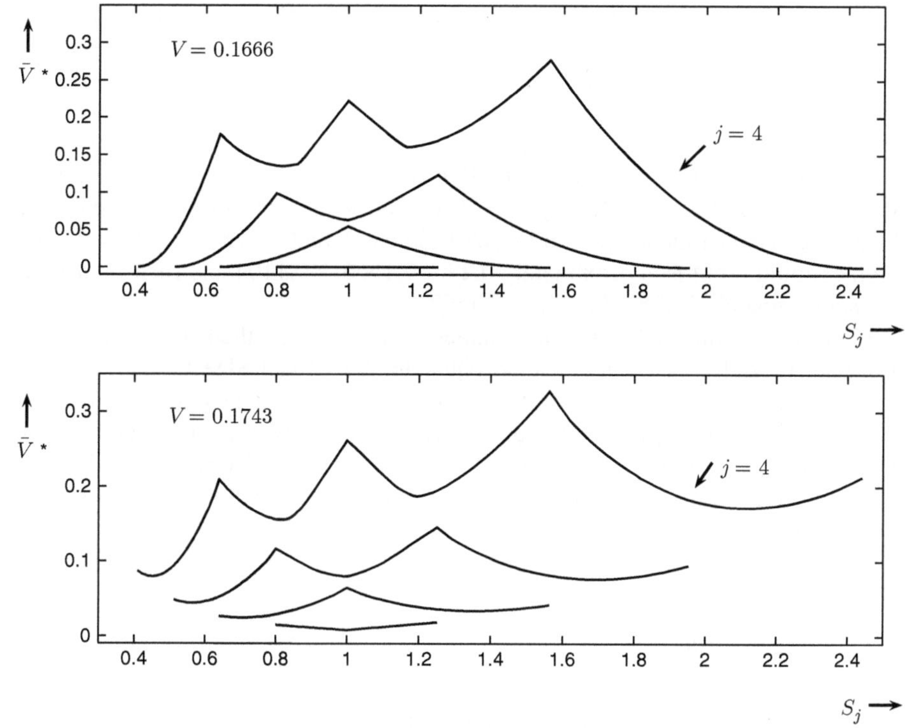

Fig. 5.1 Optimal costs under RCC restrained strategies. Both plots contain the graphs of the optimal current latitude $\bar{V}^*(j,s) := V - H^*(j,s) - f_j(s)$ with $s \in [d^j, u^j]$, for $j = 1, \ldots, 4$. In the *upper plot* the RCC limit is chosen equal to the lowest achievable cost limit, i.e., $V = f_0(S_0) = 0.1666$, while in the *lower plot* this is increased by 5% to $V = 0.1743$. Thus $\bar{V}^*(j,s)$ denotes the maximum current latitude compatible with a price $S_j = s$ under strategies that are admissible under these RCC limits. In the *upper plot* $\bar{V}^*$ is zero at the boundary points d^j and u^j because these are only achievable by a sequence of extreme jumps in prices, which keeps the latitude at the zero level, by definition of delta hedging. The fact that $\bar{V}^*(1,s) = 0$ for all $s \in [d,u]$ is somewhat coincidental because this would not be the case for exercise prices unequal to 1

The same analysis is repeated with a slighly higher RCC limit $1.05 f_0(S_0) = 0.1743$. Maximum profit turns out to be 0.1531 and is achieved for the price path

$$S^* = (S_0^*, S_1^*, S_2^*, S_3^*, S_4^*) = (1, 0.9564, 1, 5/4, 25/16). \tag{5.16}$$

To give an impression of the outcome of the optimal realized hedge cost functions $H^*(j,s)$ for $j = 1,2,3,4$ [cf. (5.9) and (5.10)], we have plotted the corresponding optimal current latitude $\bar{V}^*(j,s) = V - H^*(j,s) - f(j,s)$, both for the tight RCC limit $V = f_0(S_0)$ and for $V = 1.05 f_0(S_0)$, in Fig. 5.1. This has the following interpretation. If at time t_j the asset price is given by $S_j = s$, then the realized hedge costs H_j, defined by (5.2), are under optimal circumstances equal to $H^*(j,s)$, i.e., under optimal admissible hedging and for the best price path in $\mathbb{I}^{u,d}$ from S_0 to

$S_j = s$. Hence $\bar{V}^*(j,s)$ denotes the maximum current latitude compatible with a price $S_j = s$. In particular, $\bar{V}^*(4,S_4) = V - H^*(4,S_4) - f(4,S_4)$. By (5.12), the last price S_4^* of the optimal price path minimizes $H^*(4,S_4) + f(4,S_4)$. Thus the optimal price S_4^* is obtained from the plot as that price at which the $j=4$ curve attains its maximum value. If $V = 0.1666$, then we see that this optimal price S_4^* is approximately 1.56, whereas for $V = 0.1743$ the price S_4^* is slightly smaller. Furthermore, the maximum profit, $-\mathrm{BC}^*(S_0,V) = -H^*(4,S_4^*) - f(4,S_4^*)$, is obtained from this curve as the difference between the maximum value of the curve and V. Thus, for $V = 0.1666$ the maximum profit is approximately $0.28 - 0.1666 \approx 0.1104$ and for $V = 0.1743$ it is approximately $0.325 - 0.1743 \approx 0.1507$.

If the RCC limit is fixed at its smallest value, $f_0(S_0)$, then the initial hedge position is fixed because it must be equal to the delta hedge 0.5830. For the increased RCC limit the interval of admissible initial hedge positions is given by $[0.5369, 0.6863]$. In Fig. 5.2 we show the range of admissible strategies for time instants t_1, t_2, and t_3 under RCC limits $V = f_0(S_0)$ and $V = 1.05 f_0(S_0)$. This gives an idea of how far hedge positions may deviate from delta hedging under best-case circumstances. □

5.4 Extensions

We have described how a worst-case cost restriction can be translated to strategy limits and shown how to determine maximum profits under such a constraint. In this section we discuss extensions of this result with respect to the choice of the RCC limit and variants of the cost criterion.

First we pursue the pure interval calculus a little further in Sect. 5.4.1. It is shown how cost limits can be chosen on the basis of the maximum loss/profit ratio and how an option premium can be based on this criterion.

These criteria, which are based solely on interval limits for asset prices, have some degenerate features as a performance measure for investments, especially if the number of time steps is large. Under the RCC restriction, the downside risk is limited, by construction, but (as was already mentioned in Remark 5.5) in each step hedge volumes are driven to the maximum amount, and consequently all the gained reserves beyond the RCC limit are put at risk at each step. In particular, for a long sequence of time steps this seems odd, and it may be more desirable to secure profit, at least partially.

Therefore, in Sect. 5.4.2 we also analyze how to minimize expected costs under additional stochastic assumptions within interval models. This relates to a fairly general result that depends only on the expected growth factor, $E(S_{j+1}/S_j)$. However, despite the different nature of the criterion, we will see that the optimal hedge volumes turn out to be maximal again at each step.

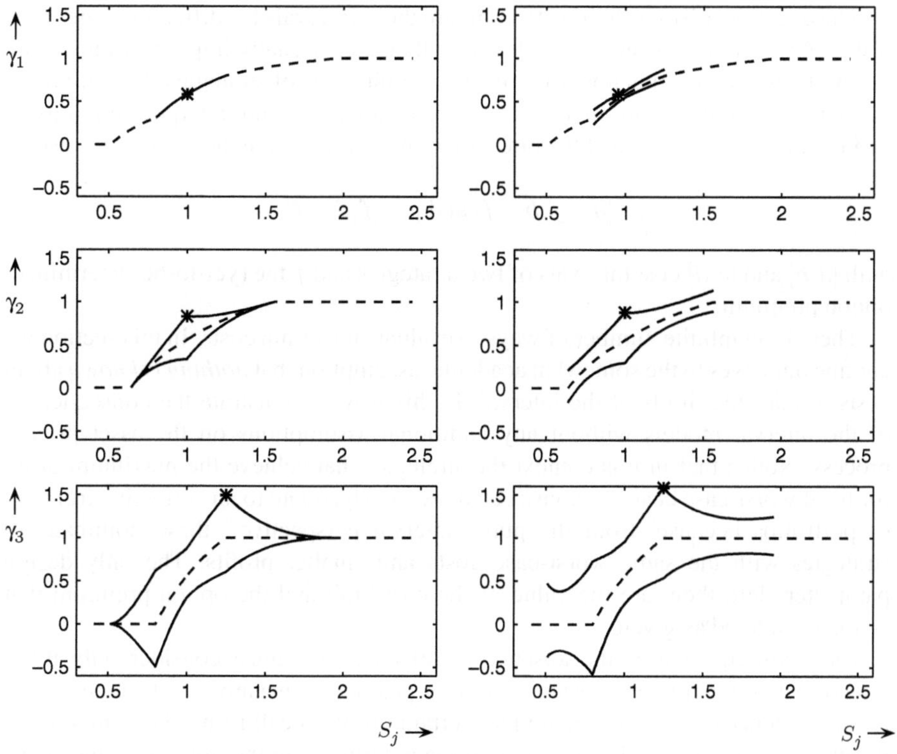

Fig. 5.2 RCC-admissible strategies. The plots on the *left-hand side* correspond to the tight RCC limit $V = f_0(S_0) = 0.1666$, on the *right-hand side* to $V = 1.05 f_0(S_0) = 0.1743$. The *dashed lines* indicate hedge positions according to delta hedging, as a function of prices S_j at time t_j, with $j = 1$ in the upper plots, $j = 2$ in the center, and $j = 3$ in the lower plots. The *solid lines* denote the graphs of $\gamma_j^{\max}$ and $\gamma_j^{\min}$, as a function of S_j, and with realized hedge costs $H^*(j, S_j)$, which are the optimal hedge costs compatible with asset price S_j at t_j, cf. (5.11). Whenever the corresponding current latitude is zero (Fig. 5.1), the strategy must coincide with delta hedging. The prices S_1^*, S_2^*, and S_3^* in the best-case price paths (5.15) (*left plots*) and (5.16) (*right plots*) are indicated by an asterisk and hence mark the actual outcome $\gamma_j^\sharp$ of the strategy for this path

5.4.1 Loss/Profit Ratio

In Sect. 5.3 we described an algorithm for solving the minimization problem (5.5). The solution consists of an optimal strategy and the construction of a corresponding price path at which profits are maximized. It is not hard to verify that worst-case costs under this strategy reach the prescribed RCC limit V, so that in fact the entire cost interval of the strategy (4.6) is known:

$$I^{g^*} = [\mathrm{BC}^*(S_0, V), V] \tag{5.17}$$

for g^* a solution of (5.5).

The question arises as to how to compare these intervals for different values of V if the RCC limit is a design variable rather than an externally imposed value. More generally, the question is how to evaluate cost intervals of strategies. By definition, the option premium is not included in the cost intervals, and incorporating it as an additional factor we arrive at the question of how to compare the interval of results

$$[a-f, b-f] \text{ and } [c-f, d-f]$$

with $[a,b]$ and $[c,d]$ cost intervals of two strategies and f the (yet-to-be-determined) option premium.

There is an infinite number of ways to evaluate uncertain costs. In this section we confine ourselves to the somewhat academic assumption that *nothing is known* about costs besides the limits of the interval. In this way we illustrate the consequences of the interval models without any additional assumptions on the asset pricing process. Notice that in this context the strategies that achieve the maximum profit for fixed worst-case costs, as constructed previously, relate to an "effective frontier" of portfolios because, from the pure "interval perspective," these dominate all strategies with the same worst-case costs and smaller profits. The only design parameters left, then, are the value of the cost limit and the option premium if it is not considered as given.

As a consequence of the absence of arbitrage, we must consider only those intervals that contain zero because entirely positive or negative cost ranges relate to the existence of strategies that yield certain profit. Recall from Theorem 4.5 that the precise bounds for arbitrage-free option premiums are the intrinsic value of the option, $[S_0 - X]^+$ (lower bound), and the Cox–Ross–Rubinstein price $f_0(S_0)$ for maximum volatility (upper bound). The *loss/profit ratio* (LPR) of such an interval $[a-f, b-f]$ is defined as $\frac{b-f}{f-a}$. There is a simple argument for considering this ratio as the main criterion. Suppose there are strategies that lead to cost intervals $[a,b]$ and $[c,d]$, so that the net results with option premium f are respectively $[a-f, b-f]$ and $[c-f, d-f]$. If we allow for portfolio rescaling, we may scale the second one by a factor $\frac{a-f}{c-f}$, which gives $[a-f, \frac{(a-f)(d-f)}{c-f}]$. Comparing this with $[a-f, b-f]$ is now simply a matter of comparing the right bounds, and their ordering is precisely determined by the LPR of the intervals. Observe that the cost interval with the smallest ratio is preferable in the absence of additional information.

From (5.17) it now follows that for a given (arbitrage-free) option premium f and an achievable RCC limit V [hence not below the Cox–Ross–Rubinstein price $f_0(S_0)$], the *optimal LPR criterion* is given by

$$\mathrm{LPR}^*(f, V) := \frac{V - f}{f - \mathrm{BC}^*(S_0, V)}.$$

Here $\mathrm{BC}^*(S_0, V)$ is given by (5.14), and it is interesting to see how this ratio depends on V and f. In particular, one might address the question of how to minimize this ratio over V for a fixed f. Notice that for the maximum arbitrage-free premium,

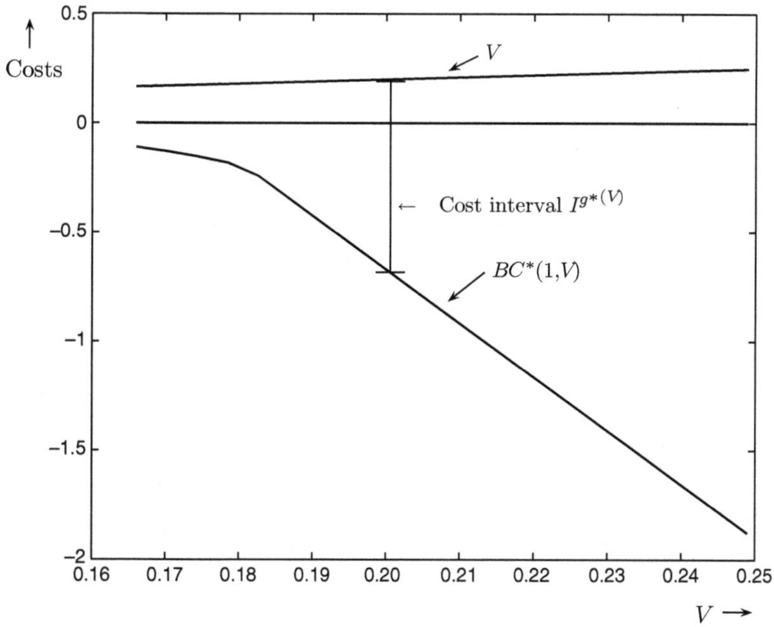

Fig. 5.3 Best-case costs as a function of RCC limit. The *lower curve* shows best-case costs $\mathrm{BC}^*(S_0,V)$ as a function of RCC limit V, with $S_0 = 1$; cf. (5.5). The *upper line* in the plot equals V. The minimally achievable RCC limit is 0.1666, which equals the Cox–Ross–Rubinstein option premium

$f = f_0(S_0)$, the RCC that minimizes this $\mathrm{LPR}^*(V)$ ratio is $V = f_0(S_0)$ because then the ratio is zero. Moreover, it is easily verified that for a fixed RCC limit V, $\mathrm{LPR}^*(f)$ is a decreasing function. So its minimum value is attained at $f = f_0(S_0)$.

In the following example we consider this dependency of the optimal LPR on its RCC limit V and option prices f in more detail.

Example 5.7. We proceed with Example 5.6. Figure 5.3 illustrates the dependence of best-case costs $\mathrm{BC}^*(S_0,V)$ on cost limit V by applying the algorithm to several values of V. It seems that $\mathrm{BC}^*(S_0,V)$ depends piecewise linearly on this RCC limit V. There seems to be a change in the slope of the best-case costs around $V = 0.18$.

In Fig. 5.4 the LPRs are shown for a range of arbitrage-free option premiums. The fair price interval is in this case given by $[0, 0.1666]$. We plotted the optimal LPR for different choices of the option price from this fair price interval. It turns out that there are two ranges of option premiums for which $\mathrm{LPR}^*(V)$ has a different minimum location. For option prices sufficiently close to the minimal RCC bound $f_0(S_0)$, the minimal $\mathrm{LPR}^*(V)$ is attained at $V = f_0(S_0)$. For option prices below a certain threshold, $\mathrm{LPR}^*(V)$ has an infimum at $V = \infty$. Thus, by relaxing in those cases the RCC constraint, we improve the LPR^*. □

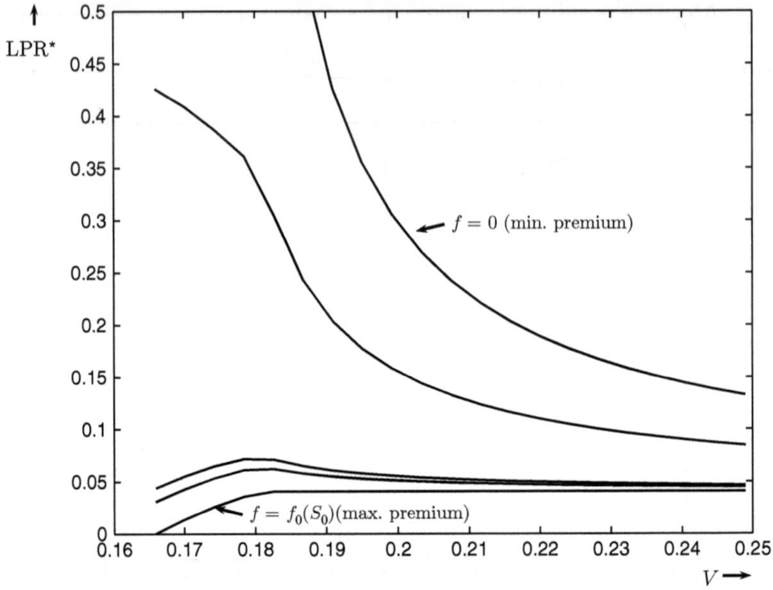

Fig. 5.4 Optimal loss/profit ratio for several option prices. The *curves* denote the LPR* for several option prices. The lowest one corresponds to the Cox–Ross–Rubinstein price $f_0(S_0) = 0.1666$; hence it is zero at $V = f_0(S_0)$. The other *curves* (*bottom* to *top*) correspond to option premiums of, respectively, 0.95, 0.93, 0.5, and 0 times $f_0(S_0)$. Prices outside this range are not arbitrage free. For factors from 0.93 to 1, the minimum is at the left, so the minimum LPR* is achieved for the lowest achievable cost bound $V = f_0(S_0)$, while for lower premiums the optimum switches to infinity

5.4.2 *Maximum Expected Profit Under a Cost Constraint*

In this section we take a step from the purely indeterministic features, as represented by interval models, to the probabilistic properties of prices. We assume that, in addition to the *limits* that an interval model induces for the growth factor of prices, its *expected value* is also given. Thus we assume that in each step

$$E_j[S_{j+1}/S_j] = 1 + e(j, S_j), \tag{5.18}$$

where $E_j[\cdot]$ denotes the expectations conditioned on past prices up to and including t_j, and e is a real-valued function denoting the expected growth rate at t_j for current price level S_j. Notice that under a risk-neutral probability measure and prices relative to a tradable asset, there are theoretical arguments for setting $e = 0$, making the measure a martingale. With e expressing "real" expectations, the value of e is typically positive and depends on the market price of risk.

It turns out that minimizing the expected costs under RCC still corresponds to hedging at the strategy limits imposed by the cost constraint.

Proposition 5.8 (Minimizing expected cost under RCC). *Assume that asset prices follow paths in $\mathbb{I}^{u,d}$ and have an expected growth rate of $1+e(j,S_j)$ for price S_j at time t_j. The strategy that minimizes the expected value of costs (4.5) under a cost constraint V,*

$$g^* := \operatorname{argmin}_{g\in\mathbb{G}^V} E[Q^g(S)],$$

is given by

$$g_j^*(S_0,\ldots,S_j) = \begin{cases} \gamma_j^{\max}(S_j,H_j) & \text{if } e(j,S_j) > 0, \\ \gamma_j^{\min}(S_j,H_j) & \text{if } e(j,S_j) < 0, \end{cases}$$

where $\gamma_j^{\min}$ and $\gamma_j^{\max}$ are the RCC strategy bounds as defined in Theorem 5.2. □

The proof of this result can be found in the appendix. This result shows that the expected-cost criterion does not restrain us from strategies that are at the cost limit at each step.

5.5 Summary

In retrospect, we showed that in interval models the price of an option is in general not uniquely determined. A fair price for an option with a convex payoff may be any price in a compact interval. The lower bound of this interval is determined by the smallest price that can occur in the interval model using, for example, a stop-loss strategy, and the upper bound is determined by the largest price that can occur in the interval model using a delta-hedging strategy. In a simulation study, we showed that in a discrete-time setting with uncertain volatility the use of binomial tree models may severely underestimate the involved cost of hedging. This applies particularly when the hedging strategy underestimates the volatility of prices. Finally, we considered the question of how to find a hedging strategy for a call option that maximizes potential profits under the restriction that costs must not exceed an a priori given bound. We derived a numerical algorithm to calculate such a strategy under the assumption that this cost bound is not too strict. This strategy has the property that gained reserves beyond this cost bound are put at risk at every step. This outcome is on the one hand quite rational given our modeling framework. On the other hand, the strategy does not seem to be in line with the basic idea behind hedging, which is to reduce risk. Clearly, the strategy satisfies the strict requirement of not crossing the cost bound. Profit maximization, however, might not be the correct specification of the hedger's objective. To find hedging strategies that are perhaps more in line with the idea of risk reduction, one could look for different objective specifications or restrict the set of admissible strategies to an a priori defined class. Together with the issue of seeing how this theory works in practice, these are challenging subjects for future research.

Chapter 6
Appendix: Proofs

Proof of Proposition 4.2.

The proof proceeds by induction with respect to the number of periods N.

For $N = 1$ the cost of a strategy is given by $F(S_1) - \gamma_0(S_1 - S_0)$ for some real number $\gamma_0 = g_0(S_0)$, so it depends continuously on S_1. Since S_1 is restricted to an interval and since continuous functions map intervals to intervals, I^g must be an interval.

Next, assume that the proposition is true for models with fewer than N steps, and consider the total cost range I^g in an N-step model for some fixed strategy g. First consider the costs of price paths $\{S_0, \ldots, S_N\}$ with $S_N = S_{N-1}$. It follows from the induction hypothesis that the cost range of strategy g over these paths forms an interval, say, I'. Take $p \in I^g$ and let $\{S_0, \ldots, S_{N-1}, S_N\}$ be the corresponding path. Consider the paths $\{S_0, \ldots, S_{N-1}, \alpha S_N + (1-\alpha)S_{N-1}\}$ for $0 < \alpha < 1$. Since the corresponding costs depend continuously on α, they form an interval that contains p and that also contains at least one point of I'. Therefore, the set I^g may be written as a union of intervals that all have at least one point in common with the interval I', and so I^g is itself an interval.

If a strategy is continuous, then the cost function associated to it is continuous in the price paths. Because the set $\mathbb{I}^{u,d} \subset \mathbb{R}^{N+1}$ is compact, the cost function then achieves both its maximum and its minimum value on $\mathbb{I}^{u,d}$. □

The proof of Theorem 4.5 requires the following two technical lemmas.

Lemma 6.1. *Let u and d be such that $d < 1 < u$. If $h : (0,\infty) \mapsto \mathbb{R}$ is convex, then the function $\tilde{h}(x)$ defined for $x > 0$ by*

$$\tilde{h}(x) = \min_{\gamma \in \mathbb{R}} \max_{dx \le y \le ux} [h(y) - \gamma(y - x)] \tag{6.1}$$

is convex as well.

P. Bernhard et al., *The Interval Market Model in Mathematical Finance*, Static & Dynamic Game Theory: Foundations & Applications, DOI 10.1007/978-0-8176-8388-7_6,

Proof. Since $h(y)-\gamma(y-x)$ is convex as a function of y, the maximum in (6.1) must be taken at the boundary of the interval $[dx, ux]$, so

$$\tilde{h}(x) = \min_{\gamma} \max[h(dx)+\gamma(1-d)x, h(ux)-\gamma(u-1)x].$$

Since the first argument in the "max" operator is increasing in γ and the second is decreasing, the minimum is achieved when both are equal, that is to say, when γ is given by

$$\gamma = \frac{h(ux)-h(dx)}{(u-d)x}.$$

Therefore, we have the following explicit expression for $\tilde{h}$ in terms of h:

$$\tilde{h}(x) = \frac{1-d}{u-d}h(ux) + \frac{u-1}{u-d}h(dx).$$

Since the property of convexity is preserved under scaling and under positive linear combinations, it is seen from the preceding expression that the function $\tilde{h}$ is convex. □

Lemma 6.2. *Let $h(\cdot)$ be a convex function, and let u and d be such that $d<1<u$. Then we have*

$$\max_{\gamma\in\mathbb{R}} \min_{dx\le y\le ux} [h(y)-\gamma(y-x)] = h(x).$$

Proof. We obviously have

$$\min_{dx\le y\le ux} [h(y)-\gamma(y-x)] \le h(x)$$

for all γ since the value on the right-hand side is achieved on the left-hand side for $y=x$. So to complete the proof it suffices to show that there exists γ such that

$$h(y) \ge h(x)+\gamma(y-x)$$

for all y. Clearly, any subgradient of h at x has this property. □

Proof of Theorem 4.5.

1. The value function for the problem of minimizing worst-case costs is given by

$$V(S,j) = \min \max_{S_j=S} \left[F(S_N) - \sum_{k=j}^{N-1} \gamma_k (S_{k+1}-S_k) \right],$$

where the minimum is taken over all strategies and the maximum is taken over all paths in the given interval model that satisfy $S_j = S$. The value function satisfies the recursion

$$V(S, j-1) = \min_{\gamma} \max_{dS \le S' \le uS} [V(S', j) - \gamma(S' - S)],$$

and of course we have

$$V(S,N) = F(S).$$

It follows from Lemma 6.1 that the functions $V(\cdot, j)$ are convex for all j. Therefore, the strategy that minimizes the maximum costs is the same as the minmax strategy for the binomial tree model with parameters u and d, and the corresponding worst-case paths are the paths of this tree model.

2. The proof is *mutatis mutandis* the same as above; use Lemma 6.2 rather than Lemma 6.1.
3. This is by definition a consequence of items 1 and 2. □

Proof of Theorem 4.7.

Items 2 and 3 are clear from Theorem 4.5. One part of item 1 follows easily from the characterization of the consistent price interval as the intersection of all cost intervals. Indeed, if $\mathbf{Q}$ is a martingale measure, then $E_{\mathbf{Q}}F(S_N)$ is in the cost interval I^g for any strategy g since the expected result from any trading strategy under the martingale measure is zero. Thus $E_{\mathbf{Q}}F(S_N)$ is in the intersection of all cost intervals. To show that every such premium can be obtained as an expected value under some martingale measure, let $\mathbf{Q}^{\alpha}$ denote the martingale measure associated to the binomial tree $\mathbb{B}^{u,d}$ with parameters $u_{\alpha} := 1 + \alpha(u-1)$ and $d_{\alpha} := 1/u_{\alpha}$. For $0 \le \alpha \le 1$ the measure $\mathbf{Q}^{\alpha}$ is also a martingale measure on $\mathbb{I}^{u,d}$. The expected option value $f_{\alpha} := E_{\mathbf{Q}^{\alpha}}F(S_N)$ is continuous in α; moreover, $f_{\alpha} = f_{\min}$ for $\alpha = 0$ and $f_{\alpha} = f_{\max}$ for $\alpha = 1$. Hence every price $f \in [f_{\min}, f_{\max}]$ occurs as an expected option value under some martingale measure. □

Proof of Proposition 4.10.

For $N = 1$ it is obvious that worst cases are at the boundary of $S_1 = [dS_0, uS_0]$ with $u = u_{\tau}$ and $d = \frac{1}{u}$ and that these worst-case costs are convex in the initial price.

Similar to the proof of Theorem 4.5, it can be proved by induction that worst cases have extreme jumps and remain convex in the initial price for any number of time steps. □

Proof of Results Algorithm 4.9.

We prove that $H^{\min}(j,S_j)$ and $H^{\max}(j,S_j)$ are respectively the minimal and maximal current hedge costs $\Sigma_{k=0}^{j-1} - g_k(S_k)(S_{k+1} - S_k)$ over all paths in $\mathbb{I}^{u,d}$ that start in S_0 and end in S_j. This is obvious for $j = 0$ and $j = 1$. For $j > 1$, observe that I_j denotes the range of all prices at S_j that are compatible with given end points S_0 and S_{j+1} in $\mathbb{I}^{u,d}$. From the continuity of the strategy it follows that the minimum (4.16) and the maximum in the definition of $H^{\max}$ are well defined and indeed denote respectively the minimal and maximal current hedge costs. In particular, $H^{\min}(N,S_N)$ and $H^{\max}(N,S_N)$ denote respectively the minimal and maximal realized hedge costs over all paths in $\mathbb{I}^{u,d}$ that end in S_N. Once $S_N^{\min}$ and $S_N^{\max}$ are determined in (4.17) and its maximum analog, the correctness of (4.19) is obvious. The backward recursions for $S^{\min}$ and $S^{\max}$ simply reconstruct the paths corresponding to the interval bounds. □

To prove Theorem 5.2, we need the next two lemmas.

Lemma 6.3. *A strategy is compatible with the restriction on worst-case costs V if and only if along all paths the current latitude is always nonnegative, i.e.,*

$$g \in \mathbb{G}^V \text{ iff } \bar{V}_j(S_j,H_j) \geq 0 \,\forall\, S \in \mathbb{I}^{u,d},\ j = 0,\ldots,N-1.$$

Proof. As soon as $\bar{V}_j$ drops below zero for some j and some price path, there is a worst-case path in the tree $\mathbb{B}^{u,d}$ that brings total costs above level V. Conversely, the condition is sufficient for g to be in $\mathbb{G}^V$, as then for all price paths $S \in \mathbb{I}^{u,d}$, $\bar{V}_N = V - H_N - \bar{V}_N(S_N,H_N) = V - Q^g(S) \geq 0$. □

Lemma 6.4. *For all $S_j \in \mathbb{R}^+$, for all $0 \leq j \leq N$,*

1. *$f_j(S_j)$ is convex in S_j;*
2. *On $S_{j+1} \in [dS_j, uS_j]$, $f_{j+1}(S_{j+1}) - \Delta_j(S_{j+1} - S_j)$ has equal boundary maxima $f_j(S_j)$.* □

Proof of Theorem 5.2.

First notice that, as $V \geq f_0(S_0)$, $\mathbb{G}^V$ is nonempty and contains delta hedging.

Now suppose we apply a strategy $g_j(S_j,H_j)$ at t_j, i.e., we choose the portfolio $C - \gamma_j S$ at t_j, with γ_j the outcome of g_j, given the past price path (which determines, in turn, the realized hedge costs H_j). Then at t_{j+1}, $H_{j+1} = H_j - \gamma_j(S_{j+1} - S_j)$, with $S_{j+1} \in [dS_j, uS_j]$, and $\bar{V}_{j+1} = V - f_{j+1}(S_{j+1}) - H_{j+1}$. From Lemma 6.3 it follows that g_j is admissible if and only if $\bar{V}_{j+1} > 0$ for all $S_{j+1} \in [dS_j, uS_j]$. So the strategy position γ_j is admissible if and only if for all $S_{j+1} \in [dS_j, uS_j]$, $V - H_j - f_{j+1}(S_{j+1}) + \gamma_j(S_{j+1} - S_j) \geq 0$. Substituting $\gamma_j =: \Delta_j + \bar{\gamma}_j$ this gives $V - H_j - f_{j+1}(S_{j+1}) + (S_{j+1} - S_j)\Delta_j - (S_{j+1} - S_j)\bar{\gamma} \geq 0$. With Lemma 6.4 we obtain that the left-hand side of this inequality is a concave function in S_{j+1}, with boundary values $V - f_j(S_j) - H_j - \bar{\gamma}(d-1)S_j$ on the left (for $S_{j+1} = dS_j$) and $V - f_j(S_j) - H_j - \bar{\gamma}(u-1)S_j$ on the right (for $S_{j+1} = uS_j$). As $d - 1 < 0$ and $u - 1 > 0$, this induces an upper and lower bound for $\bar{\gamma}$, from which the strategy bounds follow. □

Proof of Proposition 5.3.

The existence of g^* and S^* is equivalent to the existence of subsequent solutions of the minimizations in the definition of BC_j. First we write out BC_{N-1}:

$$\mathrm{BC}_{N-1}(S_{N-1}, H_{N-1}) := \min_{\{\gamma_{N-1}\in\Gamma_{N-1}, S_N\in[dS_{N-1}, uS_{N-1}]\}} [S_N - X]^+ - H_{N-1} - \gamma_{N-1}(S_N - S_{N-1}),$$

with $\Gamma_{N-1} = [\gamma_{N-1}^{\min}, \gamma_{N-1}^{\max}]$

$$= \left[\Delta_{N-1}(S_{N-1}) - \frac{V - f_{N-1}(S_{N-1}) - H_{N-1}}{(u-1)S_{N-1}}, \Delta_{N-1}(S_{N-1}) + \frac{V - f_{N-1}(S_{N-1}) - H_{N-1}}{(1-d)S_{N-1}}\right].$$

For each S_{N-1}, H_{N-1} this involves minimization of a continuous function over a compact domain, so BC_{N-1} is well defined. Further, this domain of optimization is itself a continuous function of S_j and H_j, so BC_{N-1} is continuous itself on the entire domain given in (5.7).

The existence of solutions for $j = N-2, \ldots, 0$ now follows from an obvious inductive argument. □

To prove the correctness of the statements made in the results of Algorithm 5.4 we need the next lemma.

Lemma 6.5.
For each price path $S \in \mathbb{I}^{u,d}$,

$$\min_{g\in\mathbb{G}^V} Q^g(S) = Q^{g^\sharp}(S) \tag{6.2}$$

with the (noncausal "strategy") $g^\sharp$ defined by $\gamma^\sharp$,

$$g_j^\sharp(S_0, \ldots, S_j, S_{j+1}) := \gamma_j^\sharp.$$

Proof. We have

$$\mathrm{BC}_j(S_j, H_j) = \min_{S_{j+1}\in[dS_j, uS_j]} \left(\min_{\gamma_j\in\Gamma_j} \mathrm{BC}_{j+1}(S_{j+1}, H_j - \gamma_j(S_{j+1} - S_j))\right).$$

The inner minimization has solutions at the boundary of Γ_j [see (6.4) below for a motivation]. More precisely, the minimum is at the left bound if $S_{j+1} < S_j$ and at the right bound if $S_{j+1} > S_j$. Finally, notice that if $S_{j+1} = S_j$, the value of γ_j affects neither the hedge costs over the jth time step nor future cost constraint implications, so its value may be chosen arbitrarily in Γ_j. In particular, then, $\gamma^\sharp$ can be taken, arbitrarily, at the left boundary of Γ_j, without affecting optimality.

This implies that the right-hand side in (6.2) is a lower bound for the left-hand side: price jumps cannot be amplified by larger factors with the right sign under the RCC restriction V.

Equality does not follow immediately, as $g^\sharp$ is *not* a strategy, due to the fact that it anticipates whether an increase or decrease will follow the current asset price. To derive equality, observe that for each fixed price path S' there is a (by definition *causal*) strategy $g \in \mathbb{G}^V$ that coincides with $g^\sharp$ for that particular price path, namely,

$$g_j(S_0,\ldots,S_{j+1}) = \gamma_j = \begin{cases} \gamma_j^{\max} & \text{if } S'_{j+1} > S'_j \\ \gamma_j^{\min} & \text{if } S'_{j+1} \leq S'_j \end{cases}. \tag{6.3}$$

Then $g(S') - g^\sharp(S')$, and hence the outcome of costs are the same: $Q^g(S') = Q^{g^\sharp}(S')$. By taking S' a best-case price path (which exists according to Proposition 5.3), a causal strategy is obtained with the same best-case costs as $g^\sharp$. □

Proof of Correctness of Algorithm 5.4.

First we note a specific feature of the dynamic programming problem that underlies the first step:

$$\mathrm{BC}_j(S_j,H_j) \leq \mathrm{BC}_j(S_j,H_j+h) \;\; \forall\, h \geq 0.$$

In fact, it even holds that the difference in realized hedge costs h can be maintained until the final time because any strategy that is admissible under the RCC with initial state (S_j,H_j+h) at t_j is also admissible from a state with lower accrued hedge costs, so

$$\mathrm{BC}_j(S_j,H_j+h) \geq \mathrm{BC}_j(S_j,H_j)+h \;\; \forall\, h \geq 0. \tag{6.4}$$

From Lemma 6.5 we have now that due to this monotonicity in hedge costs we can reduce the double optimization over paths and strategies to a single one over price paths. This eliminates optimization over strategies in the best-case criterion (5.5).

A further reduction in computational complexity is achieved by selecting optimal paths among all those that recombine in the same price. In view of the previous results, this is simply a matter of comparing the realized "hedge" costs under $g^\sharp$ in each step.

Let $\mathbb{I}^{u,d}(j,s)$ denote the price paths in $\mathbb{I}^{u,d}$ with price s at time t_j, and let $H^*(j,s)$ denote the minimally achievable realized hedge costs for those paths under limit V

on worst-case costs. Then the optimal realized hedge costs at t_j for given asset price S_j are given by

$$\mathbb{I}^{u,d}(j,s) := \{S \in \mathbb{I}^{u,d} | S_j = s\}, \tag{6.5}$$

$$H^*(j,s) := \min_{\{g \in \mathbb{G}^V, S \in \mathbb{I}^{u,d}(j,s)\}} -\Sigma_{k=0}^{j-1} \gamma_k (S_{k+1} - S_k), \tag{6.6}$$

with $\gamma_k = g_k(S_0, \ldots, S_k)$ the outcome of the strategy for a given price path S.

Now (5.8) is trivial because hedge costs are zero before hedging starts. The formula for $H^*(1, S_1)$ follows from

$$H^*(1,S_1) = \min_{\gamma^{\min} \leq \gamma_0 \leq \gamma^{\max}} -\gamma_0(S_1 - S_0)$$

because $\mathbb{I}^{u,d}(1,s)$ consists of at most one path. This minimum is achieved for $\gamma_0 = \gamma_0^\sharp$ [see (5.11)], from which (5.9) follows.

To prove (5.10), observe that (6.6) can be rewritten as

$$H^*(j+1, S_{j+1}) = \min_{\{\gamma_j \in \Gamma_j, s \in I^j, S \in \mathbb{I}^{u,d}(j,s)\}} -\gamma_j(S_{j+1} - s) + H^*(j,s)). \tag{6.7}$$

For each fixed value s for S_j it is optimal to take $\gamma_j = \gamma_j^\sharp$, according to Theorem 5.2, and then (5.10) follows from the fact that the domain for s is indeed given by I^j.

Hence $H^*(N, S_N)$ denotes the minimal hedge costs that are compatible with final price S_N for strategies that are admissible by the RCC restriction. Thus S_N^* does indeed occur in a best-case price path. Further, $S_0^*, \ldots, S_{N-1}^*$ is the price path to S_N^* that realizes the minimal hedge costs $H^*(N, S_N)$ if g^* is applied. Thus indeed $\mathrm{BC}^*(S_0, V) = Q^{g^*}(S^*)$. □

Proof of Proposition 5.8.

Define the value function J_j for $j = 0, \ldots, N$ by

$$J_N := [S_N - X]^+ + H_N,$$

$$J_{j-1}(S_{j-1}, H_{j-1}) := \min_{\gamma_{j-1} \in \Gamma_{j-1}} E_j[J_j(S_j, H_{j-1} - \gamma_{j-1}(S_j - S_{j-1}))].$$

Then J_j denotes the expected costs at t_j under an optimal strategy as a function of the current asset price S_j and realized hedge costs H_j. So J is indeed a value function, and in particular, J_0 denotes the expected costs under optimal hedging.

To show that g^* is indeed a solution, we first derive that

$$J_j(S_j, H_j) = \beta_j H_j + h_j(S_j), \tag{6.8}$$

with $\beta_j \in \mathbb{R}^+$, and h_j a function of S_j that is independent of H_j. This is obviously true for $j = N$, with $\beta_N = 1$ and $h_N(S_N) = [S_N - X]^+ =: f_N(S_N)$. Now take (6.8) as induction hypothesis; then for $j-1$ we have, with the assumption $e = e(j-1, S_{j-1}) > 0$ and omitting the function arguments S_{j-1} and S_j in order to avoid confusion with multiplication,

$$\begin{aligned}
J_{j-1}(S_{j-1}, H_{j-1}) :=& \min_{\gamma_{j-1}\in\Gamma_{j-1}} E_j[J_j(S_j, H_{j-1} - \gamma_{j-1}(S_j - S_{j-1}))] \\
=& \min_{\gamma_{j-1}\in\Gamma_{j-1}} E_j[\beta_j(H_{j-1} - \gamma_{j-1}(S_j - S_{j-1})) + h_j] \\
=& \beta_j H_{j-1} - \max_{\gamma_{j-1}\in\Gamma_{j-1}} \gamma_{j-1}\beta_j e S_{j-1} + E_j[h_j] \\
=& \beta_j H_{j-1} - \gamma_{j-1}^{\max}\beta_j e S_{j-1} + E_j[h_j] \\
=& \beta_j H_{j-1} - \beta_j\left(\Delta_{j-1} + \frac{V - f_{j-1} - H_{j-1}}{(1-d)S_{j-1}} e S_{j-1}\right) + E_j[h_j] \\
=& \beta_j H_{j-1} - \beta_j \frac{e}{1-d}(\Delta_{j-1}(1-d)S_{j-1} + V - f_{j-1} - H_{j-1}) + E_j[h_j].
\end{aligned}$$

Now, with function arguments included, delta hedging has the property

$$(1-d)S_{j-1}\Delta_{j-1}(S_{j-1}) - f_{j-1}(S_{j-1}) = -f_j(dS_{j-1}).$$

Substituting the left-hand side in the last formula for J_{j-1}, we obtain

$$J_{j-1}(S_{j-1}, H_{j-1}) = \beta_j\left(1 + \frac{e}{1-d}\right)H_{j-1} + E_j[h_j(S_j)] + \beta_j \frac{e}{1-d}(f_j(dS_{j-1}) - V).$$

Now take $\beta_{j-1} := \beta_j(1 + \frac{e}{1-d})$ and $h_{j-1}(S_{j-1}) := E_j[h_j(S_j)] + \beta_j \frac{e}{1-d}(f_j(dS_{j-1}) - V)$; then $J_{j-1}(S_{j-1}, H_{j-1}) = \beta_{j-1}H_{j-1} + h_{j-1}(S_{j-1})$ with h_{j-1} being indeed independent of H_{j-1}. Hence, by induction, (6.8) must be valid for all j.

For negative $e(j-1, S_{j-1})$ the computations are analogous, with $\gamma^{\max}$ replaced by $\gamma^{\min}$. The derivation of the formula immediately reveals that g^* is indeed optimal. □

Part III
Robust-Control Approach to Option Pricing

Author: Pierre Bernhard
INRIA-Sophia Antipolis-Méditerranée,
France

Most of the material presented henceforth, except for Sect. 8.3.1, was obtained while the author was a professor at École Polytech'Nice, a department of the University of Nice-Sophia Antipolis, and Director of I3S, a joint laboratory of the university and CNRS.

In this part, we develop the full theory of European option hedging in the interval market model with transaction costs, equating "hedging" with worst-case design, yielding a "seller's price" (since the market model is incomplete). The theory is comprehensive concerning vanilla options, including continuous and discrete trading and the convergence of the latter to the former, as well as providing a fast algorithm. We also develop the more difficult case of digital options. In that case, the full theory still relies on two unproven, though numerically substantiated, conjectures.

The uniqueness theorem of Sect. 8.3.1 owes much to Naïma El Farouq of Blaise Pascal University, Clermont-Ferrand, France, and to Guy Barles of François Rabelais University, Tours, France. The analysis of digital options is a joint work with Stéphane Thiery, now with ENSAM-Lille, France. Chapter 10, which relies heavily on our joint publication [39], is essentially an English translation of part of the corresponding chapter of his unpublished French dissertation [142].

Discussions with the coauthors of this volume, and particularly with Jean-Pierre Aubin, have been instrumental in developing the ideas embodied in the theory presented here in its entirety for the first time.

Notation

- a^t: For any vector or matrix a: *a transposed*

Universal constants

- $\mathbb{R}$: The real line
- $\mathbb{N}$: The set of natural (positive) integers
- $1\!\!1$: A vector of any dimension with all entries equal to 1

Constants

- T: Exercise time [time]
- $\mathscr{K}$: Exercise price [currency]
- D: Exercise digital payoff [currency]
- μ_0: Riskless interest rate [time^{-1}]
- $\tau^- < 0$, $\tau^+ > 0$: Extreme relative rates of change of the underlying stock's price [time^{-1}]
- $\tau^\sharp = \max\{-\tau^-, \tau^+\}$ [time^{-1}]
- h: Time step for the discrete trading problem [time]
- $K \in \mathbb{N}$: $Kh = T$ [dimensionless integer]
- $\mathbb{K} = \{0, 1, \ldots, K-1\}$
- $\tau_h^- = \mathrm{e}^{\tau^- h} - 1$, $\tau_h^+ = \mathrm{e}^{\tau^+ h} - 1$ [dimensionless]
- C^-, C^+: Rate of proportional transaction cost for a sale, a buy, of underlying stock [dimensionless]
- $c^- \in [C^-, 0]$, $c^+ \in [0, C^+]$: Same as C^- and C^+ for closure costs [dimensionless]
- $\tau^\varepsilon, C^\varepsilon, c^\varepsilon$ See "special convention" below

Variables (with time)

- $t \in [0, T]$: Time
- $t_k = kh$: Trading time instants in discrete trading
- $R(t) = \mathrm{e}^{\mu_0(t-T)}$: An end-time value coefficient [dimensionless]
- $S(t)$: Underlying stock's market price [currency]
- $X(t)$: Number of underlying stocks in the hedging portfolio [dimensionless]
- $Y(t)$: Number of riskless bonds in the hedging portfolio, = the normalized worth of the riskless part of the portfolio [currency]
- $u(t) = S(t)/R(t)$: Normalized underlying stock's price [currency]
- $v(t) = X(t)S(t)/R(t)$: Normalized worth of underlying stock in the hedging portfolio, or *exposure* [currency]
- $w(t) = v(t) + Y(t)$: Normalized portfolio worth [currency]
- $V = \begin{pmatrix} v \\ w \end{pmatrix}, \quad x = \begin{pmatrix} u \\ v \end{pmatrix}, \quad y = \begin{pmatrix} t \\ u \\ v \end{pmatrix}, \quad z = \begin{pmatrix} t \\ u \\ v \\ w \end{pmatrix}$
- $\tau(t) = \dot{u}(t)/u(t) \in [\tau^-, \tau^+]$: Relative rate of change of $u(t)$ [time^{-1}]

- $(u_k, v_k, w_k) = (u(t_k), v(t_k), w(t_k))$
- $\tau_k = (u_{k+1} - u_k)/u_k$: One-step relative rate of change of u
- $\xi(t)$: Seller's control, rate of sale or buy of underlying stock [a distribution]
- ξ_c: Continuous component of $\xi(\cdot)$ [currency$\times$time^{-1}]
- ξ_k: Intensity of Dirac impulse in $\xi(\cdot)$ at time t_k (jump in v at time t_k) [currency]

Functions

- $M(u,v)$: Terminal payment of seller (bank) to option holder [currency]
- $N(u,v)$: Terminal payment augmented by closing costs [currency]
- $W(t,u,v)$: Optimal hedging portfolio worth (Value of the minimax game problem) [currency]
- $W_k^h(u,v)$: Value of the discrete time minimax game at time $t_k = kh$ [currency]
- $W^h(t,u,v)$: A carefully chosen interpolation of the sequence $\{W_k^h\}$ [currency]

Sets

- $\mathscr{U}$: Set of admissible underlying stock's price trajectories $u(\cdot)$
- Ω: Set of admissible relative rate of change histories $\omega = \tau(\cdot)$
- $\mathscr{U}_h$: Set of sequences = sampled elements of $\mathscr{U}$
- Ω_h: Set of admissible sequences $\{\tau_k\}_{k\in\mathbb{K}}$ (discrete trading)
- Ξ: Set of admissible seller's controls
- Φ: Set of admissible seller's strategies (nonanticipative mappings $\mathscr{U} \to \Xi$)

Auxiliary quantities

- $\nu^t = (n\ p\ q\ r) = -(\frac{\partial W}{\partial t}\ \frac{\partial W}{\partial u}\ \frac{\partial W}{\partial v}\ 1)$: Semipermeable normal
- $\sigma = pu + (1+q)v$: Switch function whose opposite sign $\eta = \text{sign}(-\sigma)$ dictates the optimum τ as $\tau^\star = \tau^\eta$
- $q^-(t,u)$, $q^+(t,u)$ (jointly $q^\varepsilon(t,u)$): Two slopes appearing in representation formula, with different (closed-form) formulas in vanilla and digital cases [dimensionless]
- $Q^\varepsilon = (q^\varepsilon\ 1)$, $\varepsilon \in \{-,+\}$: [Dimensionless line vector]
- $Q = \begin{pmatrix} Q^+ \\ Q^- \end{pmatrix} = \begin{pmatrix} q^+ & 1 \\ q^- & 1 \end{pmatrix}$: [Dimensionless 2×2 matrix]
- $\check{v}(t,u)$, $\check{w}(t,u)$: Two functions, jointly solution of fundamental PDE, appearing in the representation formula [currency]
- $\check{V}(t,u)$: [2-vector]

$$\check{V}(t,u) = \begin{pmatrix} \check{v}(t,u) \\ \check{w}(t,u) \end{pmatrix} \qquad \text{[currency]}$$

- $\mathscr{T}(t,u)$: [2×2 matrix]

$$\mathscr{T} = \frac{1}{q^+ - q^-}\begin{pmatrix} \tau^+q^+ - \tau^-q^- & \tau^+ - \tau^- \\ -(\tau^+ - \tau^-)q^+q^- & \tau^-q^+ - \tau^+q^- \end{pmatrix}. \qquad [\text{time}^{-1}]$$

- $\mathscr{S}$: [2 × 2 constant matrix]

$$\mathscr{S} = \begin{pmatrix} 1 & 0 \\ 1 & 0 \end{pmatrix} : \qquad \text{[dimensionless]}$$

- t_ε: Last time when $q^\varepsilon = C^\varepsilon$ and when the seller would use an impulse of sign ε
- $t_m = \min\{t_-, t_+\}$, $t_M = \max\{t_-, t_+\}$ [time]

Manifolds in (t, u, v, w) space (geometric theory)

3D hypersurfaces

- $\{\tau^\varepsilon\}$, $\varepsilon \in \{-,+\}$: Regular semipermeable manifold with $\tau = \tau^\varepsilon$
- $\{\mathscr{K}\}$: Singular semipermeable manifold (digital)
- $\{\iota^\varepsilon\}$, $\varepsilon \in \{-,+\}$: Jump semipermeable manifold of sign ε

2D manifolds

- $\mathscr{B}$: "Basic" manifold $\check{V} = 0$
- $\mathscr{C}$: (Digital call) "Constant" manifold $\check{V} = (0 \;\; D)^t$
- $\mathscr{D}$: Dispersal manifold, $\{\tau^-\} \cap \{\tau^+\}$ (vanilla) or $\{\tau^-\} \cap \{\mathscr{K}\}$ (digital)
- $\mathscr{E}$, $\mathscr{E}^\varepsilon$: Envelope junction of a jump manifold of sign ε with a regular (vanilla and digital) or singular (digital) semipermeable manifold
- $\mathscr{F}$: Focal manifold, junction of two jump manifolds

Special convention

- $Z^\varepsilon\langle\text{expression}\rangle$, for $Z \in \{\tau, C, c, q, Q\}$, $\quad = Z^{\text{sign(expression)}} \times (\text{expression})$

Chapter 7
Continuous and Discrete-Time Option Pricing and Interval Market Model

7.1 Introduction

Recently, there has been a large body of literature on robust control optimization, with applications to various fields including mathematical finance. See, e.g., [27, 41, 42]. Most aim to exploit the power of modern computer tools to solve complex problems whose stochastic formulation is essentially out of reach. Instances of such works will appear in the later parts of this volume. In contrast, we concentrate in this part on the simplest problems – one underlying asset and the less sophisticated interval model – and strive to push the analytic investigation as far as possible. In some sense, being very pretentious, this may be seen as a counterpart, in the robust control option pricing literature, of the Black–Scholes theory in the stochastic literature, the representation theorem playing the role of the Black–Scholes closed-form pricing formula.

The role of probabilities in finance is further discussed in other parts of the book. But special mention must be made of the outstanding book by Shafer and Vovk [136]. They start from the same analysis of hedging as we do in terms of a game against "nature," very much in the spirit of the present book. But where we conclude that we can do without probability theory, they claim that this *is* probability theory. Or rather they claim that this can be an alternative to Kolmogorov's measure-theoretical foundation of probability theory, and they proceed to recover many results, such as asymptotic and ergodic theorems, from this new viewpoint. Our seller's price is their "upper expectation" – of which they claim, as do we implicitly, that it is the price likely to be found on the market. Yet, they are more interested in recovering classical probabilities and elaborating on the classic Black–Scholes theory than in providing alternative models and tackling the problems of transaction costs and discrete trading that we consider. But the relationship of that theory to our model deserves further consideration.

P. Bernhard et al., *The Interval Market Model in Mathematical Finance*, Static & Dynamic Game Theory: Foundations & Applications, DOI 10.1007/978-0-8176-8388-7_7,

7.1.1 A New Theory of Option Pricing?

In a series of papers [34, 38–40, 70, 143], we introduced a robust control approach to option pricing and, more specifically, to the design of a hedging portfolio and management strategy using the interval model for the market and a robust control approach to hedging. An overview of the theory available at that time appeared in [35], and the most complete account to date is in [142]. But no complete account is available in print yet.

This theory arose from a failed attempt to extend to the discrete trading case the probability-free robust control approach of Chap. 2 with no assumption on the sequence of market prices of the underlying stock. It soon became clear that unbounded admissible price increments made the buy-and-hold strategy the only hedging strategy. Incidentally, this is also what precludes a satisfactory discrete trading strategy with Samuelson's model. Hence the need for the interval model henceforth, which we also extended to the continuous trading scheme.

We feel that the assumptions of this model – that we know a bound on the possible one-step relative price increments or on the relative price rate of change in the continuous trading case – while not very satisfactory, imply much less knowledge on future prices than the assumption made in the classic Black–Scholes theory – that the process is a geometric diffusion, a process with unbounded variation and thus *not graphically representable*, and, moreover, that we know its exact volatility. This is indeed an extremely rich set of assumptions on the future price history. For instance, it implies that its total squared relative variation will be exactly $\sigma^2 t$, which fact we used in Chap. 2 to derive the Black–Scholes equation.

Admittedly, much work has been done to relax this too precise description using variable volatilities, in particular stochastic volatilities. This adds more probabilities to the model and, in so doing, in some sense more – or at least not less – information on this unknown future. There have also been attempts at an "interval" description of volatility; see, e.g., [18]. But then the worst case is systematically the highest possible value.

The interval model certainly assumes less a priori knowledge on future price histories. It can turn out to be violated, of course – and we will examine in Chap. 10 why this is not too much of a concern – but this will be easy to monitor, contrary to the Samuelson model, which will undoubtedly always prove to be violated one way or another one; even if the price history turns out to be almost as irregular as a geometric diffusion, its "volatility" will never exactly match the volatility assumed, this being difficult to monitor in real time.

A more detailed comparison of the interval model with Samuelson's model was provided in the previous part. The clear conclusion, though, is that the new theory is not a priori unnatural. Is it rich enough to be useful? Its main claims to being so are, on the one hand, that it allows for constructing a consistent theory of hedging portfolios with either a continuous- or discrete-time trading paradigm, the former being the limit of the latter for vanishing time steps, with one and the same (continuous-time) market model,[1] and, on the other hand, accommodating transaction costs and closing costs in a natural way, even if large, with a nontrivial hedging portfolio. Indeed, our theory naturally accommodates transaction costs to such an extent that, without them, in the continuous-time theory – incidentally, this is just as unrealistic a model as the total absence of transaction costs – our hedging strategy degenerates into a stop-loss strategy. Thus this model allows one to circumvent the classic impossibility [138], at the cost of being incomplete, obliging us to resort to super replication and a seller's price. (This was analyzed in more detail in the previous part.)

Let us add that our "fast algorithm" makes computing a safe approximation of the optimal hedging strategy and of the associated premium almost as fast as computing the Black–Scholes optimal strategy and premium, to the point where the power of modern personal computers hides any difference in computing time.

Perhaps we could claim as positive points the ease with which we can deal with American options – a single line of code to add in our "standard algorithm" – and the possibility of accommodating a time delay in the information available to the discrete trader. See Sect. 7.3 below.

Yet, this is not, by far, as mature a theory as Black–Scholes, of course. Among other shortcomings at this time, we do not yet have detailed and comparative validation data. A first investigation in that direction is reported in Chap. 10. And, in contrast with the theory developed in Part II, but essentially in accordance with the rest of this book, we have only investigated the *seller's price* of our model – the maximum point of Part II's fair price interval – i.e., the price that the seller must charge to make sure to always hedge his costs, that is, as long as the market does not violate the the model's hypotheses. Undoubtedly, much work remains to be done and much practice accumulated before it can challenge the established theory, if ever.

7.1.2 Related Contributions

As mentioned in Chap. 2, McEneaney may have been the first to replace the stochastic framework with a robust control approach. See the introduction there.

The previous part dealt in detail with the approach of Cox et al. [57] where probabilities seem to come in as an implicit byproduct of a binary tree approximation of the geometric diffusion. This will be more extensively commented in Part IV of the book, where Kolokoltsov, following earlier work of his, generalizes that approach.

---

[1] Notice that the quite remarkable Cox–Ross–Rubinstein model does not claim any realism with a finite step size and only aims to recover the Black–Scholes theory in the limit as the step size goes to zero, in a fashion that cannot be interpreted as the sampling of a fixed continuous-time market model.

We took the phrase "interval model" from Roorda et al. [131, 132] and from their work presented in the previous part.

Yet the contribution the closest to ours, as far as we know, is that of Aubin et al., as initiated by the Ph.D. dissertation of Dominique Pujal [128] and further developed in [17] and using computer tools. This will be presented in Part V of this volume. They have a more general model, as shown by their dealing with complex options, on which they bring to bear the powerful mathematical machinery of viability theory. In the continuous case with transaction costs, they essentially ignore jumps in the portfolio – which is of no consequence in deriving a discrete-time algorithm. More importantly, our algorithmic approaches differ: while our "fast algorithm" is specifically tailored for the problem of hedging a European claim, they use a very general "guaranteed capture basin" algorithm, which is less efficient for simple European options but can be extended to a large class of complex options.

Similar thoughts were developed by Olsder [123], although this is only a preliminary analysis, as acknowledged by the author. Barles and Soner [23] developed a different approach to option pricing that lets them deal with transaction costs, but then the premium they arrive at depends on the rest of the trader's portfolio.

More recently, systematic investigations into the use of robust control in lieu of stochastic control have led authors to consider the option pricing problem. See, e.g., [41] and the references therein.

7.2 Modeling

We develop here the fundamental model of our problem, both in terms of market model and portfolio model, and therefore how the game theoretic approach comes in instead of the more traditional stochastic approach.

7.2.1 Contingent Claims

We offer a rather general presentation of European and American contingent claims, or "derivatives"; a more precise description of the most basic ones will be given in the next chapter. Contingent claims are contracts sold by a seller to a buyer. The object of the theory is to determine the "optimum" price, or *premium*, that the seller should charge for that contract.

In a European contingent claim, a "seller" sells to a "buyer" a contract per which the latter may, at a given *exercise time* T, exert a right to buy or sell a given commodity at a predetermined price, or receive a certain amount of money. The crucial point is that the exact transaction to be exerted or not may bear a cost to the seller, which depends on the state of the market at exercise time, usually the market price $S(T)$ of an "underlying stock" prescribed by the contract. We let

$M(S(T))$ be the cost to the seller, excluding closing transaction costs, which we will cover later on.

In an American contingent claim, the buyer may choose when to opt to exercise his right, anywhere between the initial time and T. We will show in Sect. 7.3 how to deal with that case. The added difficulty with respect to a European option is much less severe than in the stochastic theory. Yet we have not yet fully fleshed it out. (The computer code implementing our "standard algorithm" solves that case easily.)

7.2.2 *Market*

7.2.2.1 Riskless Interest Rate

We assume a fixed, known, riskless interest rate μ_0 characteristic of the given economy. In a classic fashion, all monetary values will be expressed in end-time value computed at that fixed riskless rate, so that, without loss of generality, the riskless rate will be taken as (seemingly) zero. (It reappears in the theory of American options; see Sect. 7.3) For this to happen, let

$$R(t) := \mathrm{e}^{\mu_0(t-T)} \tag{7.1}$$

be our end-value discounting coefficient.

The market contains a *riskless bond*, an asset with return rate μ_0, i.e., whose monetary value at time t is $S_0(t) = R(t)S_0(T)$.[2]

Concerning the risky underlying asset, we share with the other parts of the book, particulary Part II, the view that a market model is a set of possible underlying stock price trajectories $[0,T] \to \mathbb{R}^+$: $t \mapsto S(t)$. As stated previously, we will use, instead of $S(t)$, its *normalized* price (its end-time value at rate μ_0), i.e.,

$$u(t) = \frac{S(t)}{R(t)}. \tag{7.2}$$

7.2.2.2 Continuous Trading

Let $\mathscr{U}$ be the set of possible normalized price trajectories. Our model is defined by two real numbers $\tau^- < 0$ and $\tau^+ > 0$, and $\mathscr{U}$ is the set of all absolutely continuous functions $u(\cdot)$ such that for any two time instants $t_1 \le t_2$

$$\mathrm{e}^{\tau^-(t_2-t_1)} \le \frac{u(t_2)}{u(t_1)} \le \mathrm{e}^{\tau^+(t_2-t_1)}. \tag{7.3}$$

[2] $S_0(T)$ could, without loss of generality, be taken as equal to one. We avoid that convention to keep track of the dimension of the variables. Y is dimensionless, and $S_0(T)$ is a currency value.

(Notice that the use of $u(t_2)/u(t_1)$ instead of $S(t_2)/S(t_1)$ only translates the interval $[\tau^-,\tau^+]$ by μ_0.)

The notation τ^ε will be used to handle both τ^+ and τ^- at one time. Hence, in that notation, it is understood that $\varepsilon \in \{-,+\}$, sometimes identified to $\{-1,+1\}$. Moreover, as with any other symbol with a superscript ε (C^ε, c^ε, q^ε, Q^ε), the notation

$$\tau^\varepsilon\langle expression\rangle$$

will always mean that τ^ε is to be replaced by τ^- if *expression* < 0 and by τ^+ if *expression* > 0. Finally, we will let

$$\tau^\sharp = \max\{-\tau^-,\tau^+\}.$$

In the continuous trading theory, we will use the equivalent characterization

$$\dot{u} = \tau u, \quad \tau \in [\tau^-,\tau^+]. \tag{7.4}$$

In that formulation, $\tau(\cdot)$ is a measurable function that plays the role of the "control" of the market. We will let Ω denote the set of measurable functions from $[0,T]$ to $[\tau^-,\tau^+]$. It is equivalent to specifying a $u(\cdot) \in \mathscr{U}$ or a $(u(0),\tau(\cdot)) \in \mathbb{R}^+ \times \Omega$. This is an a priori unknown time function. The concept of nonanticipative strategies embodies that fact.

7.2.2.3 Discrete Trading

In the discrete trading theory, we will call $h = T/K$ our time step, K being an integer. Hypothesis (7.3) translates into[3]

$$u(t+h) \in [e^{h\tau^-}u(t), e^{h\tau^+}u(t)].$$

For convenience, we let

$$u(t+h) = (1+\tau(t))u(t), \quad \tau(t) \in [\tau_h^-,\tau_h^+], \tag{7.5}$$

with

$$\tau_h^\varepsilon = e^{h\tau^\varepsilon} - 1, \quad \varepsilon \in \{-,+\}. \tag{7.6}$$

Alternatively, we will write, for any integer k, $u(kh) = u_k$, so that (7.5) also reads

$$u_{k+1} = (1+\tau_k)u_k, \quad \tau_k \in [\tau_h^-,\tau_h^+], \tag{7.7}$$

[3]As opposed to $u(t+h) \in \{e^{h\tau^-}u(t), e^{h\tau^+}u(t)\}$ in the Cox–Ross–Rubinstein theory.

and we let $\mathscr{U}_h$ denote the set of admissible sequences $\{u_k\}$. Note that they are exact sampled functions of $\mathscr{U}$, not Euler approximations of some kind. We also let Ω_h denote the set of admissible sequences $\{\tau_k\}$.

The case where h goes to zero will be of interest also. But, contrary to the classic limit process in the Cox–Ross–Rubinstein theory, we retain the underlying continuous-time model, hence here τ^+ and τ^-, fixed. Then τ_h^ε goes to zero as $h\tau^\varepsilon$.

7.2.3 *Portfolio*

At the initial time, when the contract is sold, the seller forms a *hedging portfolio* made of $X(0)$ shares of the underlying stock, valued at the market price $S(0)$, for a normalized end value of $v(0) = X(0)u(0) = X(0)S(0)/R(0)$, and $Y(0)$ normalized riskless *bonds*. The total value of his initial portfolio is thus, in normalized units, $w(0) = v(0) + Y(0)S_0(T)$. He then manages this portfolio in a way that will be discussed hereafter, so that $v(t)$ and $w(t)$ will vary over time.

Notice that, as usual, we allow $X(t)$, and therefore $v(t)$, or $Y(t)$ to be negative if the seller decides to "short" his portfolio to get cash or, conversely, to borrow cash to build a "long" portfolio.

Assumption 7.1. We will assume that the underlying stock, as well as the bonds, are infinitely divisible, so that $X(t)$, $v(t)$, $Y(t)$, and $w(t)$ may all be considered real variables.

7.2.3.1 Buying and Selling

Continuous Trading

The seller will want to buy or sell (parts of) shares of the underlying stock.

Assumption 7.2. We assume that the market is always liquid and can provide or absorb the amount the seller wants to trade, at a price independent of that amount.

Remark 7.3. The assumption of an always liquid market could prove to be unrealistic if the seller was to attempt very large transactions. It will be important to notice that the optimal portfolio management strategies we will find *do not* imply such large transactions. Otherwise, it could be necessary to restrict their amount, or $v(t)$.

We let $\xi(t)$ be the buying rate [a sale if $\xi(t) < 0$], which is the trader's control. Therefore, we have, in continuous time,

$$\dot{v} = \tau v + \xi. \tag{7.8}$$

However, there is no reason why one should restrict the buying/selling rate, so that there is no bound on ξ. To avoid mathematical ill-posedness, we explicitly admit an "infinite" buying/selling rate in the form of instantaneous purchase or sale of a finite amount of stock at time instants chosen by the trader together with the amount. Thus the control of the trader also involves the choice of finitely many time instants t_k and trading amounts ξ_k, and the model must be augmented with

$$v(t_k^+) = v(t_k^-) + \xi_k, \tag{7.9}$$

meaning that $v(\cdot)$ has a jump discontinuity of size ξ_k at time t_k. Equivalently, we may retain formula (7.8) but allow for impulses $\xi_k \delta(t - t_k)$ in $\xi(\cdot)$. We will therefore let $\xi(\cdot) \in \Xi$, the set of distributions defined over $[0, T]$ that are the sum of a measurable function $\xi_c(\cdot) \in \mathscr{M}([0, T], \mathbb{R})$ and a finite number of weighted translated Dirac impulses $\xi_k \delta(t - t_k)$:

$$\xi(\cdot) \in \Xi = \{\mathscr{M}([0, T], \mathbb{R}) + \text{finite sums of weighted translated Dirac impulses}\}. \tag{7.10}$$

To avoid any ambiguity, whenever needed, we will call ξ_c the "continuous trading" (or measurable) part of ξ and let

$$\xi(t) = \xi_c(t) + \sum_k \xi_k \delta(t - t_k). \tag{7.11}$$

Thus, the portfolio model can be written either as (7.8), (7.11) or as (7.9), (7.12):

$$\dot{v}(t) = \tau(t)v(t) + \xi_c(t), \quad v(t_k) = v(t_k^+) \quad \text{if} \quad t \in (t_k, t_{k+1}). \tag{7.12}$$

We emphasize that, in the continuous theory, the t_k are the choice of the seller, together with the ξ_k and $\xi_c(\cdot)$. We only require that they be isolated, hence in finite number over $[0, T]$.

Discrete Trading

The discrete trading case can be seen as a sequence of jumps at prescribed time instants $t_k = kh$, $k \in \mathbb{K} = \{0, 1, \ldots, K{-}1\} \subset \mathbb{N}$, and $h \in \mathbb{R}^+$ a prescribed time step, such that $Kh = T$. Writing u_k, v_k, w_k for $u(kh), v(kh), w(kh)$, this leads to

$$v_{k+1} = (1 + \tau_k)(v_k + \xi_k). \tag{7.13}$$

7.2.3.2 Transaction Costs

We assume that there are transaction costs, proportional to the transaction amount. But we allow for different proportionality coefficients for a purchase or sale of underlying stock. Hence let C^+ be the cost coefficient for a purchase and $-C^-$ for a sale, so that the cost of a transaction of amount ξ is $C^\varepsilon \langle \xi \rangle$ (with $\varepsilon = \text{sign}(\xi)$ according to our standing convention). We have chosen C^- negative, so that the formula always gives a positive cost, as it should.

Assumption 7.4. We will only consider cases where $|C^\varepsilon| < 1$, $\varepsilon = -, +$.

Our portfolio will always be assumed to be *self-financed*, i.e., the sale of one the commodities, underlying stock or riskless bonds, must exactly cover the purchase of the other one *and* the transaction costs. It is a simple matter to see that the worth w of the portfolio then obeys

$$\dot{w} = \tau v - C^{\varepsilon}\langle \xi \rangle\,, \tag{7.14}$$

i.e., at jump instants,

$$w(t_k^+) = w(t_k^-) - C^{\varepsilon}\langle \xi_k \rangle\,, \tag{7.15}$$

and between two jumps

$$\dot{w} = \tau v - C^{\varepsilon}\langle \xi_c \rangle\,, \quad w(t_k) = w(t_k^+) \quad \text{if} \quad t \in (t_k, t_{k+1}). \tag{7.16}$$

This is equivalent to

$$\begin{aligned} w(t) &= w(0) + \int_0^t (\tau(s)v(s) - C^{\varepsilon}\langle \xi(s) \rangle)\mathrm{d}s, \\ &= w(0) + \int_0^t (\tau(s)v(s) - C^{\varepsilon}\langle \xi_c(s) \rangle)\mathrm{d}s - \sum_{k|t_k<t} C^{\varepsilon}\langle \xi_k \rangle\,. \end{aligned} \tag{7.17}$$

In the discrete trading case, this simplifies to

$$w_{k+1} = w_k + \tau_k(v_k + \xi_k) - C^{\varepsilon}\langle \xi_k \rangle\,, \tag{7.18}$$

leading to

$$w_n = w_0 + \sum_{k=0}^{n-1} [\tau_k(v_k + \xi_k) - C^{\varepsilon}\langle \xi_k \rangle]. \tag{7.19}$$

A *dynamic portfolio* will be a pair of time functions $(v(\cdot), w(\cdot))$, whether time is continuous or discrete, also denoted $(\{v_k\}, \{w_k\})_{k\in\mathbb{K}}$ in the latter case, satisfying the preceding equations.

7.2.4 Hedging

7.2.4.1 Strategies

The initial portfolio is to be created at time 0. As a consequence, the seller's price is obtained by taking $v(0^-) = 0$. Then, formally, admissible hedging strategies will be functions $\varphi : \mathscr{U} \to \Xi$ that enjoy the property of being nonanticipative:

$$\forall (u_1(\cdot), u_2(\cdot)) \in \mathscr{U} \times \mathscr{U}\,, \quad [u_1|_{(0,t)} = u_2|_{(0,t)}] \Rightarrow [\varphi(u_1(\cdot))|_{[0,t]} = \varphi(u_2(\cdot))|_{[0,t]}].$$

(It is understood here that the restriction of $\delta(t-t_k)$ to a closed interval not containing t_k is 0, and its restriction to a closed interval containing t_k is an impulse.) This definition, classic in the dynamic game literature, embodies the fact that future prices, or price fluctuations, are not known by the seller. Therefore, nonanticipative strategies play the role that adapted strategies play in the stochastic framework.

In practice, we will find optimal hedging strategies composed of a jump at the initial time, followed by a state feedback law $\xi(t) = \phi(t, u(t), v(t))$ containing jumps when the state (t, u, v) crosses certain boundaries.

In discrete time, the situation is much simpler. We need only a nonanticipative strategy $\varphi : \mathscr{U}_h \to \mathbb{R}^{\mathrm{T}}$ giving $\xi_k = \varphi_k(u_0, u_1, \ldots, u_k)$. Again, we will find it in the form of a state feedback $\xi_k = \phi_k(u_k, v_k)$. And yet these are only nonanticipative laws, the equivalent of stochastic adapted strategies. We will see in Sect. 7.3 how to handle *strictly* nonanticipative strategies, the equivalent of stochastic predictable strategies. (But this piece of theory has yet to be investigated in full detail.)

We will call Φ the set of admissible continuous trading strategies and Φ_h the set of admissible discrete trading strategies.

7.2.4.2 Terminal Costs

At exercise time T, the seller must honor his contract with the buyer and then close his position. This terminal sale or purchase incurs transaction costs as well. But we allow for a different rate for this terminal transaction. This is possibly to represent a compensation effect that lowers the price of the transaction, and also to let us deal easily with the case where there are no closure costs, by setting the rate to zero. Let, therefore, $c^- \in [C^-, 0]$ and $c^+ \in [0, C^+]$ be the terminal transaction costs, similar to C^- and C^+ in the running phase of the contract.

Notice that as a consequence of hypothesis 7.4, $|c^\varepsilon| < 1$, $\varepsilon = -, +$.

Hence, the seller will face a terminal payment equal to

$$N(u(T), v(T)) = M(u(T)) + \text{closure transaction costs.}$$

The detailed expressions for these payments depend on the particular contracts and will be given in due time.

7.2.4.3 Hedging Portfolio

An initial portfolio $(v(0), w(0))$ and an admissible trading strategy φ, together with a price history $u(\cdot)$, generate a dynamic portfolio. We establish the following definition.

Definition 7.5. An initial portfolio $(v(0) = 0, w(0) = w_0)$ and a trading strategy φ constitute a *hedge at* u_0 if for any $u(\cdot) \in \mathscr{U}$ such that $u(0) = u_0$ (or, equivalently, for any admissible $\tau(\cdot) \in \Omega$) the dynamic portfolio thus generated satisfies

$$w(T) \geq N(u(T), v(T)). \tag{7.20}$$

Continuous Trading

Now, we may use (7.17) at time T to rewrite this as

$$\forall \tau(\cdot) \in \Omega, \quad N(u(T), v(T)) + \int_0^T \Big(-\tau(t)v(t) + C^\varepsilon \langle \xi(t) \rangle \Big) \mathrm{d}t - w_0 \leq 0.$$

This in turn is clearly equivalent to

$$w_0 \geq \sup_{\tau(\cdot)\in\Omega} \left[N(u(T), v(T)) + \int_0^T \Big(-\tau(t)v(t) + C^\varepsilon \langle \xi(t) \rangle \Big) \mathrm{d}t \right].$$

To ensure that no arbitrage possibility exists, we establish the following definition.

Definition 7.6. A *seller's price* of an option at u_0 is the worth of the cheapest hedging portfolio at u_0.

The seller's price (or *premium*) at u_0 is therefore

$$P(u_0) = \inf_{\varphi\in\Phi} \sup_{\tau(\cdot)\in\Omega} \left[N(u(T), v(T)) + \int_0^T \Big(-\tau(t)v(t) + C^\varepsilon \langle \xi(t) \rangle \Big) \mathrm{d}t \right], \tag{7.21}$$

where it is understood that $u(0) = u_0$ and $v(0) = 0$ and that $\xi(\cdot) = \varphi(u_0, \tau(\cdot))$.

Discrete Trading

In the case of discrete trading, we obtain, as the seller's price at u_0,

$$P(u_0) = \inf_{\varphi\in\Phi} \sup_{\{\tau_k\}\in\Omega_h} \left[N(u_K, v_K) + \sum_{k=0}^{K-1} \Big(-\tau_k(v_k + \xi_k) + C^\varepsilon \langle \xi_k \rangle \Big) \right]. \tag{7.22}$$

7.2.5 Conclusion: A Minimax Dynamic Game

7.2.5.1 Value Function

In conclusion, and in keeping with the standard approach of robust control theory, in that framework, Merton's principle of pricing a contingent claim according to the value of a hedging portfolio leads to dynamic minimax games, (7.21) or (7.22) depending on whether we allow continuous or discrete transactions. To investigate them with the classic tools of dynamic game theory, we define the Value functions associated with these games.

Continuous Time

The continuous-time criterion is thus

$$J(t_0,u(t_0),v(t_0);\varphi,\tau(\cdot)) = N(u(T),v(T)) + \int_{t_0}^{T}\Big(-\tau(t)v(t) + C^{\varepsilon}\langle\xi(t)\rangle\Big)\mathrm{d}t,$$

where it is understood that the functions $(u(\cdot),v(\cdot))$ are generated by the dynamics (7.4), (7.8) from $u(t_0)$ and $v(t_0)$, with $\xi(t) = \varphi(\tau(\cdot))(t) = \xi_c(t) + \sum_k \xi_k \delta(t-t_k)$. And we will use the Value function of that game:

$$W(t,u,v) = \inf_{\varphi\in\Phi}\sup_{\tau(\cdot)\in\Omega} J(t,u,v;\varphi,\tau(\cdot)).$$

Discrete Time

In the discrete-time problem, we have in the same fashion

$$J(\ell,u_\ell,v_\ell;\{\tau_k\},\varphi) = N(u_K,v_K) + \sum_{k=\ell}^{K-1}\Big(-\tau_k(v_k+\xi_k) + C^{\varepsilon}\langle\xi_k\rangle\Big),$$

where it is understood that the sequences $\{u_k\}$ and $\{v_k\}$ are generated from u_ℓ and v_ℓ by the dynamics (7.7), (7.13) and $\{\xi_k\} = \varphi(\{\tau_k\})$, and

$$W_\ell^h(u,v) = \inf_{\varphi\in\Phi}\sup_{\{\tau_k\}\in\Omega_h} J(\ell,u,v;\varphi,\{\tau_k\}).$$

Interpolation

This forms a sequence of scalar functions of (u,v), not to be mistaken for $W^h(t,u,v)$, an interpolation of the $W_\ell^h(u,v)$ obtained as follows. While $W_\ell^h(u,v)$ is the minimax value obtained from the instant $t = \ell h$ and the state (u,v) when the minimizer is restricted to impulses at the discrete time instants $t_k = kh$, $W^h(t,u,v)$ is the minimax value obtained from any time t and the state (u,v) when the minimizer is allowed to perform an impulse at time t, and then only impulses at the discrete time instants $t_k = kh \geq t$. Clearly, $W^h(t_k,u,v) = W_k^h(u,v)$, and because it will follow from standard theorems that $t \mapsto W^h(t,u,v)$ is continuous, we indeed have a continuous interpolation of the sequence of functions $\{W_k^h\}_k$. The important point is that the interpolated function $W^h(t,u,v)$ is itself the Value function of an "interpolated game".

Pricing

According to formulas (7.21) and (7.22), the lowest premium that allows a seller to hedge his risk, assuming hypothesis (7.3) is satisfied, is at time t (exercise time T is assumed given) and the current underlying stock price S:

$$P(t,S) = W(t, \mathrm{e}^{\mu_0(T-t)}S, 0), \qquad \text{or} \qquad P(t,S) = W^h(t, \mathrm{e}^{\mu_0(T-t)}S, 0), \tag{7.23}$$

depending on whether trading is assumed continuous or discrete.

7.2.5.2 Notation Shorthand

We will in some instances use the following shorthand notation:

$$x = \begin{pmatrix} u \\ v \end{pmatrix}, \qquad \dot{x} = \begin{pmatrix} \tau u \\ \tau v + \xi \end{pmatrix} \qquad =: f(x,\tau,\xi),$$

$$y = \begin{pmatrix} t \\ x \end{pmatrix} = \begin{pmatrix} t \\ u \\ v \end{pmatrix}, \quad \dot{y} = \begin{pmatrix} 1 \\ f(x,\tau,\xi) \end{pmatrix} = \begin{pmatrix} 1 \\ \tau u \\ \tau v + \xi \end{pmatrix} \qquad =: g(y,\tau,\xi),$$

$$z = \begin{pmatrix} y \\ w \end{pmatrix} = \begin{pmatrix} t \\ u \\ v \\ w \end{pmatrix}, \quad \dot{z} = \begin{pmatrix} g(y,\tau,\xi) \\ \tau v - C^{\varepsilon}\langle\xi\rangle \end{pmatrix} = \begin{pmatrix} 1 \\ \tau u \\ \tau v + \xi \\ \tau v \quad C^{\varepsilon}\langle\xi\rangle \end{pmatrix} =: h(z,\tau,\xi),$$

or its equivalent in the Joshua transform (see below) with controls (τ, ι).

7.3 Extensions

We allude here to two extensions of the preceding model. They remain to be investigated in detail, but numerical procedures are readily available.

7.3.1 American Options

In American options, the buyer is free to exercise his option at any time, but not later than T. We will not provide a detailed analysis of this case. But we show here a simple way to solve it numerically. The Isaacs equation for the discrete trading problem for a European option is (Sect. 8.4.1)

$$\forall (u,v) \in \mathbb{R}_+ \times \mathbb{R}, \quad W_K^h(u,v) = N(u,v), \quad \text{and} \quad \forall k \in \mathbb{K}, \forall (u,v) \in \mathbb{R}_+ \times \mathbb{R},$$
$$W_k^h(u,v) = \inf_{\xi \in \mathbb{R}} \max_{\tau \in [\tau_h^-, \tau_h^+]} \left[W_{k+1}^h\Big((1+\tau)u, (1+\tau)(v+\xi)\Big) - \tau(v+\xi) + C^\varepsilon\langle\xi\rangle \right].$$

We will split the minimax procedure as

$$W_{k+\frac{1}{2}}^h(u,v) = \max_{\tau \in [\tau_h^-, \tau_h^+]} [W_{k+1}^h\Big((1+\tau)u, (1+\tau)v\Big) - \tau v],$$

$$W_k^h(u,v) = \inf_{\xi \in \mathbb{R}} [W_{k+\frac{1}{2}}^h(u, v+\xi) - C^\varepsilon\langle\xi\rangle].$$

For an American option, the minimax game of the preceding section is replaced by a stopping time game. We need to recover the terminal value of an exercise at a time $t_e \in [0,T]$ as

$$\widetilde{N}(t_e,u,v) = \mathrm{e}^{\mu_0(T-t_e)} N(\mathrm{e}^{\mu_0(t_e-T)}u, \mathrm{e}^{\mu_0(t_e-T)}v),$$

and the new game is

$$W(0,u(0),v(0))$$
$$= \inf_{\varphi \in \Phi} \sup_{\tau(\cdot) \in \Omega} \max_{t_e \in [0,T]} \left\{ \widetilde{N}(t_e, u(t_e), v(t_e)) + \int_0^{t_e} [-\tau(t)v(t) + C^\varepsilon\langle\xi(t)\rangle] \,\mathrm{d}t \right\}$$

or its discrete trading equivalent:

$$W_0^h(u(0),v(0))$$
$$= \inf_{\varphi \in \Phi} \sup_{\{\tau_k\} \in \Omega_h} \max_{k_e \in \mathbb{K}} \left\{ \widetilde{N}_{k_e}(u(k_e), v(k_e)) + \sum_{k=0}^{k_e-1} [-\tau_k(v_k+\xi_k) + C^\varepsilon\langle\xi_k\rangle] \right\}.$$

Its discrete Isaacs equation is

$$\forall (u,v) \in \mathbb{R}_+ \times \mathbb{R}, \quad W_K^h(u,v) = N(u,v),$$
$$\forall k \in \mathbb{K}, \forall (u,v) \in \mathbb{R}_+ \times \mathbb{R}, \quad W_k^h(u,v) = \inf_{\xi \in \mathbb{R}} \max \Big\{ \{N(u,v),$$
$$\max_{\tau \in [\tau_h^-, \tau_h^+]} \left[W_{k+1}^h\Big((1+\tau)u, (1+\tau)(v+\xi)\Big) - \tau(v+\xi) + C^\varepsilon\langle\xi\rangle \right] \Big\}.$$

We may separate the minimization and maximization as above. In the maximization, we perform exactly the same operations as in the European case, and hence use the same computer code, and add one line of code comparing the result of these computations to $\widetilde{N}_k(u,v)$ to decide which one to write into the memory for $W_k^h(u,v)$.

Both $\widetilde{N}$ and $W^h_{k+\frac{1}{2}}$ are convex in v, and therefore their maximum is also convex. Hence the minimization procedure as described in Sect. 8.4.1 can be retained.

7.3.2 Delayed Information

The hypothesis that price information is available and used instantly may be considered adequate if the frequency of trading is low, say once per day or less. For high-frequency trading, it surely is not. In the discrete trading case, a delay of one time step, requiring a *strictly nonanticipative* strategy, may be accommodated without increasing the dimension of the state space as follows.

Let again a control be computed as $\xi_k = \phi_k(u_k, v_k)$ but be available only for use at time $k+1$. The dynamics of v are modified as

$$v_{k+1} = \tau_k v_k + \xi_k.$$

The Isaacs equation is modified accordingly. We lose the possibility of splitting it as was done previously. But the computation remains well within the capabilities of any modern PC.

Chapter 8
Vanilla Options

8.1 Introduction and Main Results

We now introduce the European "vanilla" options that will be the subject of this chapter, and we particularize the general framework to them to arrive at a rather comprehensive theory.

8.1.1 Vanilla Options

8.1.1.1 Vanilla Call

We assume that a *seller* sells a European buy option or *call* to a *buyer*, giving him the right to request the possibility of buying one share of the *underlying stock* at a prescribed *exercise price* or *strike* $\mathscr{K}$ at the prescribed *exercise time* T.

It is assumed (but not necessary) that if at time T the current underlying stock's market price $S(T)$ is less than $\mathscr{K}$, the buyer will not exercise his right but that he will do so if, to the contrary, $S(T)$ is larger than $\mathscr{K}$. Whether the buyer behaves that way or not, the contract is equivalent to a contingent debt of the seller to be repaid at time T, in the amount of at most $M(S(T))$ with

$$M(u) = \max\{0, u - \mathscr{K}\}. \tag{8.1}$$

Henceforth, the case of a call will be our standard working framework.

8.1.1.2 Vanilla Put

If the seller sells a European sell option, or *put*, the buyer may sell him one share of the underlying stock at exercise time at the prescribed price $\mathscr{K}$, thus resulting in a contingent debt of $M(S(T))$, with now

P. Bernhard et al., *The Interval Market Model in Mathematical Finance*, Static & Dynamic Game Theory: Foundations & Applications, DOI 10.1007/978-0-8176-8388-7_8,

$$M(u) = \max\{\mathscr{K} - u, 0\}. \tag{8.2}$$

We will only allude to this case occasionally, as it is in many respects symmetric to the call option, and essentially the same analysis applies.

8.1.2 *Terminal Payment*

8.1.2.1 Call (Closure in Kind)

If the buyer does not exercise his option to buy, then the terminal payment faced by the seller is just the transaction costs related to his closing of his position: reselling any underlying stock left in his portfolio (said to be "long") or buying some if he shorted the portfolio, that is, a payment $c^{\varepsilon}\langle -v(T)\rangle$.

If the buyer does exercise his right, the seller must first bring the amount of underlying stock in his portfolio to one unit, buying or selling some according to the sign of $v-u$ and bearing the associated terminal transaction costs $c^{\varepsilon}\langle u(T)-v(T)\rangle$, and then perform the promised transaction with the buyer, losing an extra $M(u(T))$ (8.1) in that process.

Since we want the portfolio to hedge the risk whatever the decision of the buyer, we set

$$N(u,v) = \max\{c^{\varepsilon}\langle -v\rangle, u - \mathscr{K} + c^{\varepsilon}\langle u - v\rangle\}. \tag{8.3}$$

It is useful to draw a graph of $v \mapsto N(u,v)$ at fixed u (Fig. 8.1). The graph of the "no exercise" part is independent of u, a simple V shape with the wedge at the origin, and branch slopes of $-c^+$ to the left and $-c^-$ to the right. The graph of the "exercise" part is a V with the same slopes, its wedge at $v = u$, $N(u,u) = u - \mathscr{K}$. The ordinate

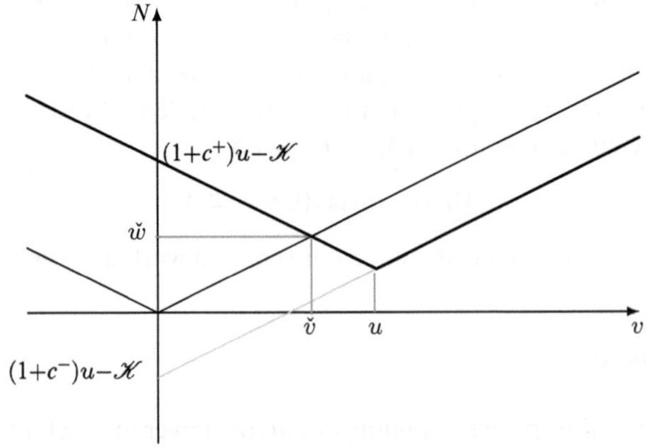

Fig. 8.1 Graph of $N(u,v)$ for $(1+c^-)u - \mathscr{K} < 0 < (1+c^+)u - \mathscr{K}$

at the origin of the left part of the V is $(1+c^+)u-\mathscr{K}$, and that of its right-hand part at $(1+c^-)u-\mathscr{K}$. Hence, the shape of the max depends on the signs of these last two quantities. If $(1+c^+)u-\mathscr{K}\le 0$, then the "no exercise" graph is always "above." If, on the contrary, $(1+c^-)u-\mathscr{K}\ge 0$, then the "exercise" graph is above the "no exercise" one. In the intermediate case, $(1+c^-)u-\mathscr{K}<0<(1+c^+)u-\mathscr{K}$, as shown in Fig. 8.1, the graph of the maximum is still a V shape with the same slopes, but its wedge where $-c^-v=u-\mathscr{K}+c^+(u-v)$, i.e.,

$$v=\check{v}(T,u):=\frac{(1+c^+)u-\mathscr{K}}{c^+-c^-},\qquad N=\check{w}(T,u):=u-\mathscr{K},$$

as shown in Fig. 8.1.

We emphasize that shape by the use of the notation

$$N(u,v)=\check{w}(T,u)+c^{\varepsilon}\langle\check{v}(T,u)-v\rangle\,, \tag{8.4}$$

with $\check{v}(T,u)$ and $\check{w}(T,u)$ given in the table thereafter.

8.1.2.2 Put (Closure in Kind)

In the case of a sell option, a similar analysis leads to a terminal payment still called $N(u(T),v(T))$ with

$$N(u,v)=\max\{c^{\varepsilon}\langle -v\rangle\,,\mathscr{K}-u+c^{\varepsilon}\langle -u-v\rangle\}. \tag{8.5}$$

It can be represented by formula (8.4), with the different formulas for $\check{v}(T,u)$ and $\check{w}(T,u)$ as given in the table.

8.1.2.3 Closure in Cash

In the case of a closure in cash, the contract is that, if the buyer wants to exercise his right, he will be paid $M(u(T))$ in cash by the buyer from the seller, who will then just close out his position. The cost incurred by the seller is therefore the sum of his loss $M(u(T))$ and the same closing cost $c^{\varepsilon}\langle -v(T)\rangle$ as in the case where the option is not exercised. The terminal payment he faces is therefore $N(u(T),v(T))$, with

$$N(u,v)=M(u)+c^{\varepsilon}\langle -v\rangle\,. \tag{8.6}$$

This admits a representation similar to (8.4), as given by Table 8.2.

Concerning all of these cases, we state the following useful fact.

Proposition 8.1. *The function N is jointly convex in (u,v).*

Table 8.1 $\check{v}(T,u)$ and $\check{w}(T,u)$ for a closure in kind

Call	$u \le \frac{\mathscr{K}}{1+c^+}$	$\frac{\mathscr{K}}{1+c^+} \le u \le \frac{\mathscr{K}}{1+c^-}$	$\frac{\mathscr{K}}{1+c^-} \le u$	Put	$u \le \frac{\mathscr{K}}{1+c^+}$	$\frac{\mathscr{K}}{1+c^+} \le u \le \frac{\mathscr{K}}{1+c^-}$	$\frac{\mathscr{K}}{1+c^-} \le u$
$\check{v}$	0	$\frac{(1+c^+)u-\mathscr{K}}{c^+-c^-}$	u	$\check{v}$	$-u$	$\frac{\mathscr{K}-(1+c^-)u}{c^--c^+}$	0
$\check{w}$	0	$-c^-\check{v}(u)$	$u-\mathscr{K}$	$\check{w}$	$\mathscr{K}-u$	$-c^+\check{v}(u)$	0

Table 8.2 $\check{v}(T,u)$ and $\check{w}(T,u)$ for closure in cash

Call	$u \le \mathscr{K}$	$u \ge \mathscr{K}$	Put	$u \le \mathscr{K}$	$u \ge \mathscr{K}$
$\check{v}(u)$	0	$\frac{u}{1+c^-}$	$\check{v}(u)$	$-\frac{u}{1+c^+}$	0
$\check{w}(u)$	0	$\frac{u}{1+c^-}-\mathscr{K}$	$\check{w}(u)$	$\mathscr{K}-\frac{u}{1+c^+}$	0
$N(u,v)=\check{w}(u)+c^-(\check{v}(u)-v)$			$N(u,v)=\check{w}(u)+c^+(\check{v}(u)-v)$		

Proof. Let us take the case of a call with closure in kind. Note that $c^{\varepsilon}\langle -v\rangle = \max\{-c^+v, -c^-v\}$ and $u+c^{\varepsilon}\langle u-v\rangle = \max\{(1+c^+)u-c^+v, (1+c^-)v-c^-v\}$. Hence, (8.3) can also be rewritten as

$$N(u,v) = \max\{-c^+v,\, -c^-v,\, (1+c^+)u-\mathscr{K}-c^+v,\, (1+c^-)u-\mathscr{K}-c^-v\}.$$

Thus, N is the supremum of four affine functions and, hence, convex.

Similar arguments apply for the other three cases. □

8.1.3 Main Results

These are the main results proved in the sequel. They hold for calls or puts alike, closure in kind or in cash, just choosing the correct $(\check{v},\check{w})$ in Table 8.1 or Table 8.2, which changes N according to formula (8.4), and changing q^- and q^+ for closure in cash: see after Eq. (8.8). The premium to be determined is $P(u) = W(0,u,0)$.

8.1.3.1 Differential Quasivariational Inequality

We introduce here the differential quasivariational inequality (DQVI) for the Value function $W(t,u,v)$:

$$\begin{aligned}
&\forall (t,u,v) \in [0,T)\times\mathbb{R}_+\times\mathbb{R},\\
&\quad \max\left\{-\frac{\partial W}{\partial t} - \tau^{\varepsilon}\left\langle \frac{\partial W}{\partial u}u + \left(\frac{\partial W}{\partial v}-1\right)v\right\rangle, -\left(\frac{\partial W}{\partial v}+C^+\right), \frac{\partial W}{\partial v}+C^-\right\} = 0,\\
&\forall (u,v) \in \mathbb{R}_+\times\mathbb{R}, \quad W(T,u,v) = N(u,v).
\end{aligned} \tag{8.7}$$

Theorem 8.2. *The Value function W is the only Lipschitz continuous viscosity solution (VS) of the DQVI (8.7).*

Furthermore, for vanilla options we have the following theorem.

Theorem 8.3. $\forall t \in [0,T]$, *the function* $(u,v) \mapsto W(t,u,v)$ *is convex.*

8.1.3.2 Representation Theorem

We need to introduce the following notation. Let, for a closure in kind,

$$\begin{aligned} q^-(t) &= \max\{(1+c^-)\exp[\tau^-(T-t)]-1, C^-\}, \\ q^+(t) &= \min\{(1+c^+)\exp[\tau^+(T-t)]-1, C^+\}. \end{aligned} \tag{8.8}$$

(For a closure in cash, both q^+ and q^- are constructed with c^- for a call and c^+ for a put instead of c^ε as above). We will refer to both at the same time as q^ε with the same convention as with C^ε and c^ε. (As a matter of fact, here, $\varepsilon = \pm$ is usually the sign of $\check{v} - v$. See below.) Note that if we define t_ε as

$$T - t_\varepsilon = \frac{1}{\tau^\varepsilon} \ln\left(\frac{1+C^\varepsilon}{1+c^\varepsilon}\right), \tag{8.9}$$

$q^\varepsilon = C^\varepsilon$ for $t \le t_\varepsilon$ and increases (for q^+) or decreases (for q^-) toward c^ε as $t \to T$. For any realistic data, for a classic stock market, t_+ and t_- are very close to T, say, 1 day or less. (This will provide a simplified algorithm.)

Let also

$$\mathscr{S} = \begin{pmatrix} 1 & 0 \\ 1 & 0 \end{pmatrix} \text{ and } \mathscr{T} = \frac{1}{q^+ - q^-}\begin{pmatrix} \tau^+q^+ - \tau^-q^- & \tau^+ - \tau^- \\ -(\tau^+ - \tau^-)q^+q^- & \tau^-q^+ \quad \tau^+q^- \end{pmatrix}. \tag{8.10}$$

We introduce a pair of functions of two variables $\check{v}(t,u)$ and $\check{w}(t,u)$ collectively called

$$\check{V}(t,u) = \begin{pmatrix} \check{v}(t,u) \\ \check{w}(t,u) \end{pmatrix} \tag{8.11}$$

and defined by the final conditions of Table 8.1 or Table 8.2, depending on which applies, and the following linear partial differential equation (PDE):

$$\frac{\partial \check{V}}{\partial t} + \mathscr{T}\left(\frac{\partial \check{V}}{\partial u}u - \mathscr{S}\check{V}\right) = 0. \tag{8.12}$$

Theorem 8.4. *The Value function W is given by*

$$W(t,u,v) = \check{w}(t,u) + q^\varepsilon \langle \check{v}(t,u) - v \rangle, \tag{8.13}$$

where $(\check{v}\ \check{w})^t = \check{V}$ *is the unique solution of (8.12) initialized as in Table 8.1 (or Table 8.2 depending on which one applies). The optimal hedging strategy is to ensure* $v(t) = \check{v}(t,u(t))$ *if* $t < t_\varepsilon$ *and not to trade if* $t \geq t_\varepsilon$*, with* $\varepsilon = \mathrm{sign}(\check{v} - v)$ *and* t_ε *as in (8.9).*

8.1.3.3 Discrete Trading and Convergence Theorem

We tackle now the discrete trading scheme. In that case, the seller is restricted to trading at predetermined time instants $t_k = kh$, $k \in \mathbb{K} = \{0,1,\ldots,K-1\}$, for a given step size $h = T/K$, with K a positive integer. Apart from that, the specification is the same as in the previous case, with the same continuous-time market model and the same hedging aim.

The mathematical problem to solve is therefore that of computing the $\inf_\varphi \sup_\tau$ in (7.22), under the dynamics (7.7), (7.13). Let $W_k^h(u,v) = W^h(t_k,u,v)$ be the Value function of the minimax game. A direct application of Isaacs' theory yields the following theorem.

Theorem 8.5. *The Value function* $W_k^h(u,v)$ *is the solution of the following Isaacs equation:*

$$\begin{aligned} W_k^h(u,v) &= \min_\xi \max_{\tau\in[\tau^-,\tau^+]} \left[W_{k+1}^h((1+\tau)u,(1+\tau)(v+\xi)) - \tau(v+\xi) + C^\varepsilon \xi \right], \\ &\forall (u,v) \in \mathbb{R}_+ \times \mathbb{R}, \quad W_K^h(u,v) = N(u,v), \end{aligned} \tag{8.14}$$

or, equivalently, of the following "fractional-step" Isaacs equation:

$$\left.\begin{aligned} W_{k+\frac{1}{2}}^h(u,v) &= \max_{\tau\in\{\tau_h^-,\tau_h^+\}} \left[W_{k+1}^h((1+\tau)u,(1+\tau)v) - \tau v \right], \\ W_k^h(u,v) &= \min_\xi \left[W_{k+\frac{1}{2}}^h(u,v+\xi) + C^\varepsilon \xi \right], \\ \forall (u,v) &\in \mathbb{R}_+ \times \mathbb{R}, \quad W_K^h(u,v) = N(u,v). \end{aligned}\right\} \tag{8.15}$$

Equations (8.15), together with a fast way to compute the $\min_\xi$ (Sect. 8.4.1) provide our *standard algorithm*, which is significantly faster than the direct implementation of Isaacs Eq. (8.14). Notice that the range of the maximization in τ has been restricted to the two end points of the interval, thanks to the following fact.

Theorem 8.6. *The functions* $(u,v) \mapsto W_k^h(u,v)$ *are convex.*

We introduce a new Value function $W^h(t,u,v)$, which is the Value of the same minimax dynamic game, but in this case the seller (minimizer here) is allowed to make one jump transaction at the initial time – even between two time instants t_k – then only at time instants $t_k = kh$, $k \in \mathbb{K}$. It continuously interpolates the sequence $\{W_k^h(\cdot,\cdot)\}_{k\in\mathbb{K}}$.

Theorem 8.7. *As the step size is subdivided and goes to zero (e.g., $h = T/2^d$, $d \to \infty \in \mathbb{N}$), for all $t \in [0,T]$, the function $W^h(t,u,v)$ decreases and converges to the function $W(t,u,v)$, uniformly on any compact in (u,v).*

8.1.3.4 Fast Algorithm

If one samples the functions q^- and q^+ [see Eq. (8.8)] at the time instants $t_k = kh$, the resulting sequences $q_k^\varepsilon := q^\varepsilon(t_k)$ can also be generated by the recursion

$$q_K^\varepsilon = c^\varepsilon \tag{8.16}$$

for a closure in kind or

$$q_K^\varepsilon = c^- \text{ for a call and } q_K^\varepsilon = c^+ \text{ for a put}$$

for a closure in cash, and

$$\left.\begin{aligned} q_{k+\frac{1}{2}}^\varepsilon &= (1+\tau_h^\varepsilon)q_{k+1}^\varepsilon + \tau_h^\varepsilon, \\ q_k^+ &= \min\{q_{k+\frac{1}{2}}^+, C^+\}, \\ q_k^- &= \max\{q_{k+\frac{1}{2}}^-, C^-\}. \end{aligned}\right\} \tag{8.17}$$

Let, for every integer ℓ,

$$Q_\ell^\varepsilon = (q_\ell^\varepsilon \;\; 1) \quad \text{and} \quad \check{V}_\ell^h(u) = \begin{pmatrix} \check{v}_\ell^h(u) \\ \check{w}_\ell^h(u) \end{pmatrix}. \tag{8.18}$$

The following system, together with (8.17), defines our "fast algorithm":

$$\begin{aligned} \check{V}_k^h(u) &= \frac{1}{q_{k+\frac{1}{2}}^+ - q_{k+\frac{1}{2}}^-} \begin{pmatrix} 1 & -1 \\ -q_{k+\frac{1}{2}}^- & q_{k+\frac{1}{2}}^+ \end{pmatrix} \begin{pmatrix} Q_{k+1}^+ \check{V}_{k+1}^h((1+\tau_h^+)u) \\ Q_{k+1}^- \check{V}_{k+1}^h((1+\tau_h^-)u) \end{pmatrix}, \\ \check{v}_K^h(u) &= \check{v}(T,u), \qquad \check{w}_K^h(u) = \check{w}(T,u), \end{aligned} \tag{8.19}$$

as given by Tables 8.1 and 8.2 depending on which one applies. [It can be verified that this is a consistent finite difference scheme for the PDE (8.12).] The following result provides a fast way of computing the discrete trading premium, which, according to the convergence Theorem 8.7, provides a safe approximation of the continuous trading premium.

Theorem 8.8. *The value functions W_k^h are given by (8.16), (8.17), (8.19), and*

$$W_k^h(u,v) = \breve{w}_k^h(u) + q_k^\varepsilon \left\langle \breve{v}_k^h(u) - v \right\rangle. \tag{8.20}$$

The optimal hedging strategy is to ensure $v_k = \breve{v}_k^h(u_k)$ if $t < t_\varepsilon$ and do not trade if $t \geq t_\varepsilon$ with t_ε as in (8.9) and $\varepsilon = \operatorname{sign}(\breve{v}_k^h(u_k) - v_k)$.

8.1.4 Joshua Transformation and the DQVI

We give here one of the two means that we will use to deal with our impulse-control problem.

8.1.4.1 Impulse Control and QVI

We are confronted with a minimax problem where the minimizer may use impulses. The natural way to deal with this, in the spirit of [32], would be to investigate the corresponding quasivariational inequality (QVI)

$$\min\left\{\frac{\partial W(t,u,v)}{\partial t} + \min_{\xi_c \in \mathbb{R}} \max_{\tau \in [\tau^-,\tau^+]} \left[\frac{\partial W(t,u,v)}{\partial u}\tau u + \frac{\partial W}{\partial v}(\tau v + \xi_c) - \tau v\right],\right.$$

$$\left.\min_{\xi_k \in \mathbb{R}} \{W(t,u,v+\xi_k) - W(t,u,v) + C^\varepsilon\langle \xi_k \rangle\}\right\} = 0,$$

$$W(T,u,v) = N(u,v).$$

However, this QVI is of no use here because it is degenerate on two counts. The least one is the absence of a second-order partial derivative, which the Itô term would bring with a diffusion model, but the really serious problem stems from the fact that the infimum of the costs of the jumps is zero. As a consequence, the second term in the QVI is always nonpositive (since $\xi = 0$ makes it zero), and therefore the first term can always be anything.

We therefore turn to other approaches. Two will be needed: the Joshua transformation and the geometric investigation of Sect. 8.2. To do so, we need the following proposition.

Proposition 8.9. *For each positive number a there exists a positive number b such that, if $|u(0)|$ and $|v(0)| \leq a$, on trajectories approaching sufficiently closely the $\inf_{\varphi \in \Phi}$, and $\tau(\cdot)$ approaching sufficiently closely the $\sup_\tau$, it holds that $|v(t)| \leq b$ for all $t \in [0,T]$.*

Proof. We noted in Sect. 7 that bounding the possible buys or short sales of an underlying stock, i.e., variations in v, might realistically be a feature of the game.

If this is not imposed a priori, then this proposition is technically a corollary of the analysis of Sect. 8.2. This analysis does not rely on the current proposition and shows it to be true. A large initial $|v|$ always leads in the optimal strategy to an initial jump toward lower values, and on optimal trajectories, $|v|$ always stays between 0 and $|u|$.

As a matter of fact, it can easily be shown that for vanilla options, a strategy leading to $v(t) \notin [0, u(t)]$, for a call, or $v(t) \notin [-u(t), 0]$, for a put, is never rational. (See [142].)

This may be understood before the complete mathematical analysis is performed. Notice that $|u(t)| \leq a\exp(\tau^\sharp T)$, so that the worst cost for $\varphi = 0$, say $J_0^\star$, is bounded. Now, large excursions in v require that the minimizer use large values of ξ, either in ξ_c or in large impulses ξ_k. These carry a cost for him. Among the possible time functions $\tau(\cdot) \in \Omega$ are those that have $\tau(t)$ of the opposite sign of v right after the large excursion, this without changing $\tau(\cdot)$ *before* the large excursion in v and, thus, without modifying $\xi(\cdot) = \varphi(\tau(\cdot))$ up to that time (since φ is nonanticipative). Now these $\tau(\cdot)$ histories result in the large v itself "costing money" to the minimizer, unless he incurs extra costs to undo at once what he has just done spending more money, which, for very large excursions, is beyond $J_0^\star$. Hence such large excursions in v cannot be part of a strategy approaching the infimum of the payoff. □

8.1.4.2 Joshua's Transformation

The technique used here has surely been proposed independently several times for other problems. The earlier reference to a similar approach of which we are aware is [91, Chap. 10, vv. 12 and 13], hence its name.

Let our original game be called $\mathscr{G}$. We introduce a new game, called $\mathscr{J}$, as follows: Its "time" variable, called *artificial time*, will be denoted by θ. Derivatives with respect to θ are denoted by a "prime" accent. Its state variables are t, u, and v. The minimizer's controls are $\xi_c \in \mathbb{R}$ and $\iota \in \{-1, 0, 1\} =: B$ (ι for "impulse".) We also use the notation $\bar{\iota} = 1 - |\iota|$. The dynamics are

$$\begin{aligned} t' &= \bar{\iota}, \\ u' &= \bar{\iota}\tau u, \\ v' &= \bar{\iota}(\tau v + \xi_c) + \iota. \end{aligned} \tag{8.21}$$

Terminal (artificial) time is defined as

$$\Theta = \inf\{\theta \mid t(\Theta) = T\}. \tag{8.22}$$

The payoff is

$$L = N(u(\Theta), v(\Theta)) + \int_0^\Theta [\bar{\iota}(-\tau v + C^\varepsilon \langle \xi_c \rangle) + \iota C^\iota] \, \mathrm{d}\theta. \tag{8.23}$$

(C^{-1} and C^1 are to be understood as C^- and C^+, respectively. C^0 need not be defined because it appears only multiplied by 0. We may think of it as being 0.) In words, when there are no impulses, the dynamics and the integral payoff run as in game $\mathscr{G}$. Impulses of $\mathscr{G}$ are replaced by (artificial) time intervals during which the natural time and classical dynamics stop, and v increases, or decreases, at rate 1. Hence the lengths of such intervals define how much v has changed while natural time was stopped, i.e., the size of the impulses.

Proposition 8.10. *Games $\mathscr{G}$ and $\mathscr{J}$ have the same value.*

Proof. To each trajectory of game $\mathscr{G}$ we can associate the corresponding trajectory of game $\mathscr{J}$ generated as follows: as a rule, $\iota = 0$, except that to each impulse of magnitude ξ_k corresponds an artificial time interval of length $|\xi_k|$ at the same natural time t_k with $\iota = \text{sign}(\xi_k)$. Clearly the payoffs J of $\mathscr{G}$ and L of the corresponding $\mathscr{J}$ are equal.

The converse requires a small argument: there are trajectories of $\mathscr{J}$ that have no natural counterpart in $\mathscr{G}$. This is when the minimizer uses two or more adjacent artificial time intervals with opposite, nonzero controls ι. This would correspond to several successive impulses, of opposite signs, at the same time. However, we may safely allow this feature in game $\mathscr{J}$ because it will never be used in an optimal strategy since it would have the same effect as a single impulse of magnitude equal to the algebraic sum of the succeeding impulses involved, but at a higher cost to the minimizer. Thus such trajectories play no role in defining the value of game $\mathscr{J}$. All other trajectories of $\mathscr{J}$ have a counterpart in $\mathscr{G}$, with the same payoff. □

8.1.4.3 Differential Quasivariational Inequality

Game $\mathscr{J}$ is a classic differential game, with no impulses. We may write its Isaacs equation (with signs adapted to the classic definition of VS), writing W for its value function since we know that it coincides with the value W of game $\mathscr{G}$. We notice that problem $\mathscr{J}$ is autonomous in artificial time, so that in that formulation, W is independent of θ. Hence we get

$$\forall(t,u,v) \in [0,T) \times \mathbb{R}_+ \times \mathbb{R}, \quad 0$$
$$= \max_{\iota \in B} \max_{\xi_c \in \mathbb{R}} \min_{\tau \in [\tau^-,\tau^+]} \left\{ -\bar{\iota} \left[\frac{\partial W}{\partial u} \tau u + \frac{\partial W}{\partial v} (\tau v + \xi_c) - \tau v + C^{\varepsilon} \xi_c \right] - \iota \left(\frac{\partial W}{\partial v} + C^{\iota} \right) \right\},$$
$$\forall(u,v) \in \mathbb{R}_+ \times \mathbb{R}, \quad W(T,u,v) = N(u,v).$$

We write separately the three possible cases $\iota = -1$, 1, and 0, making use of the notation $-\partial W/\partial v = q$.

Case $\iota = -1$:

The term $\iota = -1$ is

$$\frac{\partial W}{\partial v} + C^-.$$

It must be nonpositive, hence necessarily $q \geq C^-$.

Case $\iota = 1$:

In a similar fashion, the term $\iota = 1$ is

$$-\frac{\partial W}{\partial v} - C^+,$$

which must be nonpositive, requiring $q \leq C^+$.

Case $\iota = 0$:

The term with $\iota = 0$ is

$$\begin{aligned}
&\max_{\xi_c \in \mathbb{R}} \min_{\tau \in [\tau^-, \tau^+]} \left[-\frac{\partial W}{\partial u}\tau u - \frac{\partial W}{\partial v}(\tau v + \xi_c) + \tau v - C^{\varepsilon} \xi_c \right] \\
&= - \max_{\tau \in [\tau^-, \tau^+]} \tau \left[\frac{\partial W}{\partial u} u + \left(\frac{\partial W}{\partial v} - 1 \right) v \right] - \min_{\xi_c \in \mathbb{R}} (C^{\varepsilon} \langle \xi_c \rangle - q \xi_c).
\end{aligned}$$

It must be nonpositive. It is easy to see that

$$-\min_{\xi_c \in \mathbb{R}} (C^{\varepsilon} \langle \xi_c \rangle - q \xi_c) = \begin{cases} 0 & \text{if } C^- \leq q \leq C^+, \\ \infty & \text{if } q < C^- \text{ or } q > C^+. \end{cases}$$

According to the first two cases, it will in any case be zero. Hence ξ_c disappears from the equation, a notable fact. As for the max in τ, it is clearly reached at τ^- if the square bracket is negative and at τ^+ if it is positive. We are left with the simpler Isaacs Eq. (8.7), which we call the DQVI.

Furthermore, we need the following proposition.

Proposition 8.11. *For any $(u(0), v(0))$ there is a bound on how large Θ may be for nearly optimal strategies. Furthermore, all trajectories are transverse to the terminal manifold $t = T$.*

Proof. Note that $\Theta = T + \sum_k |\xi_k|$. Let

$$W_0(u, v) = \sup_{\tau(\cdot) \in \Omega} J(0, u, v; 0, \tau(\cdot)).$$

Also, for $v(0)$ given, let b be the bound on $|v(t)|$ according to Proposition 8.9. Then

$$\int_0^T (-\tau v(t))\,\mathrm{d}t > -b\tau^\sharp T.$$

Hence, any strategy that causes

$$\int_0^T C^\varepsilon\langle\xi(t)\rangle\,\mathrm{d}t > W_0(u,v) + b\tau^\sharp T$$

is worse than the strategy $\varphi = 0$ and, therefore, nonoptimal. It does not enter into the evaluation of $W(0,u,v)$. A fortiori, any strategy with $\sum_k \xi_k > W_0(u,v) + b\tau^\sharp T$ may be ignored.

Since t' can only be equal to zero or one, the only way in which a trajectory may fail to be transverse to the manifold $t = T$ is if an impulse occurs at $t = T$. But our definition of Θ does not allow such a terminal impulse since as soon, in artificial time, as $t = T$, the game ends.

There is no loss of optimality in that definition. As a matter of fact, if a jump in v at $t = T$ is necessary for the minimizer, it is taken care of in the terminal transaction in N and done at a cost $|c^\varepsilon| \leq |C^\varepsilon|$. □

As a corollary of this proposition we get the following classic result (see, e.g., [69]).

Theorem 8.12. *The function W is a continuous VS of the DQVI (8.7).*

At this stage, we are still left with the task of proving the uniqueness of that VS, which, as far as we know, does not follow from known results. Notice also that W is not bounded.

8.2 Geometric Approach

In an alternative way of dealing with impulses in ξ, or jumps in v, we turn to a four-dimensional representation of the problem where jumps are ordinary trajectories, just orthogonal to the time (and the u) axis, so that we do not need to distinguish ξ_c from ξ_k, and we just use $\xi \in \mathbb{R}$.

In effect, we investigate the DQVI via its characteristics, using the geometric concept of semipermeable hypersurfaces. This analysis will let us identify the singular surfaces in the candidate solution, all of which satisfy the fundamental PDE (8.12), and some of their properties in order to verify later on that they do satisfy the viscosity conditions.

8.2.1 Geometric Formulation

8.2.1.1 Notation

To ease the investigation in (t,u,v,w) space, henceforth we will make use of the shorthand notation of Sect. 7.2.5.2:

$$z = \begin{pmatrix} t \\ u \\ v \\ w \end{pmatrix} \qquad \dot{z} = h(z,\tau,\xi) = \begin{pmatrix} 1 \\ \tau u \\ \tau v + \xi \\ \tau v - C^{\varepsilon}\langle \xi \rangle \end{pmatrix}.$$

In the four-dimensional representation, we will call $v^t = (n \ p \ q \ r)$ a normal to a semipermeable hypersurface, or semipermeable normal, pointing toward the increasing w. The variables n, p, q (and r, but it remains equal to one) are also called *adjoint variables* or *costate*.

We will also need the notation

$$V = \begin{pmatrix} v \\ w \end{pmatrix}, \qquad \check{V} = \begin{pmatrix} \check{v} \\ \check{w} \end{pmatrix}. \tag{8.24}$$

8.2.1.2 The Game in (t,u,v,w) Space

We consider our game problem from Sect. 7.2.5 in (t,u,v,w) space. The goal (7.20) is now a condition on the final state we want to reach:

$$\{(t,u,v,w) \mid t = T, \quad w - N(u,v) \geq 0\}. \tag{8.25}$$

We must therefore solve the following formulation of the game (Sect. 7.2.5):

$$\sup_{\varphi \in \Phi} \inf_{\tau(\cdot) \in \Omega} [w(T) - N(u(T), v(T)) \geq 0. \tag{8.26}$$

Following a classic interpretation (see [89]) we consider the problem as being that of finding a hypersurface separating states z that can be driven to region (8.25) from those that cannot. This manifold, a *barrier* in Isaacs' parlance, is necessarily semipermeable, meaning that if we let $v = (n \ p \ q \ r)^t$ be a normal to this manifold (where it is smooth) pointing toward larger w, it should hold that

$$\max_{\xi \in \mathbb{R}} \min_{\tau \in [\tau^-, \tau^+]} \langle v, h(z,\xi,\tau) \rangle = 0. \tag{8.27}$$

This is Isaacs' *second main equation*. (See [89]).

Focusing on trajectories in z space and manifolds traversed by them allows us to consider jumps in v as ordinary trajectories orthogonal to the time axis (and also to the u axis). It easily follows from the dynamics the following fact.

Proposition 8.13. *The slopes in (v,w) space of jump trajectories are always $-C^-$ for negative jumps and $-C^+$ for positive ones.*

The following fact will be useful:

Proposition 8.14. *Let a smooth two-dimensional manifold, nowhere orthogonal to the t axis, be given, and consider the hypersurface made of jump trajectories of fixed sign reaching it. It is semipermeable.*

Proof. These trajectories are in fact orthogonal to the (t,u) plane. Hence there is no time for the ordinary dynamics to have an effect while traversing them and cause the trajectory to drift away from the hypersurface.

A formal way to see that is as follows. We *renormalize* the dynamics by the norm of ξ, say ρ, and let ρ go to infinity. Moreover, (8.27) is insensitive to the norm of the vector ν, which can be changed arbitrarily. We thus renorm it, multiplying by $1/\rho$ to keep the dot product finite. Let, therefore, $\nu^t = (1/\rho)(n' \;\; p' \;\; q' \;\; r')$ and

$$h(t,u,v,w,\xi,\tau) = \begin{pmatrix} 1 \\ \tau u \\ \tau v \\ \tau v \end{pmatrix} + \rho \begin{pmatrix} 0 \\ 0 \\ 1 \\ -C^{\varepsilon} \end{pmatrix} \left\langle \frac{\xi}{\rho} \right\rangle,$$

so that

$$\langle \nu, h \rangle = \frac{1}{\rho}[n' + \tau(p'u + (q' + r')v)] + (q' - r'C^{\varepsilon})\langle \xi/\rho \rangle.$$

It follows from the previous proposition that on jump trajectories, we have $q - rC^{\varepsilon} = 0$; hence here $q' - r'C^{\varepsilon} = 0$. Thus the limit of the above scalar product as $\rho \to \infty$ is zero. On the other hand, consider a positive jump manifold, i.e., with $q' = C^+$, $r' = 1$. A negative ξ gives $\langle \nu, h \rangle \to (C^+ - C^-)\xi < 0$, and thus $\xi > 0$ does maximize it, and similarly *mutatis mutandis* for negative jump manifolds. □

Our problem has the special feature that the dynamics are independent of w. Hence, increasing or decreasing the initial w increases or decreases by the same amount the final one, without modifying $u(T)$ and $v(T)$. Hence, there is, for each (t,u,v), a limiting w that lies on the separating hypersurface. More precisely, we may rewrite the dynamics of w as [here $\xi(\cdot)$ contains the impulses]

$$w(T) = w(t) + \int_t^T [-\tau(s)v(s) + C^{\varepsilon}\langle \xi(s) \rangle]\mathrm{d}s,$$

so that, in view of formulation (8.26), the separating hypersurface is given by

$$\sup_{\varphi \in \Phi} \inf_{\tau(\cdot) \in \Omega} \left[w(t) - N(u(T),v(T)) - \int_t^T [-\tau(s)v(s) + C^{\varepsilon}\langle \xi(s) \rangle]\mathrm{d}s \right] = 0,$$

or, equivalently,

$$w - W(t,u,v) = 0.$$

And positions that can be hedged are of the form

$$(t,u,v,w) \mid w \geq W(t,u,v). \tag{8.28}$$

The next proposition follows.

Proposition 8.15. *The barrier of the four-dimensional game is the graph of the function W.*

As a consequence, we may identify ν in (8.27) as

$$\nu^t = (n\ p\ q\ r) = \left(-\tfrac{\partial W}{\partial t}\ -\tfrac{\partial W}{\partial u}\ -\tfrac{\partial W}{\partial v}\ 1\right).$$

(We have already used the notation $q = -\partial W/\partial v$.)

8.2.2 *Primary Field and Dispersal Manifold $\mathscr{D}$*

8.2.2.1 Characteristic System

The *Hamiltonian* of the problem is

$$H = \langle \nu, h(z,\xi,\tau)\rangle = n + \tau[pu + (q+1)v] + (q - C^\varepsilon)\xi.$$

Hence, $\max_{\xi\in\mathbb{R}}\min_{\tau\in[\tau^-,\tau^+]} H$ is obtained for $\tau = \tau^\star$ and $\xi = \xi^\star$ defined as

$$\sigma = pu + (1+q)v, \qquad \xi^\star = \begin{cases} -\infty & \text{if } q < C^-, \\ \leq 0 & \text{if } q = C^-, \\ 0 & \text{if } C^- < q < C^+, \\ \geq 0 & \text{if } q = C^+, \\ +\infty & \text{if } q > C^+. \end{cases}$$
$$\tau^\star = \begin{cases} \tau^- & \text{if } \sigma > 0, \\ \text{arbitrary} & \text{if } \sigma = 0, \\ \tau^+ & \text{if } \sigma < 0, \end{cases} \tag{8.29}$$

This is consistent, of course, with the findings of the previous section, the meaning of $|\xi| = \infty$ being that an impulse is optimal.

The characteristic system associated to Eq. (8.27) is, as long as there is no jump,

$$\left|\begin{array}{l} \dot t = 1, \\ \dot u = \tau^\star u, \\ \dot v = \tau^\star v + \xi_c^\star, \\ \dot w = \tau^\star v - C^\varepsilon\langle \xi_c^\star\rangle, \end{array}\right. \qquad \left|\begin{array}{l} \dot n = 0, \\ \dot p = -\tau^\star p, \\ \dot q = -\tau^\star(q+1), \\ \dot r = 0. \end{array}\right. \tag{8.30}$$

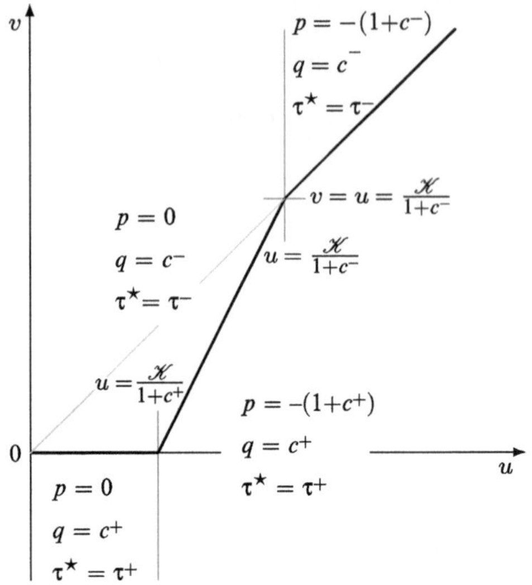

Fig. 8.2 Partition of (u,v) plane at $t = T$ according to N and its (opposite) gradient. The *thick line* is the graph $v = \check{v}(T,u)$

Terminal Conditions

We integrate the characteristic system (8.30) backward from the boundary of the target set, i.e., from all

$$(T, u(T), v(T), N(u(T), v(T))),$$

that is, a two-dimensional manifold parameterized by $(u(T), v(T))$. Integration down to t will yield the third parameter to create a three-dimensional semipermeable surface.

As previously, we will stress the case of a call with payment in kind.

Notice that, according to Table 8.1, we should partition the plane (u,v) at time T along the pecked line $v = \check{v}(T,u)$. "Above" this line, i.e., for $v \geq \check{v}(T,u)$, one has $q := -\partial N/\partial v = c^-$, "below" this line, $q = c^+$. A second dividing line is $\{u = \mathscr{K}/(1+c^+), v \leq 0\}$ followed by $\{\mathscr{K}/(1+c^+) \leq u \leq \mathscr{K}/(1+c^-), v = \check{v}(T,u)\}$ and, finally, $\{u = \mathscr{K}/(1+c^-), v \geq u\}$. To the "left" of that line (smaller u), we have $p := -\partial N/\partial u = 0$. To the "right" of that line, we have $p = -(1+c^\varepsilon)$, with $\varepsilon = \mathrm{sign}(\check{v}(T,u) - v)$ (Fig. 8.2).

The barrier of the four-dimensional game, the graph of the function W, will be obtained by backward integration from the boundary of the target set [see (8.25)], with due care for the singularities, according to the classic Isaacs–Breakwell theory [33, 116].

8.2.2.2 Sheets $\{\tau^-\}$ and $\{\tau^+\}$

We may compute $pu+(1+q)v$ at time T to find $\tau^\star$ according to (8.29). Using the remark that $0 \le \check{v}(T,u) \le u$, we find that $\tau^\star = \tau^-$ for $v(T) > \check{v}(T,u(T))$ and $\tau^\star = \tau^+$ for $v(T) < \check{v}(T,u(T))$. We therefore have two families of characteristic curves, extremal trajectories, forming two hypersurfaces of dimension 3 in the four-dimensional space. We will refer to them as the *sheet* $\{\tau^-\}$ and the *sheet* $\{\tau^+\}$.

Given the assumption that $|c^-| < |C^-|$ and $c^+ < C^+$, it follows, still according to (8.29), that at the terminal time, $\xi^\star = 0$. Notice also that if we call $\sigma = pu+(1+q)v$ the switch function such that $-\tau^\star\sigma = \tau^\varepsilon\langle -\sigma\rangle$, it follows from the characteristic Eqs. (8.30) that $\dot\sigma = 0$. Hence $\tau^\star$ is constant along these curves. They are therefore, according to (8.4), given by Table 8.1 and

$$\left| \begin{array}{ll} u(t) &= u(T)\mathrm{e}^{-\tau^\varepsilon(T-t)}, \\ v(t) &= v(T)\mathrm{e}^{-\tau^\varepsilon(T-t)}, \\ w(t) &= \check{w}(T,u(T)) + c^\varepsilon\langle \check{v}(T,u(T)) - v(T)\rangle + (v(t)-v(T)). \end{array} \right. \tag{8.31}$$

These calculations hold as long as $C^- \le q \le C^+$. Hence we need to use

$$q^\varepsilon(t) = (1+c^\varepsilon)\mathrm{e}^{\tau^\varepsilon(T-t)} - 1.$$

Accordingly, we define t_- and t_+ via $q^\varepsilon = C^\varepsilon$, i.e., (8.9), together with

$$t_m = \min\{t_-,t_-\}, \qquad t_M = \max\{t_-,t_+\}. \tag{8.32}$$

(In the case $c^\varepsilon = C^\varepsilon$, $t_\varepsilon = T$, one may skip Sect. 8.2.2.5.) We recall the complete definition (8.8) of the q^ε, $\varepsilon \in \{-,+\}$, or, alternatively,

$$q^\varepsilon = \begin{cases} (1+c^\varepsilon)\mathrm{e}^{\tau^\varepsilon(T-t)} - 1 & \text{if } t > t_\varepsilon, \\ C^\varepsilon & \text{if } t \le t_\varepsilon. \end{cases} \tag{8.33}$$

We will also use the notation

$$Q^\varepsilon = (\,q^\varepsilon \;\; 1\,). \tag{8.34}$$

8.2.2.3 Primary Field: Trivial Regions

We consider here the two regions with "small" or "large" underlying prices, where things are easy. We will need the following definitions.

Definition 8.16. Let

$$u_\ell(t) := \frac{\mathscr{K}}{1+c^+}\mathrm{e}^{-\tau^+(T-t)}, \tag{8.35}$$

$$u_r(t) := \frac{\mathscr{K}}{1+c^-} \mathrm{e}^{-\tau^-(T-t)} \tag{8.36}$$

for a call and $u_\ell = \mathscr{K} \mathrm{e}^{-\tau^-(T-t)}$ and $u_r = \mathscr{K} \mathrm{e}^{-\tau^+(T-t)}$ for a put.

Case $u(T) < \mathscr{K}/(1+c^-)$

According to Table 8.1, if $u(T) < \mathscr{K}/(1+c^-)$, then $\check{v}(T,u(T)) = \check{w}(T,u(T)) = 0$. Also according to the preceding remark, $\sigma(T) = (1+c^\varepsilon)v(T)$ with $\varepsilon = -\text{sign}(v(T))$. Trajectories ending with $v(T) < 0$ would therefore have $\tau^\star = \tau^+$ (they belong to the sheet $\{\tau^+\}$), a negative $v(t)$, and $u(t) \le u_\ell(t)$ (8.35).

A quick calculation integrating system (8.30) with $\tau = \tau^+$ yields, for $t \in [t_+, T]$, using the notation (8.33),

$$w(t) = -q^+(t)v(t).$$

On the other hand, trajectories with $v(T) > 0$ have $v(t) > 0$ and belong to the sheet $\{\tau^-\}$. The same calculation as above yields, for $t \in [t_-, T]$,

$$w(t) = -q^-(t)v(t).$$

Finally, if $v(T) = 0$, then $v(t) = w(t) = 0$ on both sheets [and $\sigma(t) = 0$, hence $\tau(t)$ can be anything, generating the same manifold $\{\tau^-\} \cap \{\tau^+\}$].

Remembering that this is the graph of W, we end up, for the region $u(t) < u_\ell(t)$, $t \ge t_M$, with

$$W(t,u,v) = q^\varepsilon(t)\langle -v \rangle . \tag{8.37}$$

We explain below that indeed, with the notation (8.33), this formula extends to smaller t, the sheets $\{\tau^\varepsilon\}$ being replaced before $t = t_\varepsilon$ by jump manifolds, with slope $-C^\varepsilon$ in the (v,w) plane.

We notice that in the region $u \le u_\ell$, for all admissible $\tau(\cdot)$ the underlying stock price ends with $u(T) < \mathscr{K}/(1+c^+)$. Below that price, while the buyer should (rationally) not exercise his right to buy the underlying stock at a price $\mathscr{K}$, should he nevertheless do so, the seller may buy this stock at that time, at a unit cost of $(1+c^+)u$ (including transaction costs) and sell it back for $\mathscr{K}$, still making a profit. Hence there is no dilemma for the seller; the empty portfolio suffices to hedge the risk.

The seller should therefore return to $v = 0$ instantly to protect himself against the worst case $\tau(\cdot)$: τ^- if $v > 0$ (he has a long portfolio) or τ^+ if $v < 0$ (a short portfolio), except if we are so close to the terminal time that the benefit of the smaller closure transaction costs overcomes the risk due to an unfavorable evolution of the underlying stock's price.

Case $u(T) > \mathscr{K}/(1+c^+)$

If $u(T) > \mathscr{K}/(1+c^+)$, then Table 8.1 tells us that $\check{v}(T,u(T)) = u(T)$, $\check{w}(T,u(T)) = u(T) - \mathscr{K}$. Now $\sigma(T) = (1+c^\varepsilon)(v(T) - u(T))$. Hence trajectories ending with $v(T) > u(T)$ have $\tau^\star = \tau^-$. (They belong to the sheet $\{\tau^-\}$.) On these trajectories, we have $u(t) \geq u_r(t)$ [see (8.36)].

Again, a quick calculation shows that, on these trajectories,

$$w(t) = (1+q^-(t))u(t) - \mathscr{K} - q^-(t)v(t).$$

On the other hand, trajectories ending with $v(T) < u(T)$ belong to a sheet $\{\tau^+\}$. The same calculation as above shows that

$$w(t) = (1+q^+(t))u(t) - \mathscr{K} - q^+(t)v(t).$$

These two surfaces intersect along $u = v$ in $w = u - \mathscr{K}$. The graph of W is given by the max of the two, leading to

$$W(t,u,v) = u - \mathscr{K} + q^\varepsilon\langle u - v\rangle. \tag{8.38}$$

Again, we argue that this formula holds for smaller t, with jump manifolds replacing the sheets $\{\tau^-\}$ and $\{\tau^+\}$.

We notice that for $u \geq u_r$, for all admissible $\tau(\cdot)$ the underlying stock price ends with $u(T) \geq \mathscr{K}/(1+c^-)$. Above that price, the buyer should (rationally) exercise his right to buy the underlying stock at a price $\mathscr{K}$. But should he nevertheless not do so, if the seller has one unit of stock in hand that he wishes to sell back, he will still get more than $\mathscr{K}$ after the closing costs have been taken into account. Therefore, he may have borrowed up to an amount $\mathscr{K}$ and still hedge against any risk. There is no dilemma for him.

The seller should therefore get to $v = u$ (holding one unit of stock) as soon as possible, except if we are so close to the terminal time that the advantage of closure transaction costs overcomes the possible rise (if $v < u$) or drop (if $v > u$) in prices.

8.2.2.4 Primary Field: Region of Interest

Definition 8.17. We call a *region of interest* the region of (t,u) space, using the notation (8.35), (8.36):

$$\Lambda = \{(t,u) \mid u_\ell(t) \leq u \leq u_r(t)\}. \tag{8.39}$$

It is more easily represented in logarithmic coordinates for u or, better, in coordinates $(\ln(u/\mathscr{K}),t)$ (Fig. 8.3).

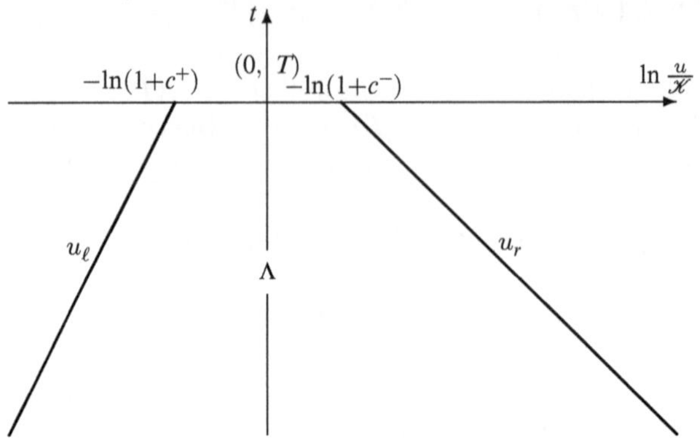

Fig. 8.3 Region of interest in logarithmic coordinates

We now integrate the characteristic system (8.30) in this region. We still have that the switch function σ is constant as long as these extremal controls hold. We therefore have two semipermeable surfaces to investigate, depending on whether $\sigma > 0$, yielding the sheet $\{\tau^-\}$, or $\sigma < 0$, yielding the sheet $\{\tau^+\}$.

Sheet $\{\tau^-\}$

If $\check{v}(u(T)) \le v(T) \le u(T)$, then $\sigma(T) = (1+c^-)v(T)$, which is positive since $v(T) \ge \check{v}(T) > 0$. Hence $\tau^\star = \tau^-$, the primary trajectories are given by

$$\left|\begin{array}{l} u(t) = u(T)\mathrm{e}^{-\tau^-(T-t)}, \\ v(t) = v(T)\mathrm{e}^{-\tau^-(T-t)}, \\ w(t) = v(T)[\mathrm{e}^{-\tau^-(T-t)} - (1+c^-)], \end{array}\right. \qquad \left|\begin{array}{l} p = 0, \\ q = (1+c^-)\mathrm{e}^{\tau^-(T-t)} - 1, \\ r = 1. \end{array}\right. \tag{8.40}$$

The sheet $\{\tau^-\}$ satisfies the equation

$$w = -q^-(t)v, \quad \text{or} \quad Q^- V = 0. \tag{8.41}$$

Sheet $\{\tau^+\}$

If $0 \le v(T)\ le\check{v}(u(T))$, then $\sigma(T) = (1+c^+)(v(T) - u(T))$, which is negative since $v(T) < \check{v}(u(T)) \le u(T)$. Hence, $\tau^* = \tau^+$, the primaries are given by

$$\left|\begin{array}{l} u(t) = u(T)\mathrm{e}^{-\tau^+(T-t)}, \\ v(t) = v(T)\mathrm{e}^{-\tau^+(T-t)}, \\ w(t) = (1+c^+)[u(T) - v(T)] - \mathscr{K} - v(T)\mathrm{e}^{-\tau^+(T-t)}, \end{array}\right. \quad \left|\begin{array}{l} p = -(1+c^+)\mathrm{e}^{\tau^+(T-t)}, \\ q = (1+c^+)\mathrm{e}^{\tau^+(T-t)} - 1, \\ r = 1. \end{array}\right. \tag{8.42}$$

The sheet $\{\tau^+\}$ satisfies the equation

$$w = (1+q^+(t))u - q^+(t)v - \mathscr{K}, \quad \text{or} \quad Q^+V = Q^+\mathbb{1}u - \mathscr{K}. \tag{8.43}$$

Time, State, and Costate

These formulas are correct as long as $C^- \le q \le C^+$, i.e., for each sheet $\{\tau^\varepsilon\}$, from time t_ε to T. We may furthermore derive the time component n of the semipermeable normal using the Isaacs "Main equation 2": $\langle \nu, h(\tau^\star, \xi^\star)\rangle = 0$, which yields

$$n = \tau^\varepsilon\langle -\sigma\rangle = -\tau^\star\sigma. \tag{8.44}$$

Singular Manifolds

The case $v(T) = 0$ for the sheet $\{\tau^-\}$ and $v(T) = u(T)$ for the sheet $\{\tau^+\}$ lead to $\sigma(t) = 0$. Therefore, $\tau(\cdot)$ is arbitrary along these trajectories. They correspond to an empty portfolio and just traverse the border of the sheet $\{\tau^-\}$ at $v = 0$, in the first case, and to the "buy and hold" portfolio, traversing the border of the sheet $\{\tau^+\}$ at $u = v$, in the second case. Geometrically, they do not add anything to these two three-dimensional sheets.

8.2.2.5 Dispersal Manifold $\mathscr{D}$

We want to investigate the intersection $\mathscr{D} = \{\tau^-\} \cap \{\tau^+\}$. It is a two-dimensional dispersal manifold. This intersection can exist only in the time range $[t_M, T]$ [Eq. (8.32)], and in the region of interest Λ [see (8.39)], where both sheets have $v \in [0, u]$.

A simple calculation using (8.41) and (8.43) shows that it can be characterized as $v = \breve{v}(t,u)$, $w = \breve{w}(t,u)$, with

$$\breve{v}(t,u) = \frac{(1+q^+(t))u - \mathscr{K}}{q^+(t) - q^-(t)}, \qquad \breve{w}(t,u) = -q^-(t)\breve{v}(t,u). \tag{8.45}$$

Notice that these expressions do coincide at $t = T$ with $\breve{v}(T,u)$ and $\breve{w}(T,u)$ as in Table 8.1. Furthermore, we stress the following fact.

Proposition 8.18. *For $(t,u) \in \Lambda$ as defined in (8.39), and for $\breve{v}$ defined as in (8.45), it holds that $0 \le \breve{v}(t,u) \le u$.*

Proof. It suffices to notice that

$$\breve{v}(t,u) = \frac{(1+c^+)e^{\tau^+(T-t)}}{q^+ - q^-}\left[u - \frac{\mathscr{K}}{(1+c^+)}e^{-\tau^+(T-t)}\right]$$

to see that $\check{v}(t,u) \geq 0$, and that

$$\check{v}(t,u) - u = \frac{(1+q^-)u - \mathscr{K}}{q^+ - q^-} = \frac{(1+c^-)\mathrm{e}^{-\tau^-(T-t)}}{q^+ - q^-}\left[u - \frac{\mathscr{K}}{1+c^+}\mathrm{e}^{-\tau^-(T-t)}\right]$$

to see that $\check{v}(t,u) - u \leq 0$. □

We may use this proposition to see the following proposition.

Proposition 8.19. *In the domain $t \in [t_M, T]$, $v \in [0,u]$, we have according to the foregoing analysis*

$$W(t,u,v) = \max\{-q^- v, (1+q^+)u - q^+ v - \mathscr{K}\}.$$

Proof. We check that trajectories of each sheet remain "above" (w-wise) the other sheet. This is seen by computing

$$\langle \nu^+, h(t,u,\check{v},\check{w},\tau^-,\xi_c^-)\rangle = (\tau^+ - \tau^-)(1+q^+)(u-\check{v}) > 0$$

and

$$\langle \nu^-, h(t,u,\check{v},\check{w},\tau^+,\xi_c^+)\rangle = (\tau^+ - \tau^-)(1+q^-)\check{v} > 0.$$

Hence, in the region where both sheets exist, the portion of each between their intersection $\mathscr{D}$ and the target at $t = T$ lies "above" the other sheet.

Remark 8.20. This can be understood in the following way. The hedging portfolio must be worth enough to hedge both risks, either $\tau = \tau^-$ or $\tau = \tau^+$, both of which are risks in the region of interest Λ. For each sheet, the states that correspond to portfolios actually hedging the risk are those that are "above" the sheet. The states that hedge both risks are therefore those that are above the maximum w.

It follows that W can be represented as the supremum of two affine functions, and hence we have the following proposition.

Proposition 8.21. *For $t \in [t_M, T]$ the function $(u,v) \mapsto W(t,u,v)$ is convex.*

We also stress the following proposition.

Proposition 8.22. *In the region $t \in [t_M, T]$, the representation formula (8.13)*

$$W(t,u,v) = \check{w}(t,u) + q^\varepsilon \langle \check{v}(t,u) - v \rangle.$$

still holds.

As a matter of fact, for $(t,u) \in \Lambda$, this is a direct consequence of the foregoing analysis. Furthermore, a direct calculation using formulas (8.35), (8.36), and (8.45) shows that

$$\check{v}(t,u_\ell) = 0, \quad \check{w}(t,u_\ell) = 0, \qquad \check{v}(t,u_r(t)) = u_r(t), \quad \check{w}(t,u_r(t)) = u_r(t) - \mathscr{K}.$$

Hence, if we extend continuously the definition of $(\check{v}(t,u),\check{w}(t,u))$ to $(0,0)$ to the left of $u_\ell(t)$ and $(u,u-\mathscr{K})$ to the right of u_r, the representation formula holds for all $(u,v)\in\mathbb{R}_+\times\mathbb{R}$.

We finally stress the following fact, using Proposition 8.18.

Proposition 8.23. *Direct calculation using formulas (8.45) yields, on $\mathscr{D}$,*

$$Q^-\check{V}_t=\tau^-(1+q^-)\check{v}\le 0,\quad Q^+\check{V}_t=-\tau^+(1+q^+)(u-\check{v})\le 0.$$

8.2.3 Equivocal Manifold $\mathscr{E}$

We now investigate the domain $t\in[t_m,t_M]$. We will concentrate on the case $t_+<t_-$, which is more realistic than the converse. The case where $t_-<t_+$ is, indeed, very similar. We will omit it here. It is dealt with in full detail in [142].

8.2.3.1 Equivocal Junction

Let, therefore, $t\in[t_+,t_-]$. The sheet $\{\tau^-\}$ does not exist in that domain. In $t=t_-$, that sheet has a slope $-C^-$ in the plane (v,w). It will be replaced before t_- by a negative jump manifold, denoted $\{\iota^-\}$ – with reference to Sect. 8.1.4 – that joins on the sheet $\{\tau^+\}$ by an *equivocal junction*. We refer the reader to [33] for a general constructive theory of such junctions and to [142] for a detailed description of how to construct them. But we give here a sufficient description of it.

We know that the jump manifold will have a slope $-C^-$ in the (v,w) plane, so that the Hamiltonian $\langle\nu,h\rangle$ for this manifold at the junction will be singular in ξ, as long as $\xi\le 0$. Therefore, it is possible to pick a control $\xi\le 0$ from other considerations without allowing the state to cross the jump manifold. The equivocal junction, denoted $\mathscr{E}$, is made of trajectories lying on the sheet $\{\tau^+\}$. We thus identify a control strategy

$$\tilde{\xi}(\tau)=\frac{(\tau^+-\tau)(1+q^+)(u-v)}{C^--q^+}\le 0$$

that has the property that

$$\forall\tau\in[\tau^-,\tau^+],\quad \Big\langle\nu^+,h\Big(u,v,\tau,\tilde{\xi}(\tau)\Big)\Big\rangle=0.$$

This is not a nonanticipative strategy because knowing instantly τ amounts to knowing the immediate future. But, on the one hand, this should be considered a technical device to compute a solution of the DQVI, of which we will verify later on that it is a VS. On the other hand, the section on discrete trading and convergence will provide a strictly causal hedging strategy providing an arbitrary good approximation of the value function thus computed.

Let $\nu_j^- = (n_j^- \; p_j^- \; C^- \; 1)^t$ denote the normal to the negative jump manifold. The scalar product

$$\left\langle \nu_j^-, h(z,\tau,\xi) \right\rangle = n_j^- + \tau[p_j^- u + (1+C^-)v] \tag{8.46}$$

is independent of ξ and linear in τ. It is therefore minimized in τ by either τ^- or τ^+. Let us assume that it is minimized by τ^-, and let us integrate both the state dynamic equations and the generalized adjoint equations below for the normal ν_j^- with the controls $\tau = \tau^-$, $\xi = \tilde{\xi}(\tau^-)$. Verifying that τ^- indeed minimizes the Hamiltonian in the field thus constructed, and therefore that no control τ can cause the state to cross the jump manifold in the negative direction, is done via the following device.

Proposition 8.24. *The control $\tau = \tau^-$ minimizes the Hamiltonian (8.46) if and only if $n_j^- \geq 0$.*

Proof. Computing state and gradient variables with the dynamic and (generalized) adjoint equations insures that the hamiltonian (8.46) remains zero. Thus it suffices to verify that $n_j^-(t) \geq 0$ to ensure that $\tau^-[p_j^- u + (1+C^-)v] \leq 0$, and therefore that τ^- is the minimizing τ in (8.46). □

8.2.3.2 Sign of n_j^- on $\mathscr{E}$

According to Proposition 8.24, we need to prove the following proposition.

Proposition 8.25. *On $\mathscr{E}$, $n_j^- \geq 0$.*

Proof. The normal ν_j^- will be obtained using the generalized adjoint equations (see [33])

$$\dot{\nu}_j^- = -\frac{\partial H}{\partial z} + \alpha(\nu_j^- - \nu^+).$$

Thus

$$\dot{n}_j^- = \alpha n_j^- - \alpha n^+.$$

We have that $n^+ > 0$. Integrating this affine ordinary differential equation (ODE) (via Lambert's formula) backward, we see that it suffices to verify that $\alpha \geq 0$ and $n_j^- \geq 0$ on the boundary, say at $t = t_b$, to ensure that for all $t \leq t_b$, indeed $n_j^- \geq 0$.

Now, α should be chosen in such a way as to guarantee that the generalized adjoint equations keep $q_j^- = C^-$, to preserve the singularity in ξ, i.e., $\dot{q} = 0$. This yields here

$$\dot{q}_j^- = -\tau^-(1+C^-) + \alpha(C^- - q^+) = 0,$$

and hence $\alpha = \tau^-(1+C^-)/(C^- - q^+) > 0$, as needed.

Let us therefore investigate the boundary $\partial\mathscr{E}$. The trajectories of $\mathscr{E}$ are generated by the dynamics

$$\dot{z} = h\left(z, \tilde{\xi}(\tau^-), \tau^-\right)$$

integrated backward in time from the boundary at $t = t_-$ and $z \in \mathscr{D}$, that is to say, $(v(t_-), w(t_-)) = (\check{v}(t_-, u(t_-)), \check{w}(t_-, u(t_-)))$ as in (8.45). But this does not fill the region $t \in [t_+, t_-]$, $(t, u) \in \Lambda$. We must also generate trajectories from $u = u_\ell(t_b)$, $t_b \in [t_-, t_+]$, with $v(t_b) = w(t_b) = 0$ to join continuously with the negative jump sheet at $u \leq u_\ell$. (From any state on this boundary, either $\tau = \tau^+$ for all t, and we just reach $\mathscr{D}$, or at some time $\tau < \tau^+$, and then the state drifts in the trivial region where $v = w = 0$ is a safe hedge.)

Let us first investigate the boundary at t_-. We denote by $\mathscr{E}_2$ the manifold generated by the trajectories of $\mathscr{E}$ ending there. The normal ν_j^- to the jump manifold is necessarily orthogonal to $\partial\mathscr{E}_2$ and, hence, to the vector $(0\ 1\ \check{v}_u\ -C^-\check{v}_u)^t$. This yields $p_j^- = 0$. It is also normal to the trajectories that traverse $\mathscr{E}$, i.e., to $h(z, \tau^-, \xi(\tau^-))$. This yields

$$n_j^- + p_j^- \tau^- u + (1 + C^-)\tau^- v = 0, \tag{8.47}$$

i.e., taking into account that $p_j^- = 0$, $n_j^-(t_-) = -(1 + C^-)\tau^- v > 0$, as needed.

Let us proceed in the same manner along the boundary on u_ℓ. We denote by $\mathscr{E}_1$ the manifold generated by the trajectories of $\mathscr{E}$ ending there. The tangent to $\partial\mathscr{E}_1$ is $(1\ \tau^+ u_\ell(t)\ 0\ 0)^t$. Therefore, orthogonality to it yields

$$n_j^-(t_b) + \tau^+ u_\ell(t_b) p_j^-(t_b) = 0,$$

while the same calculation as for $\mathscr{E}_2$ yields again (8.47), which becomes

$$n_j^-(t_b) + \tau^- u_\ell(t_b) p_j^-(t_b) = 0,$$

hence $n_j^-(t_b) = 0$. The verification is complete. □

Remark 8.26. In the case where $t_- < t_+$, the trajectories of $\mathscr{E}$ are with $\tau = \tau^+$. The boundary of $\mathscr{E}_1$ is on u_r, with $(v, w) = (u, u - \mathscr{K})$. The remainder of the verification proceeds similarly.

8.2.3.3 Synthesis Concerning $\mathscr{E}$

This generates, via backward integration in t, a piecewise smooth two-dimensional manifold, composed of the pieces $\mathscr{E}_1$ and $\mathscr{E}_2$, as shown in Fig. 8.4.

The reference [142] gives closed form solutions of these equations. We will not need them. This manifold can be parameterized by t and u. We will let $(v\ w) = (\check{v}(t,u)\ \check{w}(t,u))$, or equivalently $V = \check{V}(t,u)$, be such a parameterization. We will show later on that $\check{V}$ still satisfies the PDE (8.12).

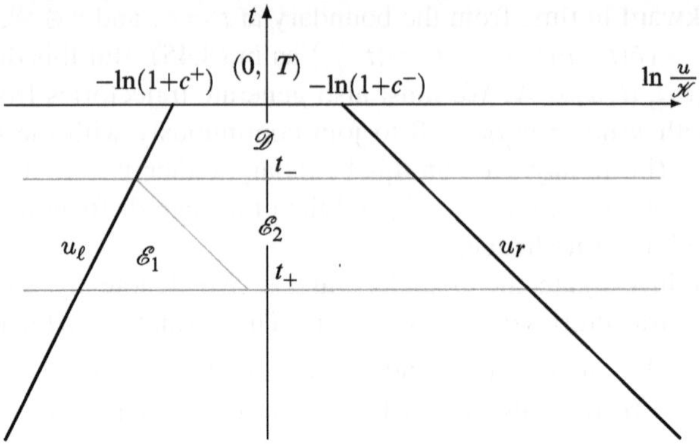

Fig. 8.4 Projection on (t,u) plane of dispersal manifold $\mathscr{D}$ and equivocal manifold $\mathscr{E}$

It is, however, useful to mention one more property.

Proposition 8.27. *On $\mathscr{E}$, we have, for all (t,u), and both $\varepsilon \in \{-,+\}$, $Q^\varepsilon \check{V}_t \leq 0$.*

Proof. The manifold $\mathscr{E}$ and two tangents are parameterized as

$$\mathscr{E} = \begin{pmatrix} t \\ u \\ \check{v}(t,u) \\ \check{w}(t,u) \end{pmatrix}, \quad \mathrm{D}_t\mathscr{E} = \begin{pmatrix} 1 \\ 0 \\ \check{v}_t(t,u) \\ \check{w}_t(t,u) \end{pmatrix}, \quad \mathrm{D}_u\mathscr{E} = \begin{pmatrix} 0 \\ 1 \\ \check{v}_u(t,u) \\ \check{w}_u(t,u) \end{pmatrix}.$$

Writing that v_j^- and v^+ are both normal to $\mathrm{D}_t\mathscr{E}$ yields

$$n_j^- + C^- \check{v}_t + \check{w}_t = 0,$$
$$n^+ + q^+ \check{v}_t + \check{w}_t = 0.$$

From these two equalities, (8.44) and Proposition (8.25), we obtain the claim. □

This last fact will be useful in proving that the function W thus constructed satisfies the viscosity condition.

Remark 8.28. The trajectory of $\mathscr{E}$ ending in $u_\ell(t_-)$ may bear a discontinuity of the gradient of W. It comes from some $u(t_+) = u_0$ of whose complicated expression we spare the reader.

Remark 8.29. In Breakwell's parlance, the condition that τ^- minimizes the scalar product $\left\langle v_j^-, h(z,\tau,\tilde{\xi})\right\rangle$ is that the corner "does not leak." It is equivalent to the viscosity condition, as seen here from the fact that both are equivalent to $n_j^- \geq 0$.

8.2.4 Focal Manifold $\mathscr{F}$

8.2.4.1 Preliminaries

For $t < t_+$, both sheets $\{\tau^-\}$ and $\{\tau^+\}$ are replaced by the jump manifolds $\{\iota^-\}$ and $\{\iota^+\}$, respectively. They join on a (singular) focal manifold $\mathscr{F}$. The general constructive theory of such manifolds can be found in [116]. The main point is Melikyan's discovery that they are traversed by two noncollinear fields of trajectories, both extremal in the nonsingular control, here τ. This fact is the crucial point in proving that the present focal manifold satisfies the fundamental Eq. (8.12). But rather than proving that only, we prove a more complete result.

We recall the notation $Q^\varepsilon = (\, q^\varepsilon \;\; 1\,)$, already introduced, and $1\!\!1$ as a vector whose entries are all ones and

$$\mathscr{S} = \begin{pmatrix} 1 & 0 \\ 1 & 0 \end{pmatrix}.$$

Lemma 8.30. *For $\varepsilon \in \{-,+\}$, if the manifold $\mathscr{V} : V = \check{V}(t,u)$ is either a submanifold of the sheet $\{\tau^\varepsilon\}$ or a submanifold of a jump manifold $\{\iota^\varepsilon\}$ traversed by trajectories generated by $\tau = \tau^\varepsilon$ and a ξ of sign ε, then it satisfies*

$$Q^\varepsilon \check{V}_t + \tau^\varepsilon Q^\varepsilon \left(\check{V}_u u - \mathscr{S}\check{V}\right) = 0. \tag{8.48}$$

Proof. If the manifold $\mathscr{V}$ is a submanifold of the sheet $\{\tau^-\}$, then its tangents

$$\mathrm{D}_t\mathscr{V} = \begin{pmatrix} 1 \\ 0 \\ \check{v}_t \\ \check{w}_t \end{pmatrix} \quad \text{and} \quad \mathrm{D}_u\mathscr{V} = \begin{pmatrix} 0 \\ 1 \\ \check{v}_u \\ \check{w}_u \end{pmatrix}$$

are orthogonal to the normal $\nu^- = (\, -\tau^- Q^- \mathscr{S}\check{V} \;\; 0 \;\; q^- \;\; 1\,)$ to the sheet $\{\tau^-\}$ [see (8.41)]. This gives two equations:

$$\begin{aligned} Q^-\check{V}_t - \tau^- Q^- \mathscr{S}\check{V} &= 0, \\ Q^-\check{V}_u &= 0. \end{aligned}$$

Multiplying the second equation by τ^- and adding to the first one yields (8.48).

If the manifold $\mathscr{V}$ is a submanifold of the sheet $\{\tau^+\}$, then these same tangents are orthogonal to the normal $\nu^+ = (\, -\tau^+ Q^+ 1\!\!1(\check{v} - u) \;\; -Q^+ 1\!\!1 \;\; q^+ \;\; 1\,)$ to the sheet $\{\tau^+\}$ [see (8.43)]. Again, this yields two equations:

$$\begin{aligned} Q^+\check{V}_t - \tau^+ Q^+ 1\!\!1(\check{v} - u) &= 0, \\ Q^+\check{V}_u - Q^+ 1\!\!1 &= 0. \end{aligned}$$

Multiplying the second equation by $\tau^+ u$ and adding to the first one yields (8.48).

Finally, if $\mathscr{V}$ is a submanifold of a jump manifold $\{\iota^\varepsilon\}$, then necessarily $t \le t_\varepsilon$ and $q^\varepsilon = C^\varepsilon$. If it is traversed by trajectories generated by τ^ε and a ξ with $\text{sign}(\xi) = \varepsilon$, then take Lagrangian derivatives along such a trajectory:

$$\begin{aligned} \dot{\check{v}} &= \check{v}_t + \tau^\varepsilon \check{v}_u u = \tau^\varepsilon \check{v} + \xi, \\ \dot{\check{w}} &= \check{w}_t + \tau^\varepsilon \check{w}_u u = \tau^\varepsilon \check{v} - C^\varepsilon \xi. \end{aligned} \tag{8.49}$$

Multiplying the first equation by C^ε and adding to the second one yields (8.48). □

8.2.4.2 State and Costate Dynamics on $\mathscr{F}$

We want to show that the focal manifold $\mathscr{F}$ is the unique solution of the PDEs (8.48) with appropriate boundary conditions. To this end, we must show two things:

1. That the singular controls ξ^- and ξ^+ that let the state traverse $\mathscr{F}$ with τ^- and τ^+, respectively, satisfy $\text{sign}(\xi^\varepsilon) = \varepsilon$;
2. That the two normals ν^- and ν^+ to the jump manifolds $\{\iota^-\}$ and $\{\iota^+\}$, respectively, both have n^ε nonnegative.

We will find that this is one and the same condition.

(Notice that we omit the index j for ν^ε as there is no possible confusion with the normals to the sheets $\{\tau^\varepsilon\}$ since they do not exist in the region investigated here.) We also need to show that the PDE we will end up with indeed has a unique solution. This will prove a classic exercise in hyperbolic PDEs.

We first investigate the controls ξ^ε. We will assume that they have the required signs, investigate the trajectory fields constructed with that hypothesis, and check a posteriori that it does give the required signs for these controls.

Each of the two dynamics $h(z, \tau^-, \xi^-)$ and $h(z, \tau^+, \xi^+)$ must be normal to both ν^- and ν^+, where we recall that, for $\varepsilon \in \{-,+\}$, $\nu^\varepsilon = (\, n^\varepsilon \;\; p^\varepsilon \;\; C^\varepsilon \;\; 1\,)$. This yields four equations:

$$\begin{aligned} n^- + \tau^-[p^- u + (1 + C^-)v] &= 0, \\ n^+ + \tau^-[p^+ u + (1 + C^+)v] + (C^+ - C^-)\xi^- &= 0, \\ n^- + \tau^+[p^- u + (1 + C^-)v] + (C^- - C^+)\xi^+ &= 0, \\ n^+ + \tau^+[p^+ u + (1 + C^+)v] &= 0. \end{aligned} \tag{8.50}$$

From these equations we easily derive

$$\xi^- = -\frac{(\tau^+ - \tau^-)n^+}{\tau^+(C^+ - C^-)}, \qquad \xi^+ = \frac{(\tau^+ - \tau^-)n^-}{-\tau^-(C^+ - C^-)}. \tag{8.51}$$

Hence, if the n^ε are both nonnegative, then $\text{sign}(\xi^\varepsilon) = \varepsilon$ for both ε.

Let $\bar{\varepsilon}$ be the opposite sign to ε. The normals ν^ε obey the generalized adjoint equations

$$\dot{\nu}^\varepsilon = -\frac{\partial H}{\partial z} + \alpha^\varepsilon(\nu^\varepsilon - \nu^{\bar{\varepsilon}}),$$

i.e., here

$$\dot{n}^\varepsilon = \alpha^\varepsilon(n^\varepsilon - n^{\bar{\varepsilon}}), \tag{8.52}$$

where, as previously, the α^ε should be chosen so as to keep q^ε constant in the adjoint equation

$$\dot{q}^\varepsilon = -\tau^\varepsilon(1 + C^\varepsilon) + \alpha^\varepsilon(C^\varepsilon - C^{\bar{\varepsilon}}),$$

i.e., here

$$\alpha^\varepsilon = \tau^\varepsilon \frac{1 + C^\varepsilon}{C^\varepsilon - C^{\bar{\varepsilon}}} > 0. \tag{8.53}$$

Yet this is not a constructive theory because these are Lagrangian derivatives along different trajectories. Adapting a proof of Gronwall's lemma to this case (see [142]) lets one prove that these equations with zero terminal conditions have $n^\varepsilon = 0$ for unique solution and, hence, via Fredholm's alternative, that they always have a unique solution. We now investigate the boundary conditions to conclude on the signs of the n^ε.

8.2.4.3 Signs of the n^ε Solutions of (8.52)

Trajectories generated by (τ^-, ξ^-) terminate on $\partial\mathscr{E}$ at $t_b^- = t_+$, $u \in [u_\ell(t_+), u_r(t_+)]$ or on $t_b^- < t_+$, $u = u_\ell(t_b^-)$. Trajectories generated by (τ^+, ξ^+) terminate on the same boundary of $\mathscr{E}$, or on $t_b^+ < t_+$, $u = u_r(t_b^+)$.

We remark first that $\mathscr{F}$ is both a submanifold of $\{\iota^-\}$ traversed by trajectories (τ^+, ξ^+) and a submanifold of $\{\iota^+\}$ traversed by trajectories (τ^-, ξ^-), with $\operatorname{sign}(\xi^\varepsilon) = \varepsilon$ if our construction succeeds. Therefore, according to Lemma 8.30, it satisfies both Eqs. (8.48). This is also true for $\mathscr{E}$, as we will emphasize in the next subsection. However, $\check{V}$ will be, by construction, continuous across $\partial\mathscr{E}$, and, hence, also $\check{V}_u$ along $t = t_+$. As a consequence, $Q^\varepsilon \check{V}_t$ is continuous across this boundary, as are the two $n^\varepsilon = -Q^\varepsilon \check{V}_t$. Hence, on that boundary, $n^- = 0$ and $n^+ > 0$.

By continuity, n^+ will remain positive in a neighborhood $t < t_+$. And this, with (8.52), shows that it is also true of n^-.

On $u = u_\ell(t)$, for the same reason as on $\mathscr{E}$, we need to have $\check{v} = \check{w} = 0$. As a consequence,

$$Q^- \dot{\check{V}} = Q^- \check{V}_t + \tau^+ Q^- \check{V}_u u = 0.$$

And, as everywhere on $\mathscr{F}$, (8.48) holds with $\varepsilon = -$, i.e.,

$$Q^- \check{V}_t + \tau^- Q^- \check{V}_u u = \tau^- Q^- \mathscr{S} \check{V} = 0.$$

Hence $Q^- \check{V}_t = 0 = n^-$.

To find the boundary conditions on $u = u_r$, we integrate backward the dynamic equations from $t = t_+$, $u = u_r(t_+)$, $v = \check{v}(t_+) = u_r(t_+)$, $w = \check{w}(t_+) = u_r(t_+) - \mathscr{K}$. Integration with $\tau = \tau^-$, $\xi = 0$, leads to $\check{v}(t) = u_r(t)$, $\check{w}(t) = u_r(t) - \mathscr{K}$. Indeed, from such a point, either for all t, $\tau = \tau^-$, and the state reaches $\partial\mathscr{E}$, or for some t, $\tau > \tau^-$, and the state drifts in the trivial region where $V = (\,u \;\; u - \mathscr{K}\,)^t$ is a safe hedge.

As previously, the fact that $\check{V} = (\,u \;\; u - \mathscr{K}\,)^t$ leads by differentiation to

$$Q^+\dot{\check{V}} = Q^+\check{V}_t + \tau^- Q^+\check{V}_u u = \tau^-(1+C^+)u,$$

and (8.48) with $\varepsilon = +$

$$Q^+\check{V}_t + \tau^+ Q^+\check{V}_u u = \tau^+(1+C^+)\check{v},$$

and since $\check{v} = u$, $Q^+\check{V}_t = 0 = -n^+$.

Proposition 8.31. *The preceding construction yields $n^\varepsilon > 0$, and as a consequence $Q^\varepsilon\check{V}_t \le 0$, $\forall\varepsilon \in \{-,+\}$ over the whole interior of $\mathscr{F}$.*

Proof. It may be useful, in investigating this question, to introduce, for $\varepsilon \in \{-,+\}$, $m^\varepsilon(t) = \exp[\alpha^\varepsilon(t_+ - t)]n^\varepsilon(t)$, which have the same signs as the corresponding n^ε, so that the two differential equations (8.52) become

$$\dot{m}^\varepsilon = -\alpha^\varepsilon e^{(\alpha^\varepsilon - \alpha^{\bar{\varepsilon}})(t_+ - t)} m^{\bar{\varepsilon}}.$$

Recall that the two α^ε are (constant and) positive. Now, at $t = t_+$, $u \in (u_\ell, u_r)$, we know that $m^+ > 0$. It is therefore true in a neighborhood $t < t_+$. It results that m^- strictly decreases toward 0, hence $m^- > 0$ for some $t < t_+$ in the field covered by the trajectories τ^+ arriving in $u \in (u_\ell, u_r)$. This in turn implies that in that field, for t close to t_+, $m^+(t) > m^+(t_+) = n^+(t_+) > 0$. Also, by continuity, on the trajectory τ^+ reaching u_r at t_+, and in a neighborhood of it, we get $m^- > 0$, and thus also m^+. And as long – backward – as in a neighborhood of the trajectory $u_r(\cdot)$, $m^- > 0$, it causes $m^+ > 0$ in that neighborhood.

We therefore have that $m^- > 0$ in a neighborhood of the trajectory $u_r(\cdot)$, and that $m^+ > 0$, decreasing (hence, larger than at t_+). A similar but simpler reasoning holds in the neighborhood of the trajectory $u_\ell(\cdot)$.

Now, let t_1 be the last instant when at an interior point one has $m^\varepsilon = 0$ for one of the two ε. This requires that at a strictly later time on the trajectory τ^ε through that point, m^ε reaches a maximum. Hence that $\dot{m}^\varepsilon = 0$, hence $m^{\bar{\varepsilon}} = 0$, which is a contradiction since t_1 is by assumption the last time when this happens.

We conclude that n^- and n^+ are both strictly positive in the interior of the region of interest for $t < t_+$. □

8.2.4.4 Synthesis Concerning $\mathscr{F}$

The two Eqs. (8.48) can be regrouped in the following way. Let

$$Q := \begin{pmatrix} Q^+ \\ Q^- \end{pmatrix} = \begin{pmatrix} q^+ & 1 \\ q^- & 1 \end{pmatrix}. \tag{8.54}$$

The two equations can be written as

$$Q\check{V}_t + \begin{pmatrix} \tau^+ & 0 \\ 0 & \tau^- \end{pmatrix} Q(\check{V}_u u - \mathscr{S}\check{V}) = 0,$$

or, multiplying to the left by Q^{-1} and noticing that

$$Q^{-1}\begin{pmatrix} \tau^+ & 0 \\ 0 & \tau^- \end{pmatrix} Q = \frac{1}{q^+ - q^-}\begin{pmatrix} \tau^+ q^+ - \tau^- q^- & \tau^+ - \tau^- \\ -(\tau^+ - \tau^-)q^+ q^- & \tau^- q^+ - \tau^+ q^- \end{pmatrix} = \mathscr{T}$$

as defined in (8.10), exactly (8.12):

$$\check{V}_t + \mathscr{T}(\check{V}_u u - \mathscr{S}\check{V}) = 0. \tag{8.12}$$

The boundary conditions are imposed by the solution in the trivial regions – or, equivalently, by the same considerations as for $\mathscr{E}$:

$$\check{V}(t, u_\ell(t)) = \begin{pmatrix} 0 \\ 0 \end{pmatrix}, \qquad \check{V}(t, u_r(t)) = \begin{pmatrix} u_r(t) \\ u_r(t) - \mathscr{K} \end{pmatrix}. \tag{8.55}$$

The existence and uniqueness of the solution of this (pair of) hyperbolic equation(s) will be checked in the next subsection.

It follows from the foregoing analysis that the manifold solution of this equation is indeed traversed by the trajectories generated by τ^- and, specifically, $\dot{w} = \tau^- v - C^- \xi^-$ and by τ^+ and $\dot{w} = \tau^+ v - C^+ \xi^+$, according to (8.51). The associated jump manifold normals are then positive according to Proposition 8.31, implying that $\mathrm{sign}(\xi^\varepsilon) = \varepsilon$, so that the trajectories generated are, as we just said, indeed those of the dynamics of the game.

Finally, if τ takes an intermediate value in (τ^-, τ^+), then it follows from (8.50) that $\mathrm{sign}([p^\varepsilon u + (1 + C^\varepsilon)v] = \bar{\varepsilon}$, so that $\langle \nu^\varepsilon, h(z, \tau, 0)\rangle > 0$ for both ε. Hence the control $\xi = 0$ suffices to ensure that the state drifts in the region $w > W(t, u, v)$. As a matter of fact, it suffices that

$$-\frac{(\tau^+ - \tau)n^+}{\tau^+(C^+ - C^-)} \le \xi \le \frac{(\tau - \tau^-)n^+}{-\tau^-(C^+ - C^-)}$$

to ensure that the states remains "above" (w-wise) the composite barrier.

8.2.4.5 Existence and Uniqueness

For the previous two singular surfaces, $\mathscr{D}$ and $\mathscr{E}$, we had a constructive theory based upon the integration of ODEs. Concerning $\mathscr{F}$, we only have the PDE (8.12). This is an evolution equation whose numerical integration requires some care to avoid numerical instabilities – typical of a hyperbolic PDE – but is not very difficult. Yet, to make it a solid theory, since this is the only definition we have for $\mathscr{F}$, we need to check the existence – and uniqueness, although this will also be a consequence of its yielding a VS of the DQVI (8.7) – of the solution in the domain Λ restricted to $t \in [0, t_+]$. We therefore prove the following theorem.

Theorem 8.32. *Let* $\Lambda_+ = \{(t,u) \mid t \in [0,t_+],\ u \in [u_\ell(t), u_r(t)]\}$, *and let* $\partial_0\Lambda_+ = \{t = t_+, u \in [u_\ell(t_+), u_r(t^+)]\} \cup \{t \in [0,t^+], u = u_\ell(t)\} \cup \{t \in [0,t^+], u = u_r(t)\}$ *be its boundary deprived of its part at* $t = 0$. *The fundamental Eq. (8.12) in the domain* Λ_+ *with continuous initial conditions given on* $\partial_0\Lambda_+$ *has a unique continuous solution.*

Proof. We use the form (8.48) of the PDE, with $q^\varepsilon = C^\varepsilon$:

$$\forall \varepsilon \in \{-,+\}, \quad C^\varepsilon \check{v}_t + \check{w}_t + \tau^\varepsilon[(q^\varepsilon \check{v}_u + \check{w}_u)u - (1+q^\varepsilon)\check{v}] = 0.$$

To make things simpler, we change the unknown functions and consider, for $\varepsilon \in \{-,+\}$ (this is a notation local to this subsection, not to be mistaken with the strategy ϕ),

$$\phi^\varepsilon(t,u) = C^\varepsilon(t)\check{v}(t,u) + \check{w}(t,u);$$

then we take the trajectories u_ℓ and u_r as new coordinate axes. First, let t_0, u_0 be their intersection:

$$t_0 = T + \frac{1}{\tau^+ - \tau^-}\ln\frac{1+c^+}{1+c^-}, \qquad u_0 = \mathscr{K}(1+c^-)^{\frac{-\tau^+}{\tau^+-\tau^-}}(1+c^+)^{\frac{\tau^-}{\tau^+-\tau^-}},$$

and let

$$\begin{cases} \lambda = -\tau^-(t_0 - t) - \ln\frac{u}{u_0}, \\ \mu = \tau^+(t_0 - t) + \ln\frac{u}{u_0}, \end{cases} \quad \text{i.e.,} \quad \begin{cases} t = t_0 - \frac{\lambda+\mu}{\tau^+-\tau^-}, \\ u = u_0 \exp\left(-\frac{\tau^+\lambda + \tau^-\mu}{\tau^+-\tau^-}\right), \end{cases}$$

Finally, let

$$\psi^\varepsilon(\lambda,\mu) = \phi^\varepsilon(t,u).$$

We introduce additional notation, the two positive constants

$$a = \frac{\tau^+(1+C^+)}{(\tau^+-\tau^-)(C^+-C^-)}, \qquad b = \frac{-\tau^-(1+C^-)}{(\tau^+-\tau^-)(C^+-C^-)}.$$

Using (8.49), we find

$$\begin{aligned}\psi^+_\lambda &= \frac{-1}{\tau^+-\tau^-}(\phi^+_t+\phi^+_u\tau^+u) = \frac{-(1+C^+)\tau^+}{\tau^+-\tau^-}\check{v} = a(\psi^- - \psi^+),\\ \psi^-_\mu &= \frac{-1}{\tau^+-\tau^-}(\phi^-_t+\phi^-_u\tau^-u) = \frac{-(1+C^-)\tau^-}{\tau^+-\tau^-}\check{v} = b(\psi^+ - \psi^-).\end{aligned} \tag{8.56}$$

We differentiate the first preceding equality with respect to μ, substitute ψ^-_μ using the second equality, and substitute again $(\psi^+ - \psi^-)$ using the first equality to get

$$\psi^+_{\lambda\mu} + a\psi^+_\mu + b\psi^+_\lambda = 0.$$

Now, let $\chi(\lambda,\mu) = \exp(a\lambda + b\mu)\psi^+(\lambda,\mu)$ and $ab = c$ to obtain, finally,

$$\forall(\lambda,\mu)\in\Lambda_+, \qquad \chi_{\lambda\mu}(\lambda,\mu) = c\chi(\lambda,\mu). \tag{8.57}$$

The domain considered, Λ_+, becomes

$$\begin{aligned}&\lambda \geq 0, \quad \mu \geq 0,\\ &\nu_+ := \ln\frac{1+c^+}{1+c^-} + \frac{\tau^+-\tau^-}{\tau^+}\ln\frac{1+C^+}{1+c^+} \leq \lambda+\mu \leq (\tau^+-\tau^-)T + \ln\frac{1+c^+}{1+c^-} =: \nu_0,\end{aligned}$$

and, using shorthand – hopefully unambiguous – notation,

$$\partial_0\Lambda_+ = \{(\lambda,\mu)\in\mathbb{R}_+\times\mathbb{R}_+ \mid \lambda+\mu = \nu\}\cup\{([\nu_+,\nu_0],0)\}\cup\{(0,[\nu_+,\nu_0])\}.$$

Both ϕ^ε, hence both ψ^ε, are known on this boundary, and thus ψ^+_λ is also known according to (8.56). For a given pair (λ,μ), let $\lambda_0 = \max\{0,\nu_+-\mu\}$ and $\mu_0 = \max\{0,\nu_+-\lambda\}$, so that (λ_0,μ) and (λ,μ_0) are boundary points in $\partial_0\Lambda_+$. Integrating (8.57) from (λ,μ_0) to (λ,μ), we obtain

$$\chi_\lambda(\lambda,\mu) = \chi_\lambda(\lambda,\mu_0) + c\int_{\mu_0}^{\mu}\chi(\lambda,\mu')\,\mathrm{d}\mu'. \tag{8.58}$$

Let, for any (α,β),

$$\Lambda_+^{\alpha,\beta} = \Lambda_+\cap\{\lambda\leq\alpha,\mu\leq\beta\}.$$

Integrating (8.58) from (λ_0,μ) to (λ,μ), we obtain

$$\chi(\lambda,\mu) = \chi(\lambda_0,\mu) + \int_{\lambda_0}^{\lambda}\chi_\lambda(\lambda',\mu_0(\lambda'))\,\mathrm{d}\lambda' + c\iint_{\Lambda_+^{\lambda,\mu}}\chi(\lambda',\mu')\,\mathrm{d}\lambda'\,\mathrm{d}\mu'.$$

The first two of the preceding terms are known boundary terms. This is a fixed-point equation for the function χ. It suffices to find a norm for which the space is complete and such that the mapping

$$\chi \mapsto \Gamma(\chi)(\lambda,\mu) = c \iint_{\Lambda_+^{\lambda,\mu}} \chi(\lambda',\mu')\,\mathrm{d}\lambda'\mathrm{d}\mu'$$

be contracting to conclude. This is performed by choosing

$$\|\chi\|_\alpha = \sup_{(\lambda,\mu)\in\Lambda_+} \mathrm{e}^{-\alpha(\lambda+\mu)}|\chi(\lambda,\mu)|.$$

This norm is defined for all continuous functions since the domain Λ_+ is bounded. Using the majoration $\chi(\lambda,\mu) \le \exp(\alpha(\lambda+\mu))\|\chi\|_\alpha$, we find that

$$\|\Gamma(\chi)\|_\alpha \le \frac{c}{\alpha^2}\|\chi\|_\alpha,$$

so that choosing $\alpha > \sqrt{c}$ gives the desired result.

This proves the existence and uniqueness of χ, and hence of ψ^+, but also, through formula (8.58), of ψ_λ^+, and hence, through the use of (8.56), of ψ^-. □

8.2.5 Synthesis: Representation Formula

At this stage, we have a constructive theory of three singular manifolds:

- $\mathscr{D}$ In the region $t \in [t_-, T]$, the dispersal manifold $\mathscr{D}$ is a submanifold of both sheets $\{\tau^+\}$ and $\{\tau^-\}$.
- $\mathscr{E}$ In the region $t \in [t_+, t_m]$, the equivocal manifold is a submanifold of the sheet $\{\tau^+\}$ and a submanifold of the manifold $\{\iota^-\}$ traversed by the trajectories $\{\tau^-\}$.
- $\mathscr{F}$ In the region $t \in [0, t_+]$, the focal manifold is a submanifold of the two jump manifolds $\{\iota^-\}$ and $\{\iota^+\}$ and traversed by the trajectories $\{\tau^-\}$ and $\{\tau^+\}$.

Therefore, all three satisfy both PDEs (8.48), and hence, as we stressed in the case of $\mathscr{F}$, all satisfy the fundamental Eq. (8.12). We have already noted that this equation is also satisfied by the solutions $\check{v} = \check{w} = 0$ and $\check{v} = \check{w} + \mathscr{K} = u$, which we identified in the trivial regions.

In all cases, these singular manifolds are the intersection of two sheets, either standard semipermeable or jump manifolds. The normals to these manifolds in the plane (v,w) are the two $(q^\varepsilon, 1)$, and therefore their slopes in that plane are $-q^\varepsilon$, as given by (8.33), with $\varepsilon = \mathrm{sign}(\check{v} - v)$. A graphical representation in the plane (v,w) for a fixed (t,u) is therefore as in Fig. 8.5.

Hence, we have the representation [see (8.13)]

$$W(t,u,v) = \check{w}(t,u) + q^\varepsilon\langle \check{v}(t,u) - v \rangle .$$

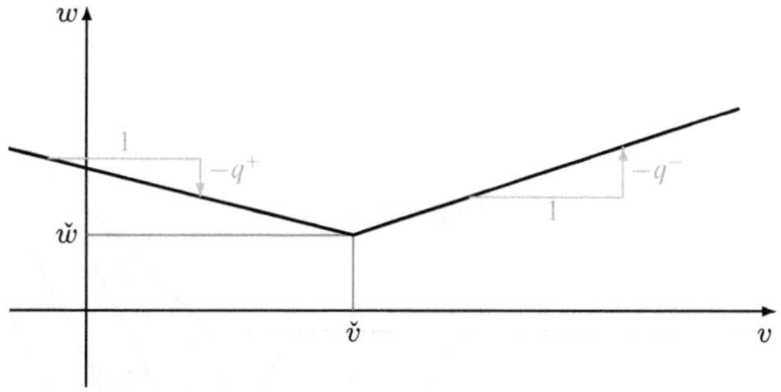

Fig. 8.5 Graph of W in the (v,w) plane for fixed (t,u)

This, together with (8.12) and the boundary conditions according to Table 8.1, gives a representation formula about which we claim that it is the VS of the DQVI (8.7). This will be verified in the next section.

It is useful to remark that, in all cases, $n^\varepsilon > 0$, whether it is the normal to a sheet $\{\tau^\varepsilon\}$, because of (8.44), or because of the specific arguments developed for jump manifolds. And as a result, since $(1\ 0\ \check{v}_t\ \check{w}_t)$ is always orthogonal to both $(n^\varepsilon\ p^\varepsilon\ q^\varepsilon\ 1)$, it follows that we always have

$$Q^\varepsilon \check{V}_t \leq 0. \tag{8.59}$$

And let us also stress that

$$\text{sign}(\check{v} - v) = \text{sign}(-\sigma). \tag{8.60}$$

Remark 8.33. We have developed here the theory in the case $t_+ < t_-$. The opposite case is not ruled out by the standing hypotheses. The analysis is very similar. The only difference comes from the fact that in that case, the envelope junction is between a positive jump manifold and the sheet $\{\tau^-\}$ (instead of a negative jump manifold joining on the sheet $\{\tau^+\}$ as here). The same representation formula holds.

Remark 8.34. As was already stressed, in the case $c^- = C^-$ and $c^+ = C^+$, $t_- = t_+ = T$, both $\mathscr{D}$ and $\mathscr{E}$ collapse, and the focal manifold $\mathscr{F}$ accounts for the whole region of interest. This simplifies the calculations because then both $q^\varepsilon = C^\varepsilon$ are constant.

8.3 Viscosity Solution

In this section, we will show that the VS of the DQVI (8.7) is unique and use this fact to confirm that the function given by the representation formula (8.12), (8.13) is indeed the value of the minimax problem, i.e., gives the best premium for the call as in (7.23).

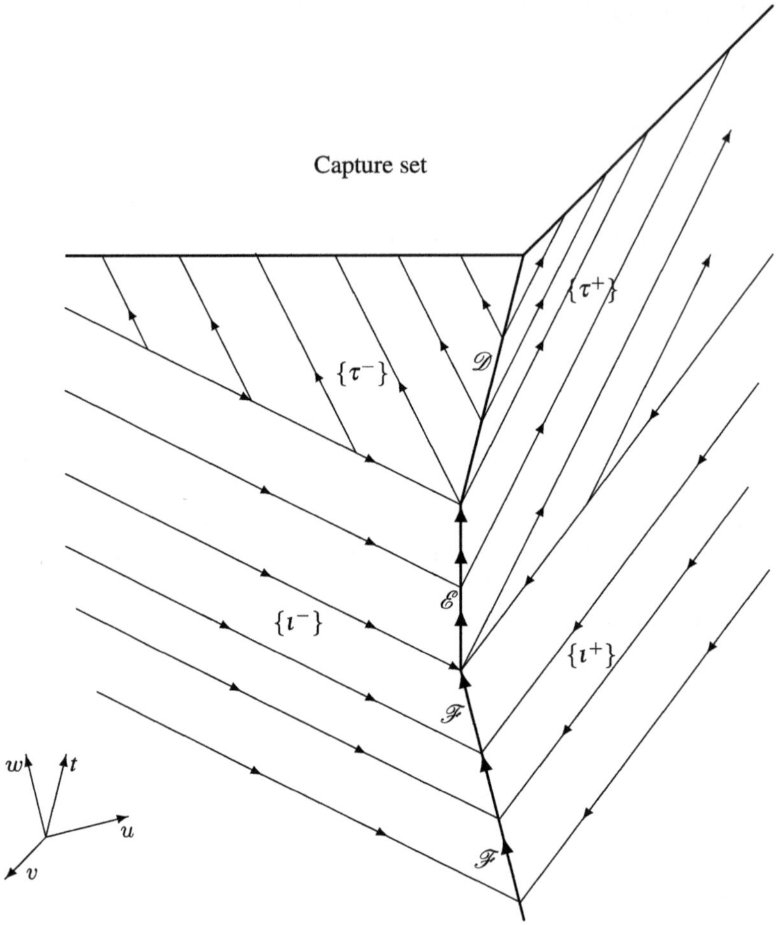

Fig. 8.6 A two-dimensional sketch of the four-dimensional geometry of the barrier, case $t_+ < t_-$

8.3.1 Uniqueness

While the value of the minimax problem is necessarily a VS – classical proofs as in [69] carry over to that case – uniqueness does not seem to be a consequence of known theorems. Even [137], which considers a problem similar to, and more general than, ours, rules out positively homogeneous impulse costs.

8.3.1.1 Viscosity Solutions and Opposite Viscosity Solutions

We first recall the definition of VSs and introduce the concept of *opposite viscosity solution* for convenience.

Definition 8.35. Let $f : \mathbb{R}^n \to \mathbb{R}$ be a uniformly continuous function. Its *subdifferential* at $x \in \mathbb{R}^n$ is the subset of $\mathbb{R}^n$

$$\partial_- f(x) := \left\{ p \in \mathbb{R}^n \mid \liminf_{y \to x} \frac{f(y) - f(x) - \langle p, y - x \rangle}{\|y - x\|} \geq 0 \right\}.$$

The *superdifferential* is the subset

$$\partial_+ f(x) := \left\{ p \in \mathbb{R}^n \mid \limsup_{y \to x} \frac{f(y) - f(x) - \langle p, y - x \rangle}{\|y - x\|} \leq 0 \right\}.$$

Only one of these two is nonempty at each x, except if f is differentiable at x, in which case both are the singleton $\{f'(x)\}$. We further have (see, e.g., [12]) the following proposition.

Proposition 8.36.

- $\partial_- f(x)$ *is the set of slopes* $g'(x)$ *of differentiable functions* $g(\cdot)$ *such that* $f - g$ *has a local minimum at* x.
- $\partial_+ f(x)$ *is the set of slopes* $g'(x)$ *of differentiable functions* $g(\cdot)$ *such that* $f - g$ *has a local maximum at* x.

Let $Q \subset \mathbb{R}^n$ be an open region with a smooth boundary ∂Q. Let $F : Q \times \mathbb{R} \times \mathbb{R}^n$ and $G : \partial Q \to \mathbb{R}$ be given continuous functions, and consider the PDE

$$\forall x \in Q \quad F(x, f(x), \mathrm{D}f(x)) = 0, \qquad \forall x \in \partial Q, \quad f(x) = G(x).$$

The following definition is classic (see, e.g., [21, 22]).

Definition 8.37.

- f is a *viscosity subsolution* if

$$\forall x \in Q, \forall p \in \partial_+ f(x), \quad F(x, f(x), p) \leq 0,$$
$$\forall x \in \partial Q, \quad f(x) \leq G(x).$$

- f is a *viscosity supersolution* if

$$\forall x \in Q, \forall p \in \partial_- f(x), \quad F(x, f(x), p) \geq 0,$$
$$\forall x \in \partial Q, \quad f(x) \geq G(x).$$

- f is a VS if it is both a viscosity subsolution and a viscosity supersolution.

It should be stressed that this very useful definition has the strange feature that a opposite viscosety solution of $F(x, f, \mathrm{D}f) = 0$ is *not* a VS of $-F(x, f, \mathrm{D}f) = 0$. Moreover, the classic choice of signs, arising from fluid mechanics and the calculus of variations, is not well suited to the Hamilton–Jacobi–Isaacs equation. Hence, we give the following definition:

Definition 8.38. f is an *opposite viscosity solution* (OVS) of $F(x,f,\mathrm{D}f)=0$ if it is a VS of $-F(x,f,\mathrm{D}f)=0$.

Consequently, Theorem 8.12 also reads as follows.

Theorem 8.39. *The value function W is a continuous OVS of the PDE*

$$\begin{aligned}
&\forall(t,u,v)\in[0,T)\times\mathbb{R}_+\times\mathbb{R},\\
&\quad\min\left\{\frac{\partial W}{\partial t}+\tau^{\varepsilon}\left\langle\frac{\partial W}{\partial u}u+\left(\frac{\partial W}{\partial v}-1\right)v\right\rangle,\frac{\partial W}{\partial v}+C^{+},-\left(\frac{\partial W}{\partial v}+C^{-}\right)\right\}=0,\\
&\forall(u,v)\in\mathbb{R}_+\times\mathbb{R},\quad W(T,u,v)=N(u,v).
\end{aligned}\tag{8.61}$$

8.3.1.2 An Auxiliary Problem with a Bounded Value

Let us fix a number a, and let b be the associated b as in Proposition 8.9. Let $\Phi_b\subset\Phi$ be the set of strategies φ that keep $|v(t)|\leq b$ for any $v(0)\leq a$. Let also $\mathscr{P}_{[c,d]}$ denote the simple projection of $\mathbb{R}$ onto $[c,d]$. Finally, let

$$N_b(u,v)=N(\mathscr{P}_{[0,ae^{\tau^{\sharp}T}]}(u),\mathscr{P}_{[-b,b]}(v)),\quad L_b(v)=\mathscr{P}_{[-b,b]}(v)$$

and

$$J_b(t,u(t),v(t);\xi,\tau(\cdot))=N_b(u(T),v(T))+\int_t^T[-\tau L_b(v(s))+C^{\varepsilon}\xi(s)]\mathrm{d}s$$

and

$$W_b(t,u,v)=\inf_{\varphi\in\Phi_b}\sup_{\tau(\cdot)\in\Omega}J_b(t,u,v;\varphi(\tau(\cdot)),\tau(\cdot)).$$

An easy result is the following proposition.

Proposition 8.40. *The function W_b is bounded uniformly continuous (BUC) over $[0,T]\times\mathbb{R}_+\times\mathbb{R}$.*

Proof. Boundedness follows from the fact that

$$W_b(t,u,v)\leq\sup_{\tau(\cdot)\in\Omega}J_b(t,u,v;0,\tau(\cdot))\leq\max_{u\in[0,a\exp(\tau^{\sharp}T)],|v|\leq b}N(u,v)+T\tau^{\sharp}b.$$

W_b is continuous. For $u(0)\geq a\exp(2\tau^{\sharp}T)$, for all $\tau(\cdot)\in\Omega$, $u(T)$ is larger than $a\exp(\tau^{\sharp}T)$, so that J_b is independent of u. But then we are left with a simple control problem whose Value is a continuous function of $v(0)=v$ alone. Moreover, if

$|v(0)|$ is so large, say $|v(0)| \geq c$, that the cost of bringing it back to b is larger than $N(a\exp(\tau^\sharp T), b) + T\tau^\sharp b$, then the best ξ can do is $\xi = 0$, so that W_b is constant for v outside a compact $[-c, c]$. Hence W_b is uniformly continuous for $u \geq a\exp(2\tau^\sharp T)$. Similarly, for $|v| \geq c$, there is nothing ξ can do, and W_b is just the Value function of a simple minimization problem depending on u alone, and constant if u is large. Therefore, W_b is a continuous function in the compact $[0, a\exp(2\tau^\sharp T] \times [-c, c]$, hence uniformly continuous there and, as we have just seen, also uniformly continuous outside this compact. □

We claim that the following proposition holds.

Proposition 8.41. *The restrictions of W and W_b to $[0, T] \times [0, a] \times [-a, a]$ coincide.*

Proof. Notice first that for trajectories with $|v(0)| \leq a$, we already know that the restriction to strategies $\varphi \in \Phi_b$ does not change the Value of the game. Additionally, it follows from the game's dynamics that if $u(0) \leq a$, then $u(t) \leq a\exp(\tau^\sharp T)$ for all $t \in [0, T]$. But in constructing J_b, we have modified J only for states (u, v) outside of the region reachable from the initial states in $[0, a] \times [-a, a]$ with the strategies of Φ_b. Hence for these initial states, and under strategies of Φ_b, the two performance indices coincide. □

The value function W_b is therefore an OVS of the modified DQVI

$$\forall (t, u, v) \in [0, T) \times \mathbb{R}_+ \times \mathbb{R},$$
$$\min\left\{\frac{\partial W_b}{\partial t} + \tau^\varepsilon\left\langle\frac{\partial W_b}{\partial u}u + \frac{\partial W_b}{\partial v}v - L_b(v)\right\rangle, \frac{\partial W_b}{\partial v} + C^+, -\left(\frac{\partial W_b}{\partial v} + C^-\right)\right\} = 0,$$
$$\forall (u, v) \in \mathbb{R}_+ \times \mathbb{R}, \quad W_b(T, u, v) = N_b(u, v). \tag{8.62}$$

The next two subsections are devoted to the proof of the following theorem.

Theorem 8.42. *The DQVI (8.62) admits a unique bounded uniformly continuous (BUC) OVS.*

From this result and Proposition 8.41 follow the next corollary, and consequently Theorem 8.2

Corollary 8.43. *The DQVI (8.7) admits a unique continuous viscosity solution.*

Proof. Assume that two different solutions W_1 and W_2 exist. There is a triplet $(\bar{t}, \bar{u}, \bar{v})$ for which $W_1(\bar{t}, \bar{u}, \bar{v}) \neq W_2(\bar{t}, \bar{u}, \bar{v})$. Choose $a > \max\{u, |\bar{v}|\}$, and let b correspond to it as in Proposition 8.9. Then, on the one hand, $W_b(\bar{t}, \bar{u}, \bar{v})$ is uniquely defined, and on the other hand, $W_1(\bar{t}, \bar{u}, \bar{v}) = W_b(\bar{t}, \bar{u}, \bar{v}) = W_2(\bar{t}, \bar{u}, \bar{v})$, a contradiction. □

8.3.1.3 Uniqueness of BUC OVS of (8.62)

We now set out to prove Theorem (8.42). We omit all indices b, but it should be understood all along that we are dealing with the modified problem.

Proof of the Theorem

We will consider the DQVI for $V = e^t W$. It satisfies another DQVI, (8.72). Assume that it has two BUC OVSs, V and V'. Choose $\varepsilon > 0$ and $\varepsilon < \|V\|_\infty$ (ε is to go to 0). Choose $\mu \in (1-\varepsilon/\|V\|_\infty, 1)$, and let $U = \mu V$. Then

$$\forall(t,u,v), \quad |V(t,u,v) - U(t,u,v)| \leq \varepsilon.$$

Let $\mathbf{M} = \sup_{t,u,v}(U(t,u,v) - V'(t,u,v))$. It follows that

$$\sup_{t,u,v}[V(t,u,v) - V'(t,u,v)] \leq \mathbf{M} + \varepsilon. \tag{8.63}$$

We now claim that the following lemma is true.

Lemma 8.44. *There exists $\mu^\star \in (1-\varepsilon/\|V\|_\infty, 1)$ and a constant $\mathsf{K} > 0$, both depending only on the data of the problem, such that, if $\mu \in (\mu^\star, 1)$, then $\mathbf{M} \leq \mathsf{K}\varepsilon$.*

As a consequence,

$$\sup_{t,u,v}[V(t,u,v) - V'(t,u,v)] \leq (\mathsf{K}+1)\varepsilon. \tag{8.64}$$

Since ε was chosen arbitrarily, it follows that for all (t,u,v), $V(t,u,v) \leq V'(t,u,v)$. But since the argument is symmetric in V and V', necessarily $V = V'$, and Theorem 8.42 is proved. □

Proof of Lemma 8.44

Notice first that if $\mathbf{M} \leq 0$, then it follows from (8.63) that, $\sup_{t,u,v}[V(t,u,v) - V'(t,u,v)] \leq \varepsilon$, and (8.64) is satisfied for any positive K. We may, therefore, from now on concentrate on the case $\mathbf{M} > 0$.

Let, thus, $0 < \varepsilon < \|V\|_\infty$ be given. Let

$$b := \max_{v\in\mathbb{R}} |L(v)| = \|L\|_\infty, \quad c := \max_{u\in\mathbb{R}_+, v\in\mathbb{R}} |N(u,v)| = \|N\|_\infty$$

(remember here that L and N stand for L_b and N_b), and choose μ such that

$$1 > \mu \geq \mu^\star = 1 - \frac{\varepsilon}{\max\{\|V\|_\infty, be^{\mathrm{T}}, ce^{\mathrm{T}}\}} < 1. \tag{8.65}$$

For three positive numbers α, β, γ (which we will choose small later on), introduce the test function $\phi_{\alpha,\beta,\gamma} : [0,T] \times \mathbb{R}^2 \times [0,T] \times \mathbb{R}^2 \to \mathbb{R}$:

$$\phi_{\alpha,\beta,\gamma}(t,u,v,t',u',v') = U(t,u,v) - V'(t',u',v') \\ -\alpha(u^2+u'^2+v^2+v'^2) - \tfrac{(u-u')^2+(v-v')^2}{\beta^2} - \tfrac{(t-t')^2}{\gamma^2}.$$

This function reaches its maximum at

$$\max \phi_{\alpha,\beta,\gamma}(t,u,v,t',u',v') = \phi_{\alpha,\beta,\gamma}(\bar{t},\bar{u},\bar{v},\bar{t}',\bar{u}',\bar{v}') =: \mathbf{M}_{\alpha,\beta,\gamma}.$$

We claim the following two lemmas, both for $\mu \in (\mu^\star, 1)$ fixed and under the hypothesis that $\mathbf{M} > 0$.

Lemma 8.45. *There exists $\alpha^\star$, $\beta^\star$, $\gamma^\star$ all positive such that for any $\alpha \leq \alpha^\star$, $\beta \leq \beta^\star$, $\gamma \leq \gamma^\star$,*

$$|U(\bar{t},\bar{u},\bar{v}) - V'(\bar{t}',\bar{u}',\bar{v}') - \mathbf{M}| \leq \varepsilon, \tag{8.66}$$

$$\alpha(\bar{u}^2+\bar{u}'^2+\bar{v}^2+\bar{v}'^2) + \frac{(\bar{u}-\bar{u}')^2+(\bar{v}-\bar{v}')^2}{\beta^2} + \frac{(\bar{t}-\bar{t}')^2}{\gamma^2} \leq 2\varepsilon. \tag{8.67}$$

Lemma 8.46. *For any $\alpha \leq \alpha^\star$, $\beta \leq \beta^\star$, $\gamma \leq \gamma^\star$,*

$$U(\bar{t},\bar{u},\bar{v}) - V'(\bar{t}',\bar{u}',\bar{v}') \leq \max\{2, 7\tau^\sharp\}\varepsilon. \tag{8.68}$$

The main lemma follows clearly, with $\mathsf{K} = \max\{3, 7\tau^\sharp + 1\}$, from inequalities (8.66) and (8.68). □

Inequality (8.67) is used in the proof of Lemma 8.46. We have split the assertions into two separate lemmas because the first one does not make use of the DQVI while the second one does.

8.3.1.4 Proof of Lemma 8.45

Assume that $\mathbf{M} > 0$. Choosing $(t,u,v) = (t',u',v')$, it follows that

$$\forall (t,u,v), \quad \mathbf{M}_{\alpha,\beta,\gamma} \geq U(t,u,v) - V'(t,u,v) - 2\alpha(u^2+v^2). \tag{8.69}$$

Pick a point $(t^\star, u^\star, v^\star)$ such that $\mathbf{M}$ is approached within $\varepsilon/2$:

$$U(t^\star,u^\star,v^\star) - V'(t^\star,u^\star,v^\star) \geq \mathbf{M} - \varepsilon/2,$$

and let $\alpha_1 = \varepsilon/[4(u^{\star 2}+v^{\star 2})]$ if $(u^{\star 2}+v^{\star 2}) \neq 0$ (and $\alpha_1 = \infty$ otherwise). It follows that for any $\alpha \le \alpha_1$,

$$U(t^\star,u^\star,v^\star) - V'(t^\star,u^\star,v^\star) - 2\alpha(u^{\star 2}+v^{\star 2}) \ge \mathbf{M} - \varepsilon,$$

and using (8.69),

$$\mathbf{M} - \varepsilon \le \mathbf{M}_{\alpha,\beta,\gamma}. \tag{8.70}$$

Hence,

$$\begin{aligned} -\varepsilon \le \mathbf{M} - \varepsilon &\le \mathbf{M}_{\alpha,\beta,\gamma} \\ &\le \|U\|_\infty + \|V'\|_\infty - \alpha(\bar{u}^2+\bar{u}'^2+\bar{v}^2+\bar{v}'^2) - \frac{(\bar{u}-\bar{u}')^2+(\bar{v}-\bar{v}')^2}{\beta^2} - \frac{(\bar{t}-\bar{t}')^2}{\gamma^2}. \end{aligned}$$

Let $r^2 := \|V\|_\infty + \|V'\|_\infty + \varepsilon$, and note that $\|U\|_\infty < \|V\|_\infty$. It follows that, for $\alpha \le \alpha_1$,

$$\alpha(\bar{u}^2+\bar{u}'^2+\bar{v}^2+\bar{v}'^2) + \frac{(\bar{u}-\bar{u}')^2+(\bar{v}-\bar{v}')^2}{\beta^2} + \frac{(\bar{t}-\bar{t}')^2}{\gamma^2} \le r^2,$$

and in particular that

$$\alpha(\bar{u}^2+\bar{u}'^2+\bar{v}^2+\bar{v}'^2) \le r^2,\ |\bar{u}-\bar{u}'| \le r\beta,\ |\bar{v}-\bar{v}'| \le r\beta,\ |\bar{t}-\bar{t}'| \le r\gamma. \tag{8.71}$$

Now, V' is uniformly continuous by hypothesis. Let, for p and q positive,

$$m(p,q) = \sup_{\substack{|t-t'|\le q \\ |u-u'|\le p \\ |v-v'|\le p}} |V'(t,u,v) - V'(t',u',v')|.$$

Clearly, m is decreasing with its arguments and decreases to 0 with $p+q$. Using (8.71), it follows that

$$U(\bar{t},\bar{u},\bar{v}) - V'(\bar{t}',\bar{u}',\bar{v}') \le U(\bar{t},\bar{u},\bar{v}) - V'(\bar{t},\bar{u},\bar{v}) + m(r\beta,r\gamma) \le \mathbf{M} + m(r\beta,r\gamma).$$

Choose β_1 and γ_1 such that for $\beta \le \beta_1$ and $\gamma \le \gamma_1$, $m(r\beta,r\gamma) \le \varepsilon$. Using again (8.70), we get

$$\mathbf{M} - \varepsilon \le \mathbf{M}_{\alpha,\beta,\gamma} \le U(\bar{t},\bar{u},\bar{v}) - V'(\bar{t}',\bar{u}',\bar{v}') \le \mathbf{M} + \varepsilon.$$

Conclusions (8.66) and (8.67) of the lemma follow. □

8.3.1.5 Proof of Lemma 8.46

We first apply a classic transformation to DQVI (8.62), introducing

$$V(t,u,v) := \mathrm{e}^t W(t,u,v),$$

which is BUC if and only if W is. Now, W is an OVS of (8.62) if and only if V is an OVS of the modified DQVI

$$\begin{aligned}
&\forall (t,u,v) \in [0,T) \times \mathbb{R}^2,\\
&0 = \min\left\{ \frac{\partial V}{\partial t} - V(t,u,v) + \tau^{\varepsilon}\left\langle \frac{\partial V}{\partial u}u + \frac{\partial V}{\partial v}v - \mathrm{e}^t L(v)\right\rangle, \right. \\
&\qquad\qquad \left. \frac{\partial V}{\partial v} + \mathrm{e}^t C^+, -\frac{\partial V}{\partial v} - \mathrm{e}^t C^- \right\}, \\
&V(T,u,v) = \mathrm{e}^{\mathrm{T}} M(u,v), \quad \forall (u,v) \in \mathbb{R}^2.
\end{aligned} \tag{8.72}$$

We will also make use of the following remark. For any positive μ, which we will take to be smaller than one, let $U(t,u,v) = \mu V(t,u,v)$. It is an OVS of the third DQVI

$$\begin{aligned}
&\forall (t,u,v) \in [0,T) \times \mathbb{R}^2,\\
&0 = \min\left\{ \frac{\partial U}{\partial t} - U(t,u,v) + \tau^{\varepsilon}\left\langle \frac{\partial U}{\partial u}u + \frac{\partial U}{\partial v}v - \mu\mathrm{e}^t L(v)\right\rangle, \right. \\
&\qquad\qquad \left. \frac{\partial U}{\partial v} + \mu\mathrm{e}^t C^+, -\frac{\partial U}{\partial v} - \mu\mathrm{e}^t C^- \right\}, \\
&U(T,u,v) = \mu\mathrm{e}^{\mathrm{T}} M(u,v), \quad \forall (u,v) \in \mathbb{R}^2.
\end{aligned}$$

As a matter of fact, the DQVI (8.72) is a particular case of this one with $\mu = 1$. We gave it separately for subsequent reference.

Case $\bar{t}$ and $\bar{t}'$ Smaller Than T

By definition of $(\bar{t},\bar{u},\bar{v})$ and $(\bar{t}',\bar{u}',\bar{v}')$, we have

$$\begin{aligned}
&U(\bar{t},\bar{u},\bar{v}) - V'(\bar{t}',\bar{u}',\bar{v}') - \alpha(\bar{u}^2+\bar{u}'^2+\bar{v}^2+\bar{v}'^2) - \tfrac{(\bar{u}-\bar{u}')^2+(\bar{v}-\bar{v}')^2}{\beta^2} - \tfrac{(\bar{t}-\bar{t}')^2}{\gamma^2}\\
&\quad \geq U(t,u,v) - V'(\bar{t}',\bar{u}',\bar{v}') - \alpha(u^2+\bar{u}'^2+v^2+\bar{v}'^2) - \tfrac{(u-\bar{u}')^2+(v-\bar{v}')^2}{\beta^2} - \tfrac{(t-\bar{t}')^2}{\gamma^2},
\end{aligned}$$

and also

$$\begin{aligned}
&U(\bar{t},\bar{u},\bar{v}) - V'(\bar{t}',\bar{u}',\bar{v}') - \alpha(\bar{u}^2+\bar{u}'^2+\bar{v}^2+\bar{v}'^2) - \tfrac{(\bar{u}-\bar{u}')^2+(\bar{v}-\bar{v}')^2}{\beta^2} - \tfrac{(\bar{t}-\bar{t}')^2}{\gamma^2}\\
&\quad \geq U(\bar{t},\bar{u},\bar{v}) - V'(t',u',v') - \alpha(\bar{u}^2+u'^2+\bar{v}^2+v'^2) - \tfrac{(\bar{u}-u')^2+(\bar{v}-v')^2}{\beta^2} - \tfrac{(\bar{t}-t')^2}{\gamma^2}.
\end{aligned}$$

Define the two test functions:

$$\phi(t,u,v) = V'(\bar{t}',\bar{u}',\bar{v}') + \alpha(u^2+\bar{u}'^2+v^2+\bar{v}'^2) + \frac{(u-\bar{u}')^2+(v-\bar{v}')^2}{\beta^2} + \frac{(t-\bar{t}')^2}{\gamma^2},$$

$$\phi'(t',u',v') = U(\bar{t},\bar{u},\bar{v}) - \alpha(\bar{u}^2+u'^2+\bar{v}^2+v'^2) - \frac{(\bar{u}-u')^2+(\bar{v}-v')^2}{\beta^2} - \frac{(\bar{t}-t')^2}{\gamma^2}.$$

The first inequality above means that $(\bar{t},\bar{u},\bar{v})$ is a maximal point of $U-\phi$, and the second that $(\bar{t}',\bar{u}',\bar{v}')$ is a minimal point of $V'-\phi'$. Using the definition of an OVS and Proposition 8.36, it follows that

$$\text{at}\quad \bar{t},\bar{u},\bar{v},\quad \min\left\{\frac{\partial\phi}{\partial t} - U + \tau^{\varepsilon}\left\langle \frac{\partial\phi}{\partial u}\bar{u} + \frac{\partial\phi}{\partial v}\bar{v} - \mu e^{\bar{t}} L\right\rangle,\right.$$
$$\left.\frac{\partial\phi}{\partial v} + \mu e^{\bar{t}} C^+, -\frac{\partial\phi}{\partial v} - \mu e^{\bar{t}} C^-\right\} \geq 0,$$

$$\text{at}\quad \bar{t}',\bar{u}',\bar{v}',\quad \min\left\{\frac{\partial\phi'}{\partial t'} - V' + \tau^{\varepsilon}\left\langle \frac{\partial\phi'}{\partial u'}\bar{u}' + \frac{\partial\phi'}{\partial v'}\bar{v}' - e^{\bar{t}'} L\right\rangle,\right.$$
$$\left.\frac{\partial\phi'}{\partial v'} + e^{\bar{t}'} C^+, -\frac{\partial\phi'}{\partial v'} - e^{\bar{t}'} C^-\right\} \leq 0.$$

The first inequality can be decomposed into three inequalities:

$$2\frac{\bar{t}-\bar{t}'}{\gamma^2} - U(\bar{t},\bar{u},\bar{v}) + 2\tau^{\varepsilon}\left\langle \alpha\bar{u}^2 + \frac{\bar{u}-\bar{u}'}{\beta^2}\bar{u} + \alpha\bar{v}^2 + \frac{\bar{v}-\bar{v}'}{\beta^2}\bar{v} - \frac{\mu}{2}e^{\bar{t}}L(\bar{v})\right\rangle \geq 0, \tag{8.73}$$

$$-\mu e^{\bar{t}} C^+ \leq 2\alpha\bar{v} + 2\frac{\bar{v}-\bar{v}'}{\beta^2} \leq -\mu e^{\bar{t}} C^-. \tag{8.74}$$

The second inequality reads

$$\min\left\{2\frac{\bar{t}-\bar{t}'}{\gamma^2} - V'(\bar{t}',\bar{u}',\bar{v}')\right.$$
$$+2\tau^{\varepsilon}\left\langle -\alpha\bar{u}'^2 + \frac{\bar{u}-\bar{u}'}{\beta^2}\bar{u}' - \alpha\bar{v}'^2 + \frac{\bar{v}-\bar{v}'}{\beta^2}\bar{v}' - \frac{1}{2}e^{\bar{t}'}L(\bar{v}')\right\rangle, \tag{8.75}$$
$$\left.-2\alpha\bar{v}' + 2\frac{\bar{v}-\bar{v}'}{\beta^2} + e^{\bar{t}'} C^+,\quad 2\alpha\bar{v}' - 2\frac{\bar{v}-\bar{v}'}{\beta^2} - e^{\bar{t}'} C^-\right\} \leq 0.$$

Now we want to use inequalities (8.74) to show that the last two terms of (8.75) can be made positive, which will imply that the first one is negative. Let us therefore write the following string of inequalities, which makes use of (8.74) between the second and third lines, then of (8.71):

$$
\begin{aligned}
&-2\alpha\bar{v}' + 2\frac{\bar{v}-\bar{v}'}{\beta^2} + e^{\bar{t}'}C^+ \\
&= 2\alpha\bar{v} + 2\frac{\bar{v}-\bar{v}'}{\beta^2} + \mu e^{\bar{t}}C^+ - 2\alpha(\bar{v}+\bar{v}') + (e^{\bar{t}'} - \mu e^{\bar{t}})C^+ \\
&\geq -2\alpha(|\bar{v}| + |\bar{v}'|) + (e^{\bar{t}'} - e^{\bar{t}})C^+ + (1-\mu)e^{\bar{t}}C^+ \\
&\geq -4r\sqrt{\alpha} - e^T r\gamma C^+ + (1-\mu)C^+.
\end{aligned}
$$

Hence, choose

$$
\alpha_2 = \min\left\{\alpha_1, \frac{(1-\mu)^2 C^{+2}}{64r^2}\right\} \quad \text{and} \quad \gamma_2 = \min\left\{\gamma_1, e^{-T}\frac{1-\mu}{2r}\right\}.
$$

The choice of $\alpha \leq \alpha_2$, $\gamma \leq \gamma_2$ ensures that this term is positive, without destroying the effects sought with the choice of α_1 and γ_1.

In a similar fashion, we have

$$
\begin{aligned}
&2\alpha\bar{v}' - 2\frac{\bar{v}-\bar{v}'}{\beta^2} - e^{\bar{t}'}C^- \\
&= -2\alpha\bar{v} - 2\frac{\bar{v}-\bar{v}'}{\beta^2} - \mu e^{\bar{t}}C^- + 2\alpha(\bar{v}+\bar{v}') - (e^{\bar{t}'} - \mu e^{\bar{t}})C^- \\
&\geq -4r\sqrt{\alpha} + e^T r\gamma C^- - (1-\mu)C^-.
\end{aligned}
$$

Again, define

$$
\alpha_3 = \min\left\{\alpha_2, \frac{(1-\mu)^2 C^{-2}}{64r^2}\right\} \quad \text{and} \quad \gamma_3 - \min\left\{\gamma_2, e^{-T}\frac{1-\mu}{2r}\right\},
$$

and the choice $\alpha \leq \alpha_3$, $\gamma \leq \gamma_3$ ensures that both terms are positive.

Therefore, with these choices of parameters α, β, γ, we have

$$
\begin{aligned}
&2\frac{\bar{t}-\bar{t}'}{\gamma^2} - V'(\bar{t}', \bar{u}', \bar{v}') \\
&+2\tau^\varepsilon\left\langle -\alpha\bar{u}'^2 + \frac{\bar{u}-\bar{u}'}{\beta^2}\bar{u}' - \alpha\bar{v}'^2 + \frac{\bar{v}-\bar{v}'}{\beta^2}\bar{v}' - \frac{1}{2}e^{\bar{t}'}L(\bar{v}')\right\rangle \leq 0.
\end{aligned} \tag{8.76}
$$

We now make the difference (8.76) – (8.73) and make use of $\tau^\sharp = \max\{\tau^+, -\tau^-\}$ and the fact that

$$
\max_{\tau\in[\tau^-,\tau^+]} \tau A - \max_{\tau\in[\tau^-,\tau^+]} \tau B \leq \max_{\tau\in[\tau^-,\tau^+]} \tau(A-B) \leq \tau^\sharp |A-B|.
$$

This yields

$$U(\bar{t},\bar{u},\bar{v})-V'(\bar{t}',\bar{u}',\bar{v}')\leq 2\tau^{\sharp}\left[\alpha(\bar{u}^2+\bar{u}'^2+\bar{v}^2+\bar{v}'^2)+\frac{(\bar{u}-\bar{u}')^2+(\bar{v}-\bar{v}')^2}{\beta^2}\right.$$
$$\left.+\frac{1}{2}(\mathrm{e}^{\bar{t}'}L(\bar{v}')-\mu\mathrm{e}^{\bar{t}}L(\bar{v}))\right].$$

Using (8.67), the first line on the right-hand side of the preceding equation is less than $4\tau^*\varepsilon$ for any $(\alpha,\beta,\gamma)\leq(\alpha_1,\beta_1,\gamma_1)$, and a fortiori for $(\alpha,\beta,\gamma)\leq(\alpha_3,\beta_3,\gamma_3)$. Also,

$$\begin{aligned}\mathrm{e}^{\bar{t}'}L(\bar{v}')-\mu\mathrm{e}^{\bar{t}}L(\bar{v})&=(\mathrm{e}^{\bar{t}'}-\mathrm{e}^{\bar{t}})L(\bar{v}')+\mathrm{e}^{\bar{t}}(L(\bar{v}')-L(\bar{v}))+(1-\mu)L(\bar{v})\\&\leq \mathrm{e}^{\mathrm{T}}[r\gamma b+|L(\bar{v}')-L(\bar{v})|+(1-\mu)b].\end{aligned}$$

According to our choice [see (8.65)] of μ, the last term on the right-hand side of the preceding equation is not larger than ε. Let β_4 be small enough so that, for any $|\bar{v}'-\bar{v}|\leq r\beta_4$, $|L(\bar{v}')-L(\bar{v})|\leq\varepsilon$, which is possible since L is uniformly continuous. Choosing $\beta\leq\beta_4$ and $\gamma\leq\gamma_4=\min\{\gamma_3,\varepsilon/(\mathrm{e}^{\mathrm{T}}rb)\}$, the first two terms are also not larger than ε. Therefore, with this choice of (α,β,γ), we have

$$0<U(\bar{t},\bar{u},\bar{v})-V'(\bar{t}',\bar{u}',\bar{v}')\leq 7\tau^{\sharp}\varepsilon.$$

It remains to use inequality (8.66) to obtain $\mathbf{M}\leq(7\tau^{\sharp}+1)\varepsilon$. This is the inequality $\mathbf{M}\leq K\varepsilon$ predicted in Sect. 8.3.1.3.

Case $\bar{t}=T$ or $\bar{t}'=T$

If $\bar{t}=T$, then it follows that $U(\bar{t},\bar{u},\bar{v})=\mu\mathrm{e}^{\mathrm{T}}N(\bar{u},\bar{v})$. It also holds that $V'(\bar{t},\bar{u},\bar{v})=\mathrm{e}^{\mathrm{T}}N(\bar{u},\bar{v})$ and $|V'(\bar{t}',\bar{u}',\bar{v}')-V'(\bar{t},\bar{u},\bar{v})|\leq m(r\beta,r\gamma)\leq\varepsilon$. (This last inequality holds as soon as $\beta\leq\beta_1$ and $\gamma\leq\gamma_1$.) Recall that $\|N\|_\infty=c$ and (8.65). Hence,

$$U(\bar{t},\bar{u},\bar{v})-V'(\bar{t}',\bar{u}',\bar{v}')\leq(1-\mu)\mathrm{e}^{\mathrm{T}}c+m(r\beta,r\gamma)\leq 2\varepsilon.$$

If $\bar{t}'=T$, then $V'(\bar{t}',\bar{u}',\bar{v}')=\mathrm{e}^{\mathrm{T}}N(\bar{u}',\bar{v}')$. Choose $\beta_5<\beta_4$ and $\gamma_5\leq\gamma_4$ such that for $|u-u'|\leq r\beta_5$, $|v-v'|\leq r\beta_5$, and $|t-t'|\leq r\gamma_5$, it results that $|U(t,u,v)-U(t',u',v')|\leq\varepsilon$. (This is possible since, like V', U is assumed to be uniformly continuous.) It results that

$$U(\bar{t},\bar{u},\bar{v})-V'(\bar{t}',\bar{u}',\bar{v}')\leq\varepsilon+(\mu-1)\mathrm{e}^{\mathrm{T}}N(\bar{u}',\bar{v}')\leq\varepsilon+(1-\mu)\mathrm{e}^{T}c\leq 2\varepsilon.$$

Finally, the case where $\bar{t}=\bar{t}'=T$ is taken care of by any of the preceding two cases.

We may now set $\alpha^\star = \alpha_3$, $\beta^\star = \beta_5$, and $\gamma^\star = \gamma_5$, and the two lemmas are proved, as are Lemma 8.44 and the theorem.

8.3.2 Verification of Representation Formula

In this subsection, we will prove the following fact.

Theorem 8.47. *The value function defined by (8.13) is the VS of the DQVI (8.7).*

The rest of this subsection is devoted to that verification. Therefore, W always stands for its representation (8.13). Again, we will be dealing with the case $t_+ < t_-$. The other case is very similar.

There are three ways in which W may have a discontinuous gradient: at $v = \check{v}$, because q^ε is discontinuous across this manifold, or at $t = t_\varepsilon$, because $\dot{q}^\varepsilon$ is discontinuous there, or, finally, along manifolds where $\check{V}$ has a discontinuous gradient. We will investigate first the regular regions, then these three cases.

Recall that we seek an *OVS* of (8.61). We use the notation

$$H(t,u,v;DW;\tau) = \frac{\partial W}{\partial t} + \tau\left[\frac{\partial W}{\partial u}u + \left(\frac{\partial W}{\partial v} - 1\right)v\right]$$

and

$$\bar{H}(t,u,v;DW) = \max_{\tau\in[\tau^-,\tau^+]} H(t,u,v;DW).$$

8.3.2.1 Regular Regions

In regions where the function W is C^1, it suffices to check that it satisfies the DQVI.

Case $t < t_+$

If $t < t_+$, then we are in the "focal" domain. Thus both $q^\varepsilon = C^\varepsilon$. It suffices to check that $\bar{H}(t,u,v;DW) \geq 0$. Here, $\varepsilon = \operatorname{sign}(\check{v} - v)$:

$$\bar{H}(t,u,v;DW) = \check{w}_t + C^\varepsilon \check{v}_t + \max_{\tau\in[\tau^-,\tau^+]} \tau[\check{w}_u u + C^\varepsilon \check{w}_u u - (C^\varepsilon + 1)v].$$

Again, we notice, using (8.48) for the third equality, that

$$\begin{aligned} H(t,u,v;DW;\tau^\varepsilon) &= \check{w}_t + C^\varepsilon \check{v}_t + \tau^\varepsilon[\check{w}_u u + C^\varepsilon \check{v}_u u - (C^\varepsilon + 1)v] \\ &= Q^\varepsilon \check{V}_t + \tau^\varepsilon[Q^\varepsilon(\check{V}_u u - \mathscr{S}\check{V}) + (1 + C^\varepsilon)(\check{v} - v)] \\ &= \tau^\varepsilon(1 + C^\varepsilon)(\check{v} - v) \geq 0. \end{aligned}$$

According to Proposition 8.31, $Q^\varepsilon \check{V}_t \leq 0$, hence

$$\tau^\varepsilon [\check{w}_u u + C^\varepsilon \check{v}_u u - (C^\varepsilon + 1)v] \geq 0,$$

and therefore $\tau = \tau^\varepsilon$ maximizes the Hamiltonian.

Case $t > t_-$

If $t > t_-$, then both $|q^\varepsilon| < |C^\varepsilon|$. The last two terms in the DQVI are both positive. We must check that $\bar{H}(t,u,v;\mathrm{D}W) = 0$. Notice also in that case

$$q_t^\varepsilon = -\tau^\varepsilon (1 + q^\varepsilon).$$

Inserting this into W_t directly yields

$$\bar{H}(t,u,v;\mathrm{D}W) = Q^\varepsilon \check{V}_t - \tau^\varepsilon (1+q^\varepsilon)(\check{v} - v) + \max_{\tau \in [\tau^-,\tau^+]} \tau [Q^\varepsilon \check{V}_u u - (1+q^\varepsilon) v].$$

We notice then that

$$H(t,u,v;\mathrm{D}W;\tau^\varepsilon) = Q^\varepsilon \check{V}_t + \tau^\varepsilon Q^\varepsilon (\check{V}_u u - \mathscr{S}\check{V}) = 0.$$

According to Proposition 8.23, $Q^\varepsilon \check{V}_t \leq 0$. A fortiori, $Q^\varepsilon \check{V}_t - \tau^\varepsilon (1+q^\varepsilon)(\check{v} - v) \leq 0$. Thus, it follows from $H(t,u,v;\mathrm{D}W;\tau^\varepsilon) = 0$ that τ^ε maximizes the Hamiltonian since $\tau^\varepsilon [Q^\varepsilon \check{V}_u u - (1+q^\varepsilon) v] \geq 0$.

Case $t_+ < t < t_-$

In the region of the equivocal manifold, $\partial W / \partial v + C^+ \geq 0$ and $\partial W / \partial v + C^- = 0$. Thus we must verify that $\bar{H}(t,u,v;\mathrm{D}W) \geq 0$. Using Proposition 8.27, if $v < \check{v}$, i.e., $\varepsilon = +$, then the same argument as we just used yields $\bar{H}(t,u,v;\mathrm{D}W) = 0$, and if $v > \check{v}$, then the same argument as for the focal region holds.

8.3.2.2 Manifold $v = \check{v}(t,u)$

On the manifold $v = \check{v}(t,u)$, W has a local minimum, as Fig. 8.5 shows. All subdifferentials are obtained by replacing q^ε in (8.13) by $\lambda q^- + (1-\lambda) q^+$, $\lambda \in [0,1]$. We have seen that, using $\varepsilon = \mathrm{sign}(\check{v} - v)$,

$$\bar{H}(t,u,v;\mathrm{D}W) = \begin{cases} 0 & \text{if } t > t_\varepsilon, \\ \tau^\varepsilon (1 + C^\varepsilon)(\check{v} - v) & \text{if } t < t_\varepsilon. \end{cases}$$

Hence, at $v = \breve{v}$, $\bar{H} = 0$. Thus, for any fixed τ, $H(t,u,v;\tau;\mathrm{D}W) \le 0$. For fixed τ, H is affine in q. Therefore, replacing q^ε by $\lambda q^- + (1-\lambda)q^+$ replaces H by a convex combination of two Hamiltonians, each nonpositive for any fixed $\tau \in [\tau^-, \tau^+]$. Thus this convex combination is itself nonpositive, and therefore so is its maximum in τ. This is the viscosity condition.

8.3.2.3 Manifold $t = t_\varepsilon$

Since $\check{V}$ and W are continuous, along a manifold at $t =$ constant, only their partial derivative in t may be discontinuous. For $\check{V}$ we have already stressed that this fact, together with the fact that it satisfies everywhere (8.48), implies that $Q^\varepsilon \check{V}_t$ is continuous. Concerning W, we have

$$W_t = Q^\varepsilon \check{V}_t + q_t^\varepsilon \langle \breve{v} - v \rangle.$$

The time derivative of q^ε is discontinuous

- At $t = t_-$ if $\varepsilon = -$, i.e., if $\breve{v} - v < 0$;
- At $t = t_+$ if $\varepsilon = +$, i.e., if $\breve{v} - v > 0$,

jumping from 0 to $-\tau^\varepsilon(1+q^\varepsilon)$ as t increases across t_ε. Thus W_t jumps from $Q^\varepsilon \check{V}_t$ to $Q^\varepsilon \check{V}_t - (1+q^\varepsilon)\tau^\varepsilon \langle \breve{v} - v \rangle$. The jump in slope is negative. Hence this is a local maximum, a place where the superdifferential is nonvoid. All elements of the superdifferential are obtained as

$$\partial_+ W = \left\{ \begin{pmatrix} Q^\varepsilon \check{V}_t + \delta \\ W_u \\ -q^\varepsilon \end{pmatrix} \middle| -(1+q^\varepsilon)\tau^\varepsilon \langle \breve{v} - v \rangle \le \delta \le 0 \right\}.$$

We have also seen that if we set $\delta = -(1+q^\varepsilon)\tau^\varepsilon \langle \breve{v} - v \rangle$, we obtain $\bar{H} = 0$. Hence for any larger δ, we obtain $\bar{H} \ge 0$, which is the viscosity condition for a superdifferential.

8.3.2.4 Discontinuities of $\mathrm{D}\check{V}$

Lemma 8.48. *Assume that on a smooth manifold $u = \tilde{u}(t)$ of (t,u) space, $\check{V}$ is continuous but $\check{V}_t$ and $\check{V}_u$ have discontinuities $\delta \check{V}_t$ and $\delta \check{V}_u$, respectively. Then, necessarily, there is an $\varepsilon \in \{-,+\}$ such that $\tilde{u}'(t) = \tau^\varepsilon \tilde{u}(t)$, $\delta \check{V}_t = -\tau^\varepsilon \check{V}_u$, and $Q^\varepsilon \delta \check{V}_t = Q^\varepsilon \delta \check{V}_u = 0$.*

Proof. The continuity of $\check{V}$ implies that the discontinuity of the gradients of $\breve{v}$ and $\breve{w}$ are orthogonal to the discontinuity manifold, i.e.,

$$\delta \check{V}_t + \tilde{u}' \delta \check{V}_u = 0.$$

Also, $\check{V}$ satisfies (8.12) in both half-spaces on both sides of the manifold. If we take the difference, it holds that

$$\delta\check{V}_t + \mathscr{T}\delta\check{V}_u u = 0.$$

Combining the preceding two equations, we obtain that

$$\left(\frac{\tilde{u}'}{u} I - \mathscr{T}\right)\delta\check{V}_u u = 0.$$

This is possible with a nonzero $\delta\check{V}_u$ only if $\tilde{u}'/u$ is an eigenvalue of $\mathscr{T}$, i.e., either τ^- or τ^+, and $\delta\check{V}_u$ must be an eigenvector, i.e., a multiple of $(-1 \;\; q^\varepsilon)$, hence orthogonal to Q^ε, as well as $\delta\check{V}_t = -\tilde{u}'\delta\check{V}_u$. □

As a corollary, we have the following proposition.

Proposition 8.49. *The gradient of W is continuous across discontinuities of* $\mathrm{D}\check{V}$.

Proof. It follows from the representation formula that

$$\frac{\partial W}{\partial t} = Q_t^\varepsilon \check{V} + Q^\varepsilon \check{V}_t - q_t^\varepsilon v, \quad \frac{\partial W}{\partial u} = Q^\varepsilon \check{V}_u, \quad \frac{\partial W}{\partial v} = -q^\varepsilon.$$

Hence $\check{V}_t$ and $\check{V}_u$ appear only premultiplied by Q^ε, causing no discontinuity.

8.4 Discrete Trading and Fast Algorithm

We now tackle the discrete-time minimax problem described in Sect. 7.2.5, with dynamics (7.7), (7.13), and its sequence of Value functions $\{W_k^h\}_{k\in\mathbb{K}}$, with $Kh = T$. This will provide us with our various algorithms to compute the premium (7.22).

8.4.1 Dynamic Programming and Algorithms

8.4.1.1 Fundamental Property

As we have stressed, the sequence of Value functions $\{W_k^h\}_k$ is the sampling at instants t_k of the Value function of a game whose underlying dynamics are the same as those of the continuous-time game, with the same disturbances $\tau(\cdot) \in \Omega$, but where the minimizer is constrained to a set Φ_h of admissible strategies that is a strict subset of the set Φ of continuous-time strategies.

As a consequence, we have the following fundamental property.

Proposition 8.50. *For any time step h, we have*

$$\forall k \in \mathbb{N}, \quad \forall (u,v) \in \mathbb{R}_+ \times \mathbb{R}, \qquad W_k^h(u,v) \geq W(kh,u,v).$$

8.4.1.2 Isaacs Equation and Standard Algorithm

We write Isaacs' main equation for the game at hand:

$$\forall (u,v) \in \mathbb{R}_+ \times \mathbb{R}, \qquad W_K^h(u,v) = N(u,v), \tag{8.77}$$

$$\forall k \in [0, K-1], \quad \forall (u,v) \in \mathbb{R}_+ \times \mathbb{R},$$
$$W_k^h(u,v) = \inf_{\xi \in \mathbb{R}} \sup_{\tau \in [\tau_h^-, \tau_h^+]} \left\{ W_{k+1}^h((1+\tau)u, (1+\tau)(v+\xi)) - \tau(v+\xi) + C^\varepsilon\langle \xi \rangle \right\}.$$

It is useful to split the last recursion into two steps, according to the following fact.

Proposition 8.51.

1. *The preceding recurrence relation can be written as follows:*

$$W_{k+\frac{1}{2}}^h(u,v) = \max_{\tau \in \{\tau_h^-, \tau_h^+\}} \left\{ W_k^h((1+\tau)u, (1+\tau)v) - \tau v \right\}, \tag{8.78}$$

$$W_k^h(u,v) = \min_{\xi \in \mathbb{R}} \left\{ W_{k+\frac{1}{2}}^h(u, v+\xi) + C^\varepsilon\langle \xi \rangle \right\}. \tag{8.79}$$

2. *The functions W_k^h and $W_{k+\frac{1}{2}}^h$ are convex for all k.*
3. *Let $\check{v}_k^\varepsilon$ be the unique value of v for which $-C^{\bar{\varepsilon}} \in \partial_- W_{k+\frac{1}{2}}^h(u,\cdot)$; it holds that*

$$W_k^h(u,v) = \begin{cases} W_{k+\frac{1}{2}}^h(u, \check{v}_k^-) + C^+(\check{v}_k^- - v) & \text{if } v \leq \check{v}_k^-, \\ W_{k+\frac{1}{2}}^h(u,v) & \text{if } \check{v}_k^- \leq v \leq \check{v}_k^+, \\ W_{k+\frac{1}{2}}^h(u, \check{v}_k^+) + C^-(\check{v}_k^+ - v) & \text{if } \check{v}_k^+ \leq v, \end{cases} \tag{8.80}$$

   as depicted graphically in Fig. 8.7.

Proof. Notice that W_K is convex. As a recurrence hypothesis, assume that W_{k+1} is convex.

1. Notice that in (8.78), we have written $\tau \in \{\tau_h^-, \tau_h^+\}$, not $\tau \in [\tau_h^-, \tau_h^+]$. If in (8.78) and (8.79) we replace $\max_{\tau \in \{\tau_h^-, \tau_h^+\}}$ by $\sup_{\tau \in [\tau_h^-, \tau_h^+]}$ and $\min_\xi$ by $\inf_\xi$, then the equivalence with Isaacs' equation is by inspection. Since W_{k+1} is convex, $\tau \mapsto W_{k+1}((1+\tau)u, (1+\tau)v) - \tau v$ is convex, hence continuous. Therefore, the max in τ is reached at either τ^- or τ^+.

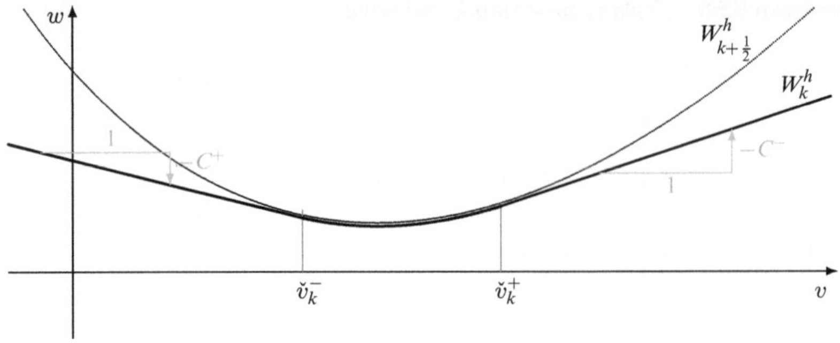

Fig. 8.7 The inf-convolution

2. As supremum of a family of convex functions $W_{k+\frac{1}{2}}$ is convex. We rewrite operation (8.79) as follows. First, let $\xi = -\xi'$. Thus

$$W_k^h(u,v) = \min_{\xi' \in \mathbb{R}} \left\{ W_{k+\frac{1}{2}}^h \left(u, v - \xi'\right) + C^\varepsilon \left\langle -\xi' \right\rangle \right\}. \tag{8.81}$$

Let

$$\Gamma(\eta,\xi) = \begin{cases} +\infty & \text{if } \eta \neq 0, \\ C^\varepsilon\langle -\xi \rangle & \text{if } \eta = 0. \end{cases}$$

This is a convex extended function. We may reinterpret (8.81) as an inf-convolution:

$$W_k^h(u,v) = \inf_{(\eta,\xi)} \{W_{k+\frac{1}{2}}^h(u-\eta, v-\xi) + \Gamma(\eta,\xi)\}.$$

The inf-convolute of two convex extended functions is convex. Hence W_k^h is convex. Finally, the particular form of this inf-convolute is best explained by Fig. 8.8, where we see that the minimum value for $W_{k+\frac{1}{2}}^h(u, v-\xi') - C^-\xi'$ is obtained when $v - \xi' = \check{v}_k^+$. In practice, because of the shape of W_{k+1}, we will have $\check{v}_k^- = \check{v}_k^+$. □

Definition 8.52. The algorithm (8.77), (8.78), (8.80) is called the *standard algorithm.*

8.4.1.3 Representation Formula and Fast Algorithm

We make use of the extension (8.18) of the notations used in the continuous trading problem. This subsubsection is devoted to proving Theorem 8.8 stated in Sect. 8.1.3.4.

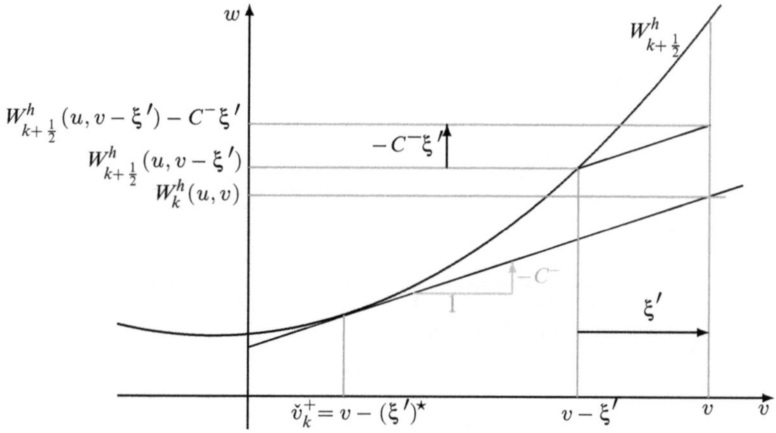

Fig. 8.8 Computing the inf-convolution

Notice first that the recursion (8.17) yields exactly $q^\varepsilon_k = q^\varepsilon(kh)$. We adopt as a recurrence hypothesis that W_{k+1} satisfies the representation formula (8.20):

$$W^h_{k+1}(u,v) = \check{w}^h_{k+1}(u) + q^\varepsilon_{k+1}\Big\langle \check{v}^h_k(u) - v)\Big\rangle.$$

We observe that this is true at the terminal time $k+1 = K$.

Consider the step (8.78). For each u, we seek the maximum of two functions, one for $\tau = \tau^-$ and one for $\tau = \tau^+$. Each is piecewise affine in v, and its graph as a function of v can be represented as a wedge with one branch sloping downward (Fig. 8.9). These can be written as

$$W^-_{k+\frac{1}{2}} := \check{w}^-_{k+\frac{1}{2}} + q^\varepsilon_{k+\frac{1}{2}}\left\langle \check{v}^-_{k+\frac{1}{2}} - v\right\rangle, \qquad W^+_{k+\frac{1}{2}} := \check{w}^+_{k+\frac{1}{2}} + q^\varepsilon_{k+\frac{1}{2}}\left\langle \check{v}^+_{k+\frac{1}{2}} - v\right\rangle,$$

where, as a simple calculation shows,

$$\begin{aligned}
\check{v}^\varepsilon_{k+\frac{1}{2}} &= \check{v}^h_{k+1}((1+\tau^\varepsilon)u)/(1+\tau^\varepsilon),\\
\check{w}^\varepsilon_{k+\frac{1}{2}} &= \check{w}^h_{k+1}((1+\tau^\varepsilon)u) - \tau^\varepsilon \check{v}^h_{k+1}((1+\tau^\varepsilon)u)/(1+\tau^\varepsilon).
\end{aligned}$$

As a result, $\check{v}_k$ is obtained as the abscissa of the intersection of the two wedges in this graph. (In the figure, $\check{v}^\varepsilon$ stands for $\check{v}^\varepsilon_{k+\frac{1}{2}}$ and $\check{v}_k$ for $\check{v}^h_k$.)

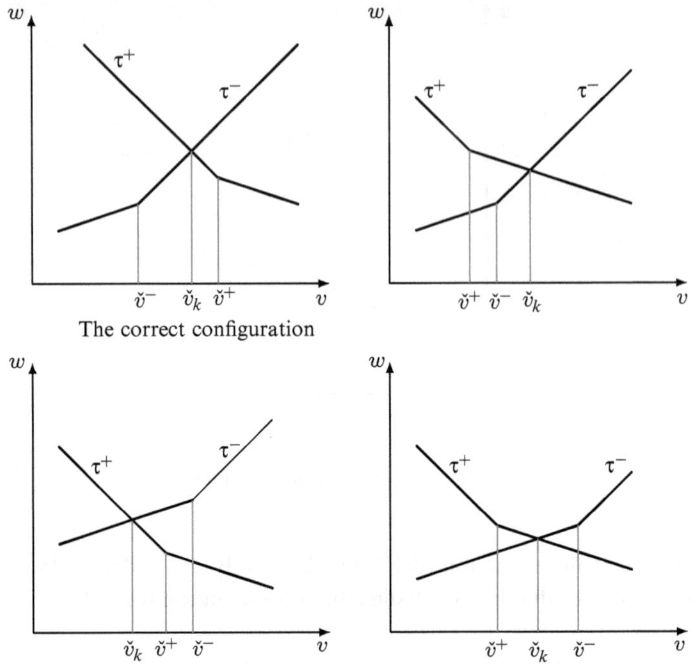

Fig. 8.9 Four possible configurations. Here $\check{v}^\varepsilon$ stands for $\check{v}^\varepsilon_{k+\frac{1}{2}}$ and $\check{v}_k$ for $\check{v}^h_k$

Now we claim the following fact.

Proposition 8.53. *We have for all* (k,u)

$$\check{v}^-_{k+\frac{1}{2}} \leq \check{v}^h_k(u) \leq \check{v}^+_{k+\frac{1}{2}}.$$

Proof. Remember that $W^h_k(u,v)$ is the worth of the cheapest hedging portfolio from the state (u,v) at time t_k. Assume that the left-hand inequality does not hold. Then a *decrease* in the price of the underlying stock (by a factor $1+\tau^-$) would result in the cheapest hedging portfolio having a *larger* content (in number of shares) in this stock than the previous one, a contradiction for a call (and for any option with an increasing payment function), and similarly, *mutatis mutandis*, for the right-hand inequality. □

As a consequence, the point $(\check{v}^h_k, \check{w}^h_k)$ is the intersection of the two upward sloping parts, ending up in formula (8.8).

This makes the step (8.78) explicit. Then the step (8.80) directly leads to the correct representation formula (8.20) for W^h_k. □

Several remarks are in order, first concerning the algorithm itself.

Remark 8.54. • While the standard algorithm is already much faster than a "naive" dynamic programming algorithm implementing Isaacs' equation, this one is two orders of magnitude faster, since the functions to be computed, $\breve{v}_k^h$ and $\breve{w}_k^h$, are functions of one scalar variable instead of two. If the range of possible v is discretized, typically, with 200 points, this is about 100 times faster. As a result, with a discretization of u and v in several hundred points, the computer time on a modern PC is negligible.

- A first-order expansion lets one verify that formula (8.19) is a finite difference discretization of the fundamental PDE (8.12). One could undertake a direct analysis of its convergence along the lines of the numerical analysis of a finite difference scheme. The direct argument that $W^h \to W$ saves us that analysis.

Then, some additional important remarks concerning its use:

Remark 8.55.

- This provides a sure – as long as hypothesis (7.3) is not violated, but see Chap. 10 – discrete-time hedging strategy, an impossibility in the Black–Scholes model.
- Because of the wedge shape of the functions W_k^h, we will have $\breve{v}_k^+ = \breve{v}_k^- = \breve{v}_k^h(u_k)$. It follows from our analysis of the minimization step in ξ that the optimal ξ is such that $v_k + \xi_k^\star = \breve{v}_k^h(u_k)$. Hence, the sequence $\{\breve{v}_k^h(u_k)\}_{k\in\mathbb{N}}$ has the status of the sequence of optimal portfolio compositions, which the trader should strive to achieve at each step in his trading.
- This defines a causal strategy: the optimal ξ_k does *not* depend on τ_k.
- For realistic data on a classic stock market, one finds that $T - t_+$ and $T - t_-$ are both less than 1 day. Hence, except perhaps for modern high-frequency trading, the algorithm simplifies even further, with $q_k^\varepsilon = C^\varepsilon$. Other types of markets, with large transaction costs and no closing costs, could require the use of variable q_k^ε near exercise time. Notice that for $t > t_M$, no trading should occur.
- A consequence of the convergence theorem of the next subsection is that this is also an algorithm by which to approach the continuous trading premium.

8.4.2 Convergence

8.4.2.1 Discretized Games

We already know (Proposition 8.10) that the original game $\mathscr{G}$ and the game transformed by the Joshua transformation $\mathscr{J}$ have the same Value. It will nevertheless be convenient to call V the Value of the game $\mathscr{J}$. It is independent of the artificial time but has t as one of its arguments. Proposition 8.10 therefore translates into

$$\forall t \in [0,T], \forall (u,v) \in \mathbb{R}_+ \times \mathbb{R}, \quad V(t,u,v) = W(t,u,v). \tag{8.82}$$

Let us furthermore consider the game $\mathscr{J}'$ derived from the game $\mathscr{J}$ by forbidding the use of the control ξ_c. We have the following fact.

Proposition 8.56. *The games $\mathscr{J}$ and $\mathscr{J}'$ have the same value.*

Proof. This is a direct consequence of the uniqueness of the VS of the DQVI. As a matter of fact, ξ does not appear in the DQVI, so that it is also the Isaacs equation of the game $\mathscr{J}'$. □

We introduce two other games. Let $\mathscr{G}^h$ be the discrete-time game of the previous subsection with a discretization step h and a Value $\{W_k^h\}_k$ and, similarly, let $\mathscr{J}^h$ be the discrete (artificial)-time game derived from $\mathscr{J}'$ in the same fashion, with Value V^h. Finally, let ℓ be a (small) positive number, define $\mathscr{G}^{h,\ell}$, with Value $\{W_k^{h,\ell}\}_k$, as the game $\mathscr{G}^h$ but where a further restriction is applied, that ξ_k itself is discretized in $\xi_k \in \ell\mathbb{Z}$. We claim that the following proposition holds.

Proposition 8.57.

$$\forall k \le K, k \in \mathbb{N}, \forall (u,v) \in \mathbb{R}_+ \times \mathbb{R}, \quad W(kh,u,v) \le W_k^h(u,v) \le W_k^{h,\ell}(u,v). \quad (8.83)$$

Proof. The left inequality is just Proposition 8.50. In going from $\mathscr{G}^h$ to $\mathscr{G}^{h,l}$, we further constrain the minimizer's controls, hence the right-hand inequality. □

The dynamics of the game $\mathscr{J}^h$ are, using the same conventions $t_k = t(kh)$, $u_k = u(kh)$, $v_k = v(kh)$, $\iota_k = \iota(kh)$, and $\bar{\iota}_k = 1 - |\iota_k|$ as previously,

$$\begin{aligned} t_{k+1} &= t_k + \bar{\iota}_k h, \\ u_{k+1} &= u_k + \bar{\iota}_k \tau_k u_k, \\ v_{k+1} &= v_k + \bar{\iota}_k \tau_k v_k + \iota_k h, \end{aligned}$$

with $\iota_k \in B = \{-1,0,1\}$, $\tau_k \in [\tau_h^-, \tau_h^+]$ as given by (7.6), and the payoff defined by $t_{\mathbb{K}} = T$ and

$$J^h(0,v_0,u_0;\{\tau_k\},\{\iota_k\}) = N(u_{\mathbb{K}}, v_{\mathbb{K}}) + \sum_{k=0}^{\mathbb{K}-1} (-\bar{\iota}_k \tau_k v_k + \iota_k h C^{\iota_k}).$$

Let us write Isaacs' equation for this game: $\forall (t,u,v)$

$$V^h(t,u,v) = \min_{\iota \in B} \max_{\tau \in [\tau_h^-, \tau_h^+]} \{V^h(t + \bar{\iota}h, u + \bar{\iota}\tau u, v + \bar{\iota}\tau v + \iota h) - \bar{\iota}\tau v + \iota h C^\iota\}. \quad (8.84)$$

As in the original Joshua transformation, splitting the $\min_\iota$ into its three possible cases, we obtain

$$\begin{aligned} &\forall (t,u,v),\; V^h(t,u,v) = \min\Big\{ \max_{\tau \in [\tau_h^-, \tau_h^+]} \Big[V^h(t+h,(1+\tau)u,(1+\tau)v) - \tau v\Big], \\ &\qquad\qquad V^h(t,u,v-h) - C^- h,\, V^h(t,u,v+h) + C^+ h\Big\}, \\ &\forall (u,v), \quad V^h(\mathbb{K}h,u,v) = N(u,v). \end{aligned}$$

The critical remark is that, as could be expected given the identity of the games $\mathscr{G}$ and $\mathscr{J}$, and according to our analysis of the inf-convolution in Proposition 8.51, this equation would be satisfied by W^h if the possible values of ξ were quantified with $\xi \in h\mathbb{Z}$. Hence the following fact.

Proposition 8.58. $\forall k \le K$, $k \in \mathbb{N}$, $\forall (u,v) \in \mathbb{R}_+ \times \mathbb{R}$, $V^h(kh,u,v) = W_k^{h,h}(u,v)$.

In the next subsubsection, we prove the following theorem.

Theorem 8.59. *Let* $h = 2^{-d}T$, $d \in \mathbb{N}$. *As* $d \to \infty$, $V^h(t,u,v)$ *converges uniformly on any compact toward* $V(t,u,v)$.

The convergence theorem, Theorem 8.7, is for discretization time points $t = kh$, $k \in \mathbb{N}$, a corollary of this theorem, together with (8.82) and (8.83). For intermediate points, the definition of the interpolation $W^h(t,u,v)$ that we have chosen is such that the proof is the same because all the foregoing arguments apply at any fixed time where an impulse is permitted.

8.4.2.2 Proof of Theorem 8.59

We first claim that the following proposition is true.

Proposition 8.60. *As* $d \to \infty$, V^h *decreases monotonously and converges uniformly on every compact toward a function* $\widehat{V}$.

Proof. As the subdivision of the interval $[0,T]$ is refined, the restrictions on the admissible controls $\xi(\cdot)$ are relaxed: if $d' > d$, then any minimizing control admissible in the discrete-time problem with $h = 2^{-d}T$ is also admissible in the problem with $h = 2^{-d'}T$, playing $\xi = 0$ at time instants that are not multiples of $2^{-d}T$. Hence, V^h decreases monotonously as d increases. And since it is bounded below by 0, it converges monotonously toward some function $\widehat{V}$. Moreover, V^h coincides with $W^{h,h}$, which is easily seen to be continuous. Hence, by Dini's theorem, its limit is uniform on any compact. □

Now we only need to prove the following fact.

Proposition 8.61. *The limit function* $\widehat{V}$ *is a VS of the DQVI (8.7).*

Then the theorem follows from the uniqueness of this solution (Theorem 8.42).

Proof. This proof is, in its essence, due to [51]. For the sake of readability of the proof, we introduce some short-hand notation:

$$\tau_h = \frac{\tau}{h} \in \left[\frac{\tau_h^-}{h}, \frac{\tau_h^+}{h}\right].$$

We recall that from formula (7.6) it follows that $\tau_h^\varepsilon/h \to \tau^\varepsilon$ as $h \to 0$. Let also

$$y = \begin{pmatrix} t \\ u \\ v \end{pmatrix}, \qquad g(y,\tau,\iota) = \begin{pmatrix} \bar{\iota} \\ \bar{\iota}\tau_h u \\ \bar{\iota}\tau_h v + \iota \end{pmatrix}, \qquad L(y,\tau,\iota) = -\bar{\iota}\tau_h v + \iota C^\iota$$

(meaning C^- for C^{-1} and C^+ for C^{+1}). With this notation, Isaacs' Eq. (8.84) also reads

$$\min_{\iota\in B} \max_{\tau_h \in \left[\frac{\tau_h^-}{h}, \frac{\tau_h^+}{h}\right]} [V^h(y + hg(y,\iota,\tau_h)) - V^h(y) + hL(y,\iota,\tau_h)] = 0. \tag{8.85}$$

Let $\Psi : \mathbb{R}^3 \to \mathbb{R}$ be a C^1 function, $y^\star \in \mathscr{N}$ a point where $\Delta := \widehat{V} - \Psi$ has a strict local maximum, and $\mathscr{N} \subset \mathbb{R}^3$ a compact neighborhood of $y^\star$ within which $y^\star$ is a strict maximum. Let also $\Delta^h := V^h - \Psi$, and let $y^h \in \mathscr{N}$ be a point where Δ^h has a maximum over $\mathscr{N}$. (All the functions involved are continuous.)

We need the following lemma, which can be found in [51].[1]

Lemma 8.62. *Let $\mathscr{N}$ be a compact neighborhood of $\mathbb{R}^n$ and $\{\Delta^h\}$ a sequence of continuous functions from $\mathscr{N}$ to $\mathbb{R}$ converging uniformly over $\mathscr{N}$ to a continuous function Δ as h converges to 0. Let Δ have a strict minimum over $\mathscr{N}$ [respectively a strict maximum] at $x^\star$. Let x^h be a minimum of Δ^h over $\mathscr{N}$ [respectively a maximum]. Then $x^h \to x^\star$ as $h \to 0$.*

Therefore, $y^h \to y^\star$ as $h \to 0$. Hence, for h small enough, $y^h + hg(y^h,\iota,\tau_h) \in \mathscr{N}$. Using (8.85) at y^h and the fact that the operator min max is nondecreasing, it follows that

$$\min_{\iota\in B} \max_{\tau_h \in \left[\frac{\tau_h^-}{h}, \frac{\tau_h^+}{h}\right]} [\Psi(y + hg(y,\iota,\tau_h)) - \Psi(y) + hL(y,\iota,\tau_h)] \geq 0.$$

The function Ψ was chosen of class C^1. Using the finite increment theorem, there exists a point $\tilde{y}$ on the segment $[y^h, y^h + hg(y^h,\iota,\tau_h)]$ such that this can also be written

$$\min_{\iota\in B} \max_{\tau_h \in \left[\frac{\tau_h^-}{h}, \frac{\tau_h^+}{h}\right]} h\left[\frac{\mathrm{d}\Psi}{\mathrm{d}y}(\tilde{y},\iota,\tau_h)g(y^h,\iota,\tau_h) + L(y,\iota,\tau_h)\right] \geq 0.$$

Divide the right-hand side above by h (which is positive), and take the limit as $h \to 0$. It follows that

$$\min_{\iota\in B} \max_{\tau\in[\tau^-,\tau^+]} \left[\frac{\mathrm{d}\Psi}{\mathrm{d}y}(y^\star,\iota,\tau)g(y^\star,\iota,\tau) + L(y^\star,\iota,\tau)\right] \geq 0.$$

[1] It is a consequence of Lemma 9.22, proved in the next chapter.

This shows that $\widehat{V}$ is an opposite viscosity subsolution of the DQVI (8.61). Clearly, the same analysis starting with a local minimum of $\widehat{V} - \Psi$ will yield the other inequality, so that $\widehat{V}$ is indeed an OVS of the DQVI (8.61), i.e., a VS of the DQVI (8.7), this being unique according to Theorem 8.42, $\hat{V} = W$. □

Chapter 9
Digital Options

9.1 Introduction and Main Results

9.1.1 Digital Options

9.1.1.1 Definition

A "cash-or-nothing," or *digital*, option is a contract by which a *seller* agrees to pay a *buyer* a fixed amount D if, at a given *exercise time* T, the market price $S(T)$ of a share of the *underlying stock* is higher – resp. lower – than an agreed *strike* $\mathscr{K}$, leading to a digital call – resp. put.

It should be noted that one could, without loss of generality, let $D = 1$ and, for an option with a different D, assume that D options have been traded. Yet, we retain D in our development to emphasize that it is not dimensionless but a currency value.

As a consequence, the payment function $M(s)$ can be expressed with the help of the Heaviside function

$$\Upsilon(s) = \begin{cases} 0 & \text{if } s < 0, \\ 1 & \text{if } s \geq 0, \end{cases}$$

as

$$M(s) = \begin{cases} D\Upsilon(s - \mathscr{K}) & \text{for a digital call,} \\ D\Upsilon(\mathscr{K} - s) & \text{for a digital put.} \end{cases} \tag{9.1}$$

In our definition, we have ruled out payments "in kind." Such "one-stock-or-nothing" options exist. We will not consider them here. And, as in the vanilla case, we will concentrate on call options. The case of put options is very similar and symmetrical.

P. Bernhard et al., *The Interval Market Model in Mathematical Finance*, Static & Dynamic Game Theory: Foundations & Applications, DOI 10.1007/978-0-8176-8388-7_9,

9.1.1.2 Terminal Payment

At exercise time, the seller must close out his position and pay $M(S(T))$ to the buyer. Hence his total terminal cost is $N(u(T),v(T))$ with

$$N(u,v) = \begin{cases} M(u) - c^- v & \text{if } v > 0, \\ M(u) - c^+ v & \text{if } v < 0. \end{cases} \tag{9.2}$$

We can write this as in (8.4)

$$N(u,v) = \check{w}(T,u) + c^{\varepsilon}\langle \check{v}(T,u) - v \rangle , \tag{9.3}$$

with

$$\check{v}(T,u) = 0, \qquad \check{w}(T,u) = M(u), \tag{9.4}$$

and $M(u)$ still given by (9.1).

Proposition 9.1. *The function $v \mapsto N(u,v)$ is convex for all $u \in \mathbb{R}_+$.*

Note, however, that N is *no longer* convex in u, or even continuous.

9.1.2 Main Results

These are the main results for digital options. They are on several counts less complete and more complicated (less elegant?) than for vanilla options. To avoid complicated statements, we limit ourselves here to call options. Put options are very similar, the symmetry with call options leading to a reversal of the sign of v.

9.1.2.1 DQVI

We consider the same differential quasivariational inequality (DQVI) (8.7) as previously:

$$\forall (t,u,v) \in [0,T) \times \mathbb{R}_+ \times \mathbb{R},$$
$$\max\left\{ -\frac{\partial W}{\partial t} - \tau^{\varepsilon}\left\langle \frac{\partial W}{\partial u}u + \left(\frac{\partial W}{\partial v} - 1\right)v \right\rangle, -\left(\frac{\partial W}{\partial v} + C^+\right), \frac{\partial W}{\partial v} + C^- \right\} = 0,$$
$$\forall (u,v) \in \mathbb{R}_+ \times \mathbb{R}, \quad W(T,u,v) = N(u,v),$$

but where the function N is now given equivalently by (9.2) or its representation (9.3) and (9.4).

All of Sect. 8.1.4 applies to the current problem. In particular, the same proof via the Joshua transformation yields the following theorem.

Theorem 9.2. *The Value function W is a (discontinuous) viscosity solution (VS) (see [22]) of the DQVI.*

Its uniqueness is not proved, however, and we will need a *conjecture*.

Conjecture 9.1. *The discontinuous VS (in the sense of Barles [22]) of the DQVI is unique.*

For digital options, we can only state the following theorem.

Theorem 9.4. *For all $(t,u) \in [0,T] \times \mathbb{R}_+$, the function $v \mapsto W(t,u,v)$ is convex.*

9.1.2.2 Representation Theorem

We introduce two special u-trajectories:

$$u_\ell(t) := \mathscr{K} \mathrm{e}^{-\tau^+(T-t)},$$
$$u_r(t) := \mathscr{K} \mathrm{e}^{-\tau^-(T-t)},$$

and as previously we let the following definition hold.

Definition 9.5. The *region of interest* Λ of the (t,u) plane is

$$\Lambda = \{(t,u) \in [0,T] \times \mathbb{R}_+ \mid u \in [u_\ell(t), u_r(t)]\}.$$

Next, we define two positive values $u_+ < u_-$,

$$u_\varepsilon = \frac{(1+c^-)\mathscr{K}}{1+C^\varepsilon}, \qquad \varepsilon \in \{-,+\}, \tag{9.5}$$

and two time instants t_- and t_+, the latter being different from its counterpart in the vanilla case, defined by

$$T - t_\varepsilon = \frac{1}{\tau^\varepsilon} \ln \frac{1+C^\varepsilon}{1+c^-}, \qquad \varepsilon \in \{-,+\}, \tag{9.6}$$

so that $u_\ell(t_+) = u_+$ and $u_r(t_-) = u_-$.

We now define q^- – the same as for vanilla options – and q^+, different from, and more complicated than, its counterpart:

$$q^-(t) = \max\{(1+c^-)\mathrm{e}^{\tau^-(T-t)} - 1, C^-\}, \tag{9.7}$$

or, equivalently,

$$q^-(t) = \begin{cases} C^- & \text{if } t \le t_-, \\ (1+c^-)\mathrm{e}^{\tau^-(T-t)} - 1 & \text{if } t \ge t_- \end{cases} \tag{9.8}$$

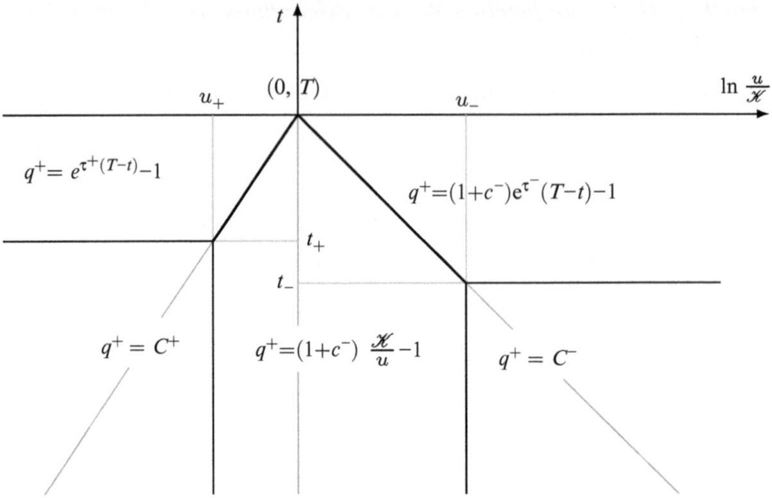

Fig. 9.1 Five regions defining q^+

Concerning q^+, we first set aside a case we will hardly consider: If $v < 0$, then we have as in (8.8)

$$q^+ = \min\{(1+c^+)e^{\tau^+(T-t)} - 1, C^+\}.$$

If $v > 0$, then we set

$$q^+(t,u) = \begin{cases} \min\{(1+c^+)e^{\tau^+(T-t)} - 1, C^+\} & \text{if } u \le u_\ell(t), \\ \min\{\max\{(1+c^-)\frac{\mathscr{K}}{u} - 1, C^-\}, C^+\} & \text{if } (t,u) \in \Lambda, \\ \max\{(1+c^-)e^{\tau^-(T-t)} - 1, C^-\} & \text{if } u \ge u_\ell(t), \end{cases} \tag{9.9}$$

or, equivalently,

$$q^+(t,u) = \begin{cases} C^+ & \text{if } t \le t_+ \text{ and } u \le u_+, \\ (1+c^-)e^{\tau^+(T-t)} - 1 & \text{if } t \ge t_+ \text{ and } u \le u_\ell(t), \\ (1+c^-)\frac{\mathscr{K}}{u} - 1 & \text{if } (t,u) \in \Lambda \cap \{[0,T] \times [u_+, u_-]\}, \\ C^- & \text{if } t \le t_- \text{ and } u \ge u_-, \\ (1+c^-)e^{\tau^-(T-t)} - 1 & \text{if } t \ge t_- \text{ and } u \ge u_r(t). \end{cases} \tag{9.10}$$

The five regions thus defined are depicted in Fig. 9.1.

Remark 9.6.

1. q^+ thus defined is continuous.
2. For $u \ge \min\{u_r(t), u_-\}$ we have $q^+ = q^-$.
3. We will see that q^+ is not really needed outside the region Λ. We defined it for convenience in the subsequent analysis.

Finally, we recall the *fundamental equation* (8.12), a pair of linear PDEs for $\check{V}^t = (\,\check{v}(t,u)\;\; \check{w}(t,u)\,)$ initialized as in (9.4), using the notation (8.10):

$$\frac{\partial \check{V}}{\partial t} + \mathcal{T}\left(\frac{\partial \check{V}}{\partial u}u - \mathcal{S}\check{V}\right) = 0, \qquad \check{V}(T,u) = \begin{pmatrix} 0 \\ M(u) \end{pmatrix}. \tag{9.11}$$

Theorem 9.7. *The Value function is given by the formula similar to (8.13):*

$$W(t,u,v) = \check{w}(t,u) + q^{\varepsilon}(t,u)\langle \check{v}(t,u) - v\rangle\,, \tag{9.12}$$

with:

1. *For $v < 0$: q^+ given by (8.8), $\check{v} = 0$, $\check{w}(t,u) = M(e^{\tau^+(T-t)}u)$ [which is a solution of (9.11)];*
2. *For $v \geq 0$: q^- given by (8.8), q^+ given by (9.9), and $\check{v}$ and $\check{w}$ are given*
   a. *In the region where $q^- \neq q^+$ (i.e., $u < \min\{u_r(t), u_-\}$) by the fundamental equation (9.11) with the initial conditions (9.4);*
   b. *In the region where $q^- = q^+$, by $\check{v}(t,u) = 0$, $\check{w}(t,u) = D$.*

9.1.2.3 Discrete Trading and Convergence Theorem

The discrete trading problem considered here is as in the vanilla case, i.e., keeping the market model unchanged, but restricting the seller to impulse trading at given time instants $t_k = kh$, $h = T/K$ a given step size, $K \in \mathbb{N}$. We let $W_k^h(u,v)$ be the Value function of the discrete-time problem (Sect. 7.2.5).

Theorem 9.8. *The Value function is given by Isaacs' Eq. (8.14) or, equivalently, by the* standard algorithm*:* $\forall (u,v) \in \mathbb{R}_+ \times \mathbb{R}$, $\forall k \in \mathbb{K}$,

$$\left.\begin{aligned} W^h_{k+\frac{1}{2}}(u,v) &= \max_{\tau \in [\tau_h^-,\tau_h^+]} \left[W^h_{k+1}((1+\tau)u,(1+\tau)v) - \tau v\right],\\ W_k^h(u,v) &= \min_{\xi}\left[W^h_{k+\frac{1}{2}}(u,v+\xi) + C^{\varepsilon}\xi\right],\\ W_K^h(u,v) &= N(u,v). \end{aligned}\right\} \tag{9.13}$$

where N is given by (9.2)

(Notice that, as compared with the vanilla case (8.15), we are obliged, in the $\max_\tau$ operation, to let τ range over the interval $[\tau_h^-, \tau_h^+]$ as opposed to its end points $\{\tau^-, \tau^+\}$.)

The next two theorems are weaker than their vanilla counterpart due to the loss of convexity in u.

Theorem 9.9. *The functions $v \mapsto W_k^h(u,v)$ are convex for all $k \in \mathbb{K}$ and $u \in \mathbb{R}_+$.*

We use the same interpolation technique as in the vanilla case, and we get the following theorem.

Theorem 9.10. *If Conjecture 9.1 is correct, as the step size is subdivided and goes to zero (e.g., $h = T/2^d$, $d \to \infty \in \mathbb{N}$), for all $t \in [0,T]$, then the function $W^h(t,u,v)$ decreases and converges pointwise to the function $W(t,u,v)$.*

9.1.2.4 Fast Algorithm

To state the fast algorithm, we need a second conjecture.

Conjecture 9.2. *In the domain $(t,u,v) \in [0,T] \times \mathbb{R}_+ \times \mathbb{R}_+$, the function $W^h_{k+\frac{1}{2}}$ of formula (9.13) is piecewise affine in v with two branches.*

This conjecture is well substantiated by numerical computations using the standard algorithm.

Theorem 9.11. *If Conjectures 9.1 and 9.2 hold, then the Value function $W_k^h(u,v)$ is given by the same representation formula (8.20),*

$$W_k^h(u,v) = \check{w}_k^h(u) + q_k^{\varepsilon}(u)\left\langle \check{v}_k^h(u) - v \right\rangle,$$

as in the vanilla case, with, for q^-, the same algorithm

$$q_K^- = c^-, \qquad q^-_{k+\frac{1}{2}} = (1+\tau_h^-)q^-_{k+1} + \tau_h^-, \qquad q_k^- = \max\left\{ q^-_{k+\frac{1}{2}}, C^- \right\}$$

and

1. *Outside of the region Λ, $\check{v}_k^h(u) = 0$, $\check{w}_k^h(u) = M(u)$,*
2. *In the region Λ,*

$$\forall (k,u) \mid (kh,u) \in \Lambda, \quad q_k^+(u) = \min\left\{ (1+c^-)\frac{\mathscr{K}}{u} - 1, C^+ \right\},$$

and $\check{v}_k^h(u)$ and $\check{w}_k^h(u)$ given by the same recursion (8.19)

$$\check{V}_k^h(u) = \frac{1}{q^+_{k+\frac{1}{2}} - q^-_{k+\frac{1}{2}}} \begin{pmatrix} 1 & -1 \\ -q^-_{k+\frac{1}{2}} & q^+_{k+\frac{1}{2}} \end{pmatrix} \begin{pmatrix} Q^+_{k+1}\check{V}^h_{k+1}\left((1+\tau_h^+)u\right) \\ Q^-_{k+1}\check{V}^h_{k+1}\left((1+\tau_h^-)u\right) \end{pmatrix},$$

$$\check{v}_K^h(u) = \check{v}(T,u), \qquad \check{w}_K^h(u) = \check{w}(T,u)$$

as in the vanilla case.

9.2 Geometric Approach

The principle of the geometric investigation is as in the vanilla case (Sect. 8.2.1). In particular, Propositions 8.13, 8.14, and 8.15 still hold. Also, the equations of the characteristics (8.30) are unchanged, as is the conclusion (8.29).

9.2.1 *Trivial Regions*

We first set aside some elementary cases.

9.2.1.1 Regular Terminal Conditions

If $u(T) \neq \mathscr{K}$, then the terminal payoff N is differentiable at $(u(T), v(T))$, giving the terminal conditions for the adjoint equations (with the same sign conventions as in Sect. 8.2.2.1):

$$(n \ p \ q \ r) = (n \ 0 \ c^{\varepsilon} \ 1), \qquad \varepsilon = \mathrm{sign}(-v),$$

and using

$$H^{\star} = n + \min_{\tau \in [\tau^-, \tau^+]} [pu + (q+1)v] + \max_{\xi}(q - C^{\varepsilon})\langle \xi \rangle = 0,$$

$$n = \tau^{\varepsilon}(1 + c^{\varepsilon})\langle -v \rangle.$$

We will see later that for $u(T) = \mathscr{K}$, necessarily $p(T) \leq 0$.

9.2.1.2 Negative v, Sheets $\{\tau^+\}$

If $v(T) < 0$, then it follows that $\sigma(T) = p(T)u(T) - (1 + c^{\varepsilon})\langle -v(T) \rangle < 0$. The optimal τ is $\tau^{\star} = \tau^+$. Also, $\xi^{\star}(T) = 0$. Integrating backward the characteristic system with these controls, σ is constant, and we obtain a sheet $\{\tau^+\}$ similar, in the domain $v < 0$, to that of the vanilla case, but with the new (discontinuous) N:

$$w(t) = M(\mathrm{e}^{-\tau^+(T-t)}u) - q(t)v(t),$$

with

$$q(t) = [(1 + c^+)\mathrm{e}^{\tau^+(T-t)} - 1]$$

(the q^+ of the vanilla case), which is smaller than C^+ only when

$$T - t \leq \frac{1}{\tau^+} \ln \frac{1 + C^+}{1 + c^+}.$$

For smaller t, the sheet $\{\tau^+\}$ is replaced by a positive jump manifold of the form $w = M(e^{\tau^+(T-t)}u) - C^+v$, i.e., the same formula as if we set, as in the vanilla case,

$$q(t) = \min\{(1+c^+)e^{\tau^+(T-t)} - 1, C^+\} =: q_V^+. \tag{9.14}$$

These are actually two disjoint sheets because of the discontinuity along the trajectory $u_\ell(t)$.

This yields formula (9.12) with $\check{v}(t,u) = 0$, $\check{w}(t,u) = M(\exp[\tau^+(T-t)]u)$, and q given by (9.14), identical to q^+ in (8.8), proving the corresponding statement of the representation Theorem 9.7.

Indeed, there is no reason for the seller to have a short portfolio since the payment is increasing in u. If he does have one, he should close it out immediately, going back to the position $v = 0$, unless the remaining time before exercise time is so short that the worst extra loss due to a possible increase in his debt is less than the difference between running and closing transaction costs.

From now on, we consider only cases with positive v. (The signs would be reversed for a put.)

9.2.1.3 Positive v, Sheets $\{\tau^-\}$

If $u(T) \neq \mathscr{K}$, then we have $\nu^-(T) = (n\ \ p\ \ q\ \ r) = (n\ \ 0\ \ c^-\ \ 1)$; therefore $\sigma = (1+c^-)v(T) > 0$, hence $\tau^\star = \tau^-$. If $u(T) < \mathscr{K}$, then we obtain exactly the same combination of a negative jump manifold followed for time t_- to T by a sheet $\{\tau^-\}$ as in the vanilla case. We will call that sheet $\{\tau_\ell^-\}$. It exists in the (t,u) plane in the region $u \leq u_r$. Its cartesian equation is as previously

$$w = -q^-v, \quad \Leftrightarrow \quad Q^-V = 0, \tag{9.15}$$

and its (semipermeable) normal is

$$\nu^- = (-\tau^-(1+q^-)v\ \ 0\ \ q^-\ \ 1) = (-\tau^-Q^-\mathscr{S}V\ \ 0\ \ q^-\ \ 1). \tag{9.16}$$

Notice also that $u_r(t_-) = u_-$.

Region $u < u_\ell(t)$, the Manifold $\mathscr{B}$

In the region $u < u_\ell$, the underlying asset will end up out of the money: $u(T) < \mathscr{K}$ whatever the market evolution, $\tau(\cdot)$, is. Hence there is no need for the seller to hedge anything. If he has some of the underlying asset in his portfolio, he should sell it back before it loses value, unless the remaining time before exercise time is so short that the worst possible loss from a decrease in u is less than the difference between the running and the closing transaction costs. (We will see that no other semipermeable hypersurface intersects the sheet $\{\tau_\ell^-\}$ in that region.)

Therefore, in that region, we have again representation (9.12) with q^- as in (8.8), $\check{v} = 0$, $\check{w} = 0$. We refer to this as the *basic* manifold $\mathscr{B}$.

Region $u > u_r(t)$, the Manifold $\mathscr{C}$

If $u(T) > \mathscr{K}$, then we have a second combination of a negative jump manifold followed from time t_- to T by a sheet $\{\tau^-\}$, which exists for $u > u_r$. We will call it $\{\tau_r^-\}$; it has $w(T) = D - c^- v(T)$. Hence it is disjoint from the sheet $\{\tau_\ell^-\}$.

Again, in that region, the asset will end up in the money – $u(T) > \mathscr{K}$, for every possible market evolution. The final payment is therefore equal to D and independent of the market evolution. Hence, there is no point in having some underlying asset in the portfolio, which might lose value but is not needed to help hedge against the final payment. Therefore, the most sensible policy is to sell all assets in the portfolio, again unless the remaining time is so short that waiting until exercise time is cheaper even if $\tau = \tau^-$. But this time, the seller needs to have an amount D in cash (or riskless bonds). The cartesian equation of this sheet $\{\tau_r^-\}$ is

$$w = D - q^- v, \quad \Leftrightarrow \quad Q^- V = D. \tag{9.17}$$

Therefore, W is again given by formula (9.12), with $\check{v} = 0$, $\check{w} = D$, a manifold we call $\mathscr{C}$ (for *constant*). There, q^- is as in formula (8.8).

The two trivial regions can be covered by the formula $\check{v} = 0$, $\check{w} = M(u)$.

9.2.2 *Region of Interest*

The problem is now solved everywhere *except* in the region of interest $u \in [u_\ell, u_r]$, i.e., for prices from which the underlying asset may end up either in the money or out of the money. As in the vanilla case, this is the region where there is a dilemma for the seller and where analysis is really needed.

We will carry out a detailed investigation of that region in the case where $t_- < t_+$. The other case, which is as plausible, is not very different, and we will give brief indications of how it is solved. (It presents no envelope junction of the type $\mathscr{E}^+$ but a third envelope $\mathscr{E}^-$.) Notice, however, that in the case $c^\varepsilon = C^\varepsilon$, contrary to the vanilla case, $t_+ < T$ while, as in the vanilla case, $t_- = T$. This is therefore a case where $t_- > t_+$, and two equivocal junctions $\mathscr{E}^-$ survive.

For this whole subsection, the reader is referred to Fig. 9.3, and to Fig. 9.4 for the case $t_- > t_+$.

9.2.2.1 Singular Sheet $\{\mathscr{K}\}$

At $u(T) = \mathscr{K}$, N is not differentiable. We chose it as $N(\mathscr{K}, v) = D + c^\varepsilon \langle -v \rangle$, hence upper semicontinuous. Therefore, it has a superdifferential in the $v > 0$ half-space

$$\partial^+ N(\mathscr{K}, v) = \{(p \quad c^-) \mid p \leq 0\}.$$

We recall that the switch function determining the optimal (worst) τ is $\sigma = pu + (1+q)v$. Hence, at exercise time for $u(T) = \mathscr{K}$, we have

$$\sigma = p\mathscr{K} + (1+c^-)v.$$

Because the component p is negative, this may be of either sign.

Remember also that as long (backward) as the characteristic system holds, i.e., as long as $C^- < q < C^+$, the switch function σ is constant. In the final period, all extremal trajectories ending with $\sigma < 0$ at $u(T) = \mathscr{K}$ have $\tau = \tau^+$, $u(t) = u_\ell(t)$, $v(t) = \exp(-\tau^+(T-t))v(T)$, and $q(t) = (1+c^-)\exp(\tau^+(T-t)) - 1$. This is valid from time t_+ as in (9.6) and $u(t_+) = u_+$ as in (9.5). In a similar fashion, all extremal trajectories ending with $\sigma > 0$ at $u = \mathscr{K}$ have $\tau = \tau^-$, $u(t) = u_r(t)$, $v(t) = \exp(-\tau^-(T-t))v(T)$, and $q(t) = (1+c^-)\exp(\tau^-(T-t)) - 1$. This is valid from time t_- as in (9.6) and $u(t_-) = u_-$ as in (9.5).

However, for $p(T) = -(1+c^-)v(T)/\mathscr{K}$, we get $\sigma(T) = 0$, hence $\sigma(t) = 0$ as long as the characteristic trajectories hold. This gives rise to singular characteristics where all controls $\tau \in [\tau^-, \tau^+]$ are extremal. Let

$$\theta(t) = \int_t^T \tau(s)\,\mathrm{d}s.$$

We now have characteristic trajectories of the form

$$\left|\begin{array}{l} u(t) = \mathscr{K}\mathrm{e}^{-\theta(t)}, \\ v(t) = v(T)\mathrm{e}^{-\theta(t)}, \\ w(t) = D + (\mathrm{e}^{-\theta(t)} - 1 - c^-)v(T), \end{array}\right. \qquad \left|\begin{array}{l} p(t) = -(1+c^-)v(T)\mathrm{e}^{\theta}(t)/\mathscr{K}, \\ q(t) = (1+c^-)\mathrm{e}^{\theta}(t) - 1, \\ r(t) = 1. \end{array}\right.$$

This defines a three-dimensional manifold in (t,u,v,w) space, with parameters $t \in [0,T]$, $v(T) > 0$, $\theta(t) \in [\tau^-(T-t), \tau^+(T-t)]$. We call it the sheet $\{\mathscr{K}\}$. However, the characteristics that make up this manifold are only valid if $C^- \le q \le C^+$; hence the above formulas only hold for

$$\frac{1+C^-}{1+c^-} \le \mathrm{e}^{\theta} \le \frac{1+C^+}{1+c^-} \Longleftrightarrow u_+ \le u \le u_-,$$

whence the definition of q^+ given in (9.10). In that region, the semipermeable normal $\nu_{\mathscr{K}}$ to the sheet $\{\mathscr{K}\}$ can be expressed in terms of q^+ as

$$\nu_{\mathscr{K}} = \left(0 \;\; -\frac{(1+q^+(u))v}{u} \;\; q^+(u) \;\; 1\right) = \left(0 \;\; -\tfrac{1}{u}Q^+\mathscr{S}V \;\; q^+ \;\; 1\right). \tag{9.18}$$

In the same region, a cartesian equation for this hypersurface is

$$w = D - q^+(u)v \quad \Leftrightarrow \quad Q^+V = D. \tag{9.19}$$

9.2.2.2 Dispersal Manifold $\mathscr{D}$

Recall that we stress the case $t_- < t_+$. In the region $\Lambda \cap \{t \geq t_-\} \cap \{u \geq u_+\}$, the sheets $\{\tau_\ell^-\}$ and $\{\mathscr{K}\}$ intersect along a two-dimensional manifold $v = \check{v}(t,u)$, $w = \check{w}(t,u)$ defined by

$$\check{w} = -q^-(t)\check{v} = D - q^+(u)\check{v};$$

hence, as long as $q^- \neq q^+$,

$$\check{v}(t,u) = \frac{D}{q^+(u) - q^-(t)}, \qquad \check{w}(t,u) = -q^-(t)\check{v}(t,u). \tag{9.20}$$

This is a dispersal manifold, which we call $\mathscr{D}$. As a matter of fact, it is a somewhat degenerate dispersal: all trajectories of $\{\mathscr{K}\}$ (except those with $\tau = \tau^-$) leave $\{\tau_r^-\}$ "above" it since, using (9.16) and $\min H = 0$, we obtain

$$\langle \nu^-, \dot{z} \rangle = n + (c^- + 1)\tau v > n + (c^- + 1)\tau^- v = 0,$$

but trajectories of $\{\tau_r^-\}$ stay on $\mathscr{D}$ since, by construction, for any τ, $\langle \nu_{\mathscr{K}}, \dot{z} \rangle = 0$.

If $u > (1 + c^-)\mathscr{K}$, then the slope in the (v,w) plane of the sheet $\{\mathscr{K}\}$ is positive, as is that of the sheet $\{\tau_\ell^-\}$. Hence, in that region, the cheaper hedging portfolio is not $v = \check{v}(t,u)$, but $v = 0$. This is because in that case, both sheets are limit portfolio values to hedge against a decrease in u. Therefore, having some of the underlying asset in the portfolio is harmful.

Formula (9.20) shows that $\check{v}$ and $\check{w}$ are positive on u_ℓ, hence discontinuous there as they are null for $u < u_\ell$, and diverge to infinity in the neighborhood of the boundary $u = u_r(t)$, where $q^- = q^+$. Yet we have the following proposition.

Proposition 9.12.

1. *The function W is bounded in the neighborhood of the manifold $u = u_r(t)$ and continuous and smooth across that manifold.*
2. *The function W has a jump discontinuity across the manifold $u = u_\ell(t)$ for $v < \check{v}(t,u)$ and is continuous for $v > \check{v}(t,u)$.*

Proof.

1. As formula (9.20) shows, in the neighborhood of $u = u_r$, both $\check{v}$ and $\check{w}$ go to infinity. Hence $\varepsilon = +$ in the representation formula, and

$$W(t,u,v) = (q^+ \;\; 1)\begin{pmatrix} 1 \\ -q^- \end{pmatrix} \frac{D}{q^+ - q^-} - q^+ v = D - q^+ v$$

remains bounded, and even continuous and smooth as u crosses the boundary $u_r(t)$, according to formula (9.17) and remembering that at $u = u_r$, $q^- = q^+$.

2. In the neighborhood of $u = u_\ell(t)$, for $v \geq \check{v}$, we have, on the dispersal manifold, using (9.20)

$$W(t,u,v) = -q^- \check{v}(t,u) + q^-(\check{v}(t,u) - v) = -q^- v,$$

which coincides with (9.15).

For $v < \check{v}$, in the closed Λ region, we get

$$W(t,u,v) = -q^- \check{v}(t,u) + q^+(u)(\check{v}(t,u) - v) = D - q^+ v.$$

The difference with W in the adjacent open "trivial" region is thus

$$D - q^+ v + q^- v = (q^+ - q^-)(\check{v} - v) > 0.$$

Therefore, we have a positive jump as u increases across u_ℓ. We may notice that, due to our choice for M, W is u.s.c. □

A last important remark is that on $\mathscr{D}$, by direct calculation,

$$Q^- \check{V}_t(t,u) = \tau^-(1 + q^-(t))\check{V}(t) \leq 0, \quad Q^+ \check{V}_t = 0.$$

9.2.2.3 Envelope Junction $\mathscr{E}^+$

At $u = u_+$, $q^+ = C^+$. Thus for smaller u, the sheet $\{\mathscr{K}\}$ is replaced by a positive jump manifold $\{\iota^+\}$. As a consequence, in the region $t \in [t_-, t_+]$, $u \in [u_\ell(t), u_+]$, we have an envelope junction of a jump manifold $\{\iota^+\}$ with the sheet $\{\tau_\ell^-\}$. The construction is similar to that of the envelope junction of Sect. 8.2.3.

The control $\xi = \tilde{\xi}(\tau)$ that keeps a trajectory on $\{\tau_\ell^-\}$ is

$$\tilde{\xi}(\tau) = \frac{(\tau - \tau^-)(1 + q^-)v}{C^+ - q^-} \geq 0.$$

We integrate both the dynamics and the generalized adjoint equations with that control and $\tau = \tau^+$, and we check a posteriori that this yields a normal ν_j^+ of the jump manifold with $n_j^+ > 0$, so that $\tau = \tau^+$ indeed minimizes the Hamiltonian $\langle \nu_j^+, \dot{z} \rangle$.

The terminal conditions for $\mathscr{E}^+$ are on the boundary $u = u_+$ of $\mathscr{D}$, where $q^+ = C^+$, as on the jump manifold, i.e., at a time $t = t_b \in [t_-, t_+]$. The boundary is therefore given, together with its tangent, as

$$\partial\mathscr{E}^+ = \begin{pmatrix} t \\ u \\ v \\ w \end{pmatrix} = \begin{pmatrix} t_b \\ u_+ \\ \check{v}(t_b, u_+) \\ -q^-(t_b)\check{v}(t_b, u_+) \end{pmatrix}, \quad \mathrm{D}_t(\partial\mathscr{E}^+) = \begin{pmatrix} 1 \\ 0 \\ \check{v}_t(t_b, u_+) \\ -C^+ \check{v}_t(t_b, u_+) \end{pmatrix}.$$

The requirement that its tangent in (t,u,v,w) space must be orthogonal to the semipermeable normal $\nu_j^+ = (\, n_j^+ \;\, p_j^+ \;\, C^+ \;\, 1\,)$ simply yields $n_j^+(t_b) = 0$.

We must integrate backward the generalized adjoint equation

$$\dot{\nu}_j^+ = -\frac{\partial H}{\partial z} + \alpha\left(\nu_j^+ - \nu^-\right),$$

i.e., here

$$\dot{n}_j^+ = \alpha\left(n_j^+ - n^-\right)$$

from $n_j^+(t_b) = 0$, with α chosen such that q_j^+ is constant, equal to C^+, i.e.,

$$\dot{q}_j^+ = -\tau^+(C^+ + 1) + \alpha(C^+ - q^-) = 0;$$

hence

$$\alpha = \frac{\tau^+(C^+ + 1)}{C^+ - q^-} > 0,$$

yielding, via Lambert's formula, a positive n_j^+ as needed.

Integrating the dynamics backward from $\partial\mathscr{E}$ with $\tau = \tau^+$ and $\xi = \tilde{\xi}(\tau^+)$ yields a field $\check{v}(t,u)$, $\check{w}(t,u)$. Closed-form formulas are given in [142], but they are complicated and not needed for the sequel of the analysis. We only need the following fact.

Proposition 9.13. *For $t \in [t_-, t_+]$, it holds that $\check{v}(t, u_\ell(t)) > 0$, and $\check{w}(t, u_\ell(t)) = -q^-(t)\check{v}(t, u_\ell(t) > 0$.*

Proof. Formula (9.20) shows that $\check{v}(t_+, u_\ell(t_+)) > D/(C^+ - C^-) > 0$. Then, we integrate backward with $\tau = \tau^+$ and

$$\xi = \tilde{\xi}(\tau^+) = \frac{(\tau^+ - \tau^-)(1 + q^-)}{C^+ - q^-}\check{v},$$

hence we have $\dot{\check{v}} = a(t)\check{v}$ for some finite $a(t)$; hence $\check{v}$ cannot change sign or reach zero. Finally, $\mathscr{E}^+$ lies on the sheet $\{\tau^-\}$, hence $\check{w} = -q^-\check{v}$. □

Of course, the Value function in the region covered here is still given by the representation formula (9.12); note that Proposition 8.27 remains unchanged.

9.2.2.4 Envelope Junction $\mathscr{E}_1^-$

At $t = t_-$, $q^-(t_-) = C^-$. Hence, for smaller t, the sheet $\{\tau^-\}$ is replaced by a negative jump manifold $\{\iota_1^-\}$. In the region $t \le t_-$, $u \in [u_+, u_-]$, the sheet $\{\mathscr{K}\}$ still exists. The jump manifold joins onto it via an envelope junction $\{\mathscr{E}_1^-\}$.

Due to the singular character (in τ) of the sheet $\mathscr{K}$, the control ξ that keeps a trajectory with an arbitrary τ on $\{\mathscr{K}\}$ is just $\tilde{\xi} = 0$. We therefore integrate both the dynamics and the generalized adjoint equations with the controls $(\xi, \tau) = (0, \tau^-)$.

The terminal conditions are found on the boundary $t = t_-$ of $\mathscr{D}$ for $u = u_b \in [u_+, u_-]$, together with its tangent:

$$\partial\mathscr{E}_1^- = \begin{pmatrix} t \\ u \\ v \\ w \end{pmatrix} = \begin{pmatrix} t_- \\ u_b \\ \check{v}(t_-, u_b) \\ -C^-\check{v}(t_-, u_b) \end{pmatrix}, \qquad \mathrm{D}_u(\partial\mathscr{E}_1^-) = \begin{pmatrix} 0 \\ 1 \\ \check{v}_u(t_-, u_b) \\ -C^-\check{v}_u(t_-, u_b) \end{pmatrix}.$$

The requirement that this tangent in (t, u, v, w) space must be orthogonal to the semipermeable jump normal $\nu_j^- = (n_j^- \ p_j^- \ C^- \ 1)$ yields $p_j^-(t_-) = 0$, and the equation $H^\star = 0$ yields $n_j^-(t_-) = -\tau^-(1 + C^-)\check{v}(t_-, u_b) > 0$.

Since $n_{\mathscr{K}} = 0$, it follows that the generalized adjoint equation is now simply

$$\dot{n}_j^- = \alpha n_j^-,$$

so that n_j^- does not change sign on a trajectory, and hence $n_j^-(t) > 0$ for all $t \leq t_-$.

Since on the trajectories of $\mathscr{E}_1^-$ one has $\xi = 0$, $\tau = \tau^-$, the dynamics integrate trivially as for the primaries in terms of $\check{v}(t_-, u_b)$ and $\check{w}(t_-, u_b)$. An elementary calculation shows that they can be expressed in terms of $\tilde{q}^- := (1 + c^-)e^{\tau^-(T-t)} - 1$ (which is *not* q^- in that region since $t < t_-$) as

$$\check{V}(t, u) = \begin{pmatrix} 1 \\ -\tilde{q}^-(t) \end{pmatrix} \frac{D}{q^+(u) - \tilde{q}^-(t)}.$$

Typical sections of the graph of W on the (v, w) plane for a given (t, u) are therefore as in Fig. 9.2.

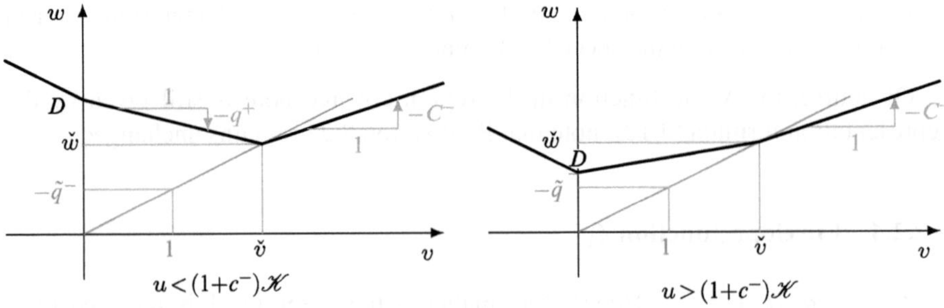

Fig. 9.2 Region $\mathscr{E}_1^-$: two sections of the graph of W for different values of u (and of D)

This whole construction holds as long as $q^+ \geq C^-$, i.e., $u \leq u_-$. The leftmost trajectory of this junction is $u = u_+ \exp(-\tau^+(t_- - t)) =: u_1(t)$. It only exists for $t \in [t_1, t_-]$, with

$$t_- - t_1 = \frac{1}{-\tau^-} \ln\left(\frac{1+C^+}{1+C^-}\right), \qquad T - t_1 = \frac{1}{-\tau^-} \ln\left(\frac{(1+c^-)(1+C^+)}{(1+C^-)^2}\right).$$

For many realistic sets of parameters, this is a short time, say on the order of 1 day. For $t < t_1$, only the next subsection holds.

The same remarks that were made at the end of Sect. 9.2.2.3 hold.

9.2.2.5 Envelope Junction $\mathscr{E}_3^-$ and Focal Manifold $\mathscr{F}$

Preliminaries

We investigate now the region $t < t_-$, $u \in [u_\ell(t), u_1(t)]$, which decomposes into the following two regions:

1. $u \in [u_+, u_1(t)]$. The situation is analogous to that of the adjacent region where we constructed the envelope junction $\mathscr{E}_1^-$. But the boundary reached by trajectories τ^+ is $u = u_+$, where $\check{v}$ and $\check{w}$ are not known, nor are the semipermeable normals.
2. $u \in [u_\ell, u_+]$. In that region, none of the sheets $\{\tau^-\}$ nor $\{\mathscr{K}\}$ exists. We must have a focal junction of two jump manifolds of opposite signs. We know (Sect. 8.2.4) that such a junction is traversed by two fields of trajectories: one with $\tau = \tau^-$, which takes its terminal values either on the boundary of $\mathscr{E}^{|}$ at $t = t_-$ or on $u = u_\ell(t)$, and one with $\tau = \tau^+$, which takes its terminal values either on the boundary of $\mathscr{E}^+$ at $t = t_-$ or on the boundary of $\mathscr{E}_3^-$ at $u = u_+$.

Consequently, we cannot construct $\mathscr{E}_3^-$ and $\mathscr{F}$ separately.

We extend Lemma 8.30 to cases where the singular sheet $\{\mathscr{K}\}$ is involved.

Lemma 9.14. *If a manifold $\mathscr{V} : V = \check{V}(t,u)$ is either*

1. *a submanifold of the sheet $\{\mathscr{K}\}$ and of a sheet $\{\tau^-\}$ (case of $\mathscr{D}$) or*
2. *a submanifold of a jump manifold $\{\iota_-\}$ traversed by trajectories τ^- and a submanifold of the sheet $\{\mathscr{K}\}$ (case of both $\mathscr{E}^-$),*

then it satisfies the two Eqs. (8.48):

$$Q^\varepsilon \check{V}_t + \tau^\varepsilon Q^\varepsilon (\check{V}_u u - \mathscr{S} V) = 0, \qquad \forall \varepsilon \in \{-,+\},$$

or, equivalently, the fundamental Eq. (9.11):

$$\check{V}_t + \mathscr{T}(\check{V}_u u - \mathscr{S}\check{V}) = 0.$$

Proof.

1. If the manifold $\mathscr{V}$ is a submanifold of a sheet $\{\tau^-\}$, then its two normals $(\,1\;0\;\check{v}_t\;\check{w}_t\,)$ and $(\,0\;1\;\check{v}_u\;\check{w}_u\,)$ are orthogonal to ν^- given by (9.16), which yields

$$\begin{aligned} Q^-\check{V}_t - \tau^- Q^- \mathscr{S}\check{V} &= 0,\\ Q^-\check{V}_u &= 0. \end{aligned}$$

   Multiply the second line by $\tau^- u$ and add the product to the first one to obtain Eq. (8.48), with $\varepsilon = -$.

   If it is a submanifold of the sheet $\{\mathscr{K}\}$, then its two normals are orthogonal to $\nu_{\mathscr{K}}$, given by (9.18), yielding

$$\begin{aligned} Q^+\check{V}_t &= 0,\\ Q^+\check{V}_u - \frac{1}{u}Q^+\mathscr{S}\check{V} &= 0. \end{aligned}$$

   Multiply the second line by $\tau^+ u$ and add the product to the first one to obtain (8.48), with $\varepsilon = +$.
2. If the manifold $\check{V}$ is a submanifold of a negative jump manifold traversed by trajectories τ^-, then it satisfies (8.48) with $\varepsilon = -$ for the same reason as in Lemma 8.30. If it is a submanifold of the sheet $\{\mathscr{K}\}$, then it satisfies this equation with $\varepsilon = +$ for the same reason as above. □

Of course, Lemma 8.30 still holds, so that $\mathscr{E}^+$ and $\mathscr{F}$ also satisfy (9.11). The manifolds $\mathscr{F}$ and $\mathscr{E}_3^-$ are therefore to be obtained as the (continuous) solution of this equation, with the boundary conditions on the three-piece boundary of the combined region: $\{t = t_-, u \in [u_\ell, u_+]\} \cup \{t \le t_-, u = u_\ell(t)\} \cup \{t \le t_-, u = u_1(t)\}$.

Boundary Conditions for $\check{V}$

Concerning the boundary $t = t_-$, $u \in [u_\ell(t_-), u_+]$, $\check{v}$ and $\check{w}$ are known on $\mathscr{E}^+$. Concerning $t \le t_-$, $u = u_1(t)$, $\check{v}$ and $\check{w}$ are known on $\mathscr{E}_1^+$.

Concerning $t \le t_-$, $u = u_\ell(t)$, we know that if at any time τ is different from τ^+, then the state drifts out of Λ, and since we are before t_-, $q^- = C^-$, i.e., the optimum behavior for the seller is to sell all the underlying asset in his portfolio at once, at a cost $-C^- v$. And since the seller needs to be left with a nonnegative portfolio, he must have $w + C^- v = Q^- V \ge 0$. The cheapest such portfolio thus satisfies $Q^- V = 0$. We therefore impose $Q^- \check{V} = 0$ on this "left" boundary of $\mathscr{F}$. This imposes

$$\xi = \tau^+ \frac{1 + C^-}{C^+ - C^-}\check{v}, \tag{9.21}$$

which is positive (as assumed to get that formula). Integrating the dynamics backward from $\check{V}(t_-, u_\ell(t_-))$ obtained on $\mathscr{E}^+$ yields $\check{V}$. We may also note that both components will be strictly positive, for the same reason as on $\mathscr{E}^+$.

Normals

As previously, we need to check that the normals ν_j^ε to the jump manifolds have their time components n_j^ε nonnegative. According to formula (8.51), which still holds here, this also serves to check that the signs hypothesized for the controls ξ of both fields traversing $\mathscr{F}$ are correct.

On $\mathscr{E}_3^-$, the only jump manifold present is with a negative jump. As on $\mathscr{E}_1^-$, n_j^- has a constant sign, given by its terminal condition, here on the boundary $u = u_+$, as usual according to $n_j^- = -Q^-\check{V}_t$ (orthogonality of $\partial\mathscr{E}_3^-$ and ν_j^-). If the normal to the manifold $\check{V}$ has a discontinuity on $u = u_+$, then the continuity of $\check{V}$ implies that its u component is also continuous. But Eq. (9.11), where $\mathscr{T}$ is invertible, implies then that $Q^-\check{V}_t$ is continuous as well. Therefore, $\mathscr{E}_3^-$ will inherit continuously n_j^- on $\mathscr{F}$.

Concerning $\mathscr{F}$ it is the junction of two jump manifolds $\{\iota^\varepsilon\}$ and traversed by two fields of trajectories, generated by $(\tau, \xi) = (\tau^\varepsilon, \xi^\varepsilon)$, for $\varepsilon = -, +$. The components n_j^ε again satisfy (8.52), (8.53).

The terminal conditions for the components n_j^- on the trajectories τ^- are either on the boundary $t = t_-$, where the jump manifold inherits the normal of the sheet $\{\tau^-\}$, again continuously because the u component is necessarily continuous and (9.11) imposes the continuity of $\check{V}_t$, or on $u_\ell(t)$. On that last boundary, either comparing (8.51) and (9.21) or using both

$$\frac{\mathrm{d}}{\mathrm{d}t}Q^-\check{V} = Q^-\check{V}_t + \tau^+ Q^-\check{V}_u u = 0$$

and (8.48) with $\varepsilon = -$, we obtain

$$n_j^- = \frac{-\tau^-\tau^+}{\tau^+ - \tau^-}(1 + C^-)\check{v} > 0.$$

The terminal conditions for the components n_j^+ on the trajectories τ^+ on the boundary $t = t_-$ are taken from the normal to $\mathscr{E}^+$, again positive, as seen in Sect. 9.2.2.3. On the boundary $u = u_+$, the normal sought is common with that of the sheet $\{\mathscr{K}\}$ (9.18) known to have $n_{\mathscr{K}} = 0$.

In the rest of the domain of $\mathscr{F}$, the reasoning goes as in the vanilla case, leading to the conclusion that, indeed, both n_j^- and n_j^+ so constructed are positive all over the domain of $\mathscr{F}$, validating the construction and yielding a positive n_j^- for $\mathscr{E}_3^-$.

9.2.2.6 Extending $\mathscr{C}$

The last part of the region of interest, and of the plane (t,u), not accounted for so far is the region $t \le t_-$, $u \in [u_-, u_r(t)]$. There, $(1+c^-)\exp(\tau^-(T-t)) - 1 < C^-$ and $(1+c^-)\mathscr{K}/u - 1 < C^-$. Hence $q^- = q^+ = C^-$, and both sheets $\{\mathscr{K}\}$ and $\{\tau^-\}$ are replaced by negative jump manifolds of slope $-C^-$ in the plane (v,w). Their equations are $Q^-V = 0$ for the first one and $Q^-V = D$ for the second one. Hence the latter is "above" the former, and is therefore the boundary of the capturable states, the graph of W.

Therefore, in that region, we may still set $\check{v} = 0$, $\check{w} = D$. Hence the manifold $\mathscr{C}$ extends beyond the line $u = u_r(t)$ to $u = u_-$. The value function W and its gradient extend continuously across $u = u_-$.

9.2.3 *Synthesis*

9.2.3.1 Case $t_- < t_+$

The same representation formula (8.13) as in the vanilla case holds for $(t,u,v) \in [0,T] \times \mathbb{R}_+ \times \mathbb{R}_+$, i.e., restricted to $v > 0$. The pair of functions $\check{v}(t,u)$, $\check{w}(t,u)$ satisfies the fundamental equation (9.11) in the region where $q^+ \neq q^-$, i.e., for $u \le \min\{u_r(t), u_-\}$. The form (8.48) can be extended to the region this excludes, but since Q is no longer invertible, this is of little help. The composite manifold $V = \check{V}(t,u)$ corresponds to singular manifolds of Isaacs' equation of various types defined in various regions according to Fig. 9.3.

The construction of $\mathscr{F}$ and $\mathscr{E}_3^-$ requires the integration of the fundamental equation over a domain where q^+ is constant on $\mathscr{F}$ and variable on $\mathscr{E}_3^-$. We did not investigate the properties of existence and uniqueness of this more complex case (as compared to the vanilla case). Our experience is that numerical integration poses no difficulty – provided one makes sure to have a small enough ratio of the time step to the u step, a classic feature of hyperbolic equations – and, moreover, the Value function thus computed agrees to a high precision with that obtained by the "standard algorithm," making no use of the representation formula.

Remark 9.15.

1. The discontinuity of the value function W along $u = u_\ell(t)$ for $v < \check{v}$, stressed in Proposition 9.12, propagates through regions $\mathscr{E}^+$ and $\mathscr{F}$ down to $t = 0$, while W is, as emphasized previously, continuous across $u = u_r(t)$.
2. Additionally, $v \mapsto W(t,u,v)$ is indeed continuous and convex over $\mathbb{R}$. This is so because, at $v = 0$, the slope in the positive v half-space is either $-q^-$ (in the "trivial regions" of (t,u) space) or $-q^+$ (in the region of interest) and $-q_V^+$ in the negative v half-space. Clearly, $-q_V^+ < 0 < -q^-$, but also, in the region of interest, $\mathscr{K}/u < \exp(\tau^+(T-t))$, hence $q^+ < q_V^+$, as one easily sees. We have used this fact in the sketch of Fig. 9.3.

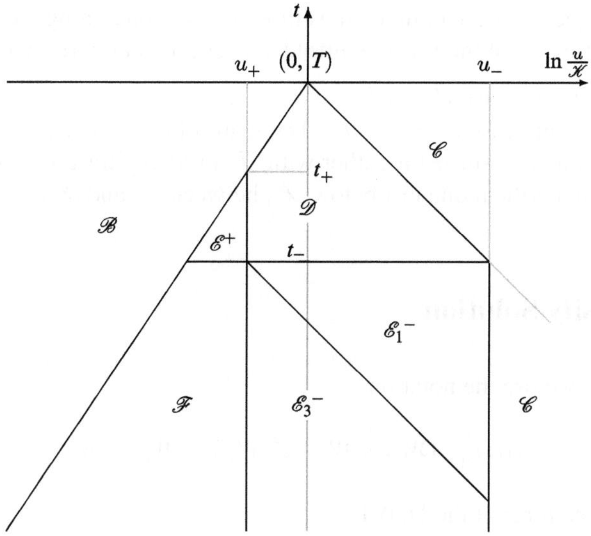

Fig. 9.3 Regions of definition of singular surfaces, case $t_- < t_+$

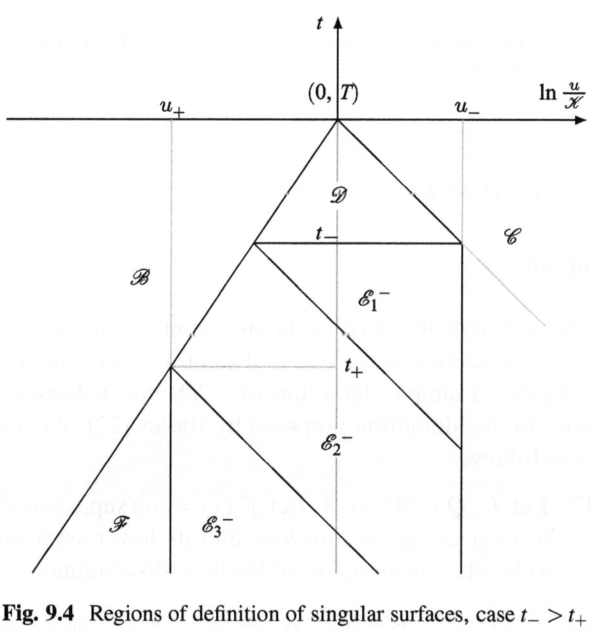

Fig. 9.4 Regions of definition of singular surfaces, case $t_- > t_+$

9.2.3.2 Case $t_- > t_+$

The case $t_- > t_+$ has not been described in detail here. It is very similar to the former one, except that $\mathscr{E}^+$ no longer exists, and another piece $\mathscr{E}_2^-$ of envelope junction with a negative jump manifold appears between $\mathscr{E}_1^-$ and $\mathscr{E}_3^-$, as shown by

Fig. 9.4. Its trajectories are computed as those of $\mathscr{E}_1^-$ concerning the dynamics and as the trajectories τ^- of the focal manifold concerning their terminal values.

Remark 9.16. In the case $c^\varepsilon = C^\varepsilon$, for $\varepsilon = -,+$, we get $u_- = \mathscr{K}$ but $u_+ < \mathscr{K}$. Hence we are in the case $t_- = T > t_+$. The manifolds $\mathscr{D}$ and $\mathscr{E}_1^-$ collapse, but the manifolds $\mathscr{E}_2^-$ and $\mathscr{E}_3^-$ subsist together with $\mathscr{F}$, probably for a very small range of values of u, but a critical one just below $\mathscr{K}$, between u_+ and $\mathscr{K}$.

9.3 Viscosity Solution

In this section, we use the notation

$$\bar{H}(t,u,v\,\mathrm{D}W) = W_t + \tau^\varepsilon \langle W_u u + (W_v - 1)v\rangle\,,$$

and the alternate form of the DQVI:

$$\min\{\bar{H}(t,u,v;\mathrm{D}W),\, W_v + C^+,\, -W_v - C^-\} = 0,$$

and we want to verify that the representation formula (8.13) does yield an opposite viscosity solution (OVS).

9.3.1 *The Discontinuity*

9.3.1.1 Definition

The Value function W calculated so far is discontinuous along the boundary $u = u_\ell(t)$ of the region of interest Λ for $v < \check{v}$. Therefore, as compared to the vanilla case, we have neither a simple definition of a VS nor, a fortiori, a uniqueness theorem. We will use the definition proposed by Barles [22]. We therefore modify Definition 8.37 as follows.

Definition 9.17. Let $f : Q \subset \mathbb{R}^n \to \mathbb{R}$. Let $f^\star(y) = \limsup_{y' \to y} f(y')$ and $f_\star(y) = \liminf_{y' \to y} f(y')$ be its u.s.c. *upper envelope* and its lower semicontinuous *lower envelope*, respectively. The function f is said to be a (discontinuous) VS of the PDE

$$\forall y \in Q,\quad F(y, f(y), \mathrm{D}f(y)) = 0, \qquad \forall y \in \partial Q,\quad f(y) = G(y),$$

if $f^\star$ which is u.s.c., is a subsolution (with $f^\star \le G^\star$ on ∂Q) and $f_\star$ which is l.s.c., is a super solution (with $f_\star \ge G_\star$ on ∂Q).

The rest of this section is devoted to proving the following fact.

Theorem 9.18. *The Value function W is a discontinuous VS of the DQVI (8.7).*

However, as the theory stands, this is proved to be neither a necessary nor a sufficient condition for the Value function. Yet we strongly conjecture that it is both. As a matter of fact, in any open region where W is continuous, it must satisfy the viscosity inequalities. The proof, applied to the Joshua transform, proceeds as in the classic case because it is local. As for the discontinuity, we claim that the following proposition is true ($\mathscr{B}$ in the proposition is a barrier not to be mistaken with the two-dimensional manifold $\mathscr{B}$ of the previous section).

Proposition 9.19. *If the Value function W has a simple discontinuity on a smooth manifold $\mathscr{B} : B(y) = 0$, with smooth restrictions W^- and W^+ in the open regions $B(y) < 0$ and $B(y) > 0$, admitting smooth extensions across $\mathscr{B}$ and with $W^- < W^+$ on $\mathscr{B}$, then the viscosity condition implies that $\mathscr{B}$ is a semipermeable manifold.*

Proof. We need yet another fact: let ∇W^- and ∇W^+ be the gradients of the restrictions of W to the two half-spaces.

Proposition 9.20. *The superdifferential of $W^\star$ and the subdifferential of $W_\star$ on $\mathscr{B}$ are given by*

$$\partial_+ W^\star(y) = \{\nabla W^+(y) + \alpha \nabla B(y), \quad \forall \alpha > 0\},$$
$$\partial_- W_\star(y) = \{\nabla W^-(y) + \alpha \nabla B(y), \quad \forall \alpha > 0\}.$$

Proof of the proposition. We prove the first statement; the second one proceeds by changing signs. Let, therefore, $y \in \mathscr{B}$, i.e., $B(y) = 0$, a test function $\psi(\cdot)$ be such that $W^\star(y) - \psi(y) = 0$ and is a local maximum. For any η such that $\langle \nabla B(y), \eta \rangle > 0$, and for small enough $\theta > 0$, $B(y + \theta\eta) > 0$, and therefore $W^\star(y + \theta\eta) = W^+(y + \theta\eta)$. Since y is a local maximum of $W^\star - \psi$, we also get $W^+(y + \theta\eta) - \psi(y + \theta\eta) < 0$, and thus $\langle (\nabla W^+ - \nabla\psi), \eta \rangle < 0$. Therefore, we find that

$$\forall \eta : \langle \nabla B(y), \eta \rangle > 0, \qquad \langle (\nabla W^+(y) - \nabla\psi(y)), \eta \rangle < 0.$$

Therefore, $(\nabla W^+(y) - \nabla\psi(y))$ must be collinear and opposite to $\nabla B(y)$, i.e., there exists a positive α such that $\nabla W^+(y) - \nabla\psi(y) = -\alpha \nabla B(y)$.

Conversely, let $\alpha > 0$. Let $\psi(y') = W^+(y') + \alpha \langle \nabla B(y), y' - y \rangle + \|y' - y\|^2$. Clearly $\psi(y) = W^+(y)$, and $\nabla\psi(y) = \nabla W^+(y) + \alpha \nabla B(y)$. For y' close enough to y with $\langle \nabla B(y), y' - y \rangle < 0$, $B(y') < 0$, so that $W^\star(y') = W^-(y') < W^+(y')$, and ψ being continuous, for y' close enough to y, $W^\star(y') < \psi(y')$. For y' such that $\langle \nabla B(y), y' - y \rangle > 0$, $W^\star(y') = W^+(y') < \psi(y')$. Finally, if $\langle \nabla B(y), y' - y \rangle = 0$, then either $B(y') < 0$, and the argument of continuity of ψ applies, or $B(y') \geq 0$, and the last argument applies. □

We use that proposition in the viscosity statement, using the Joshua transform to deal with impulses as well as ordinary trajectories, and the shorthand notation of Sect. 7.2.5.2

$$\forall \delta \in \partial_+ W^\star, \qquad \min_\iota \max_\tau [\langle \delta, g(y, \tau, \iota) \rangle + L(y, \tau, \iota)] \leq 0,$$

to get

$$\forall \alpha > 0, \qquad \min_{\iota}\max_{\tau}[\langle(\nabla W^{+}(y) + \alpha \nabla B(y)), g(y,\tau,\iota)\rangle + L(y,\tau,\iota)] \le 0.$$

Clearly, this implies

$$\min_{\iota}\max_{\tau}[\langle \nabla B(y), g(y,\tau,\iota)\rangle + L(y,\tau,\iota)] \le 0.$$

And the other viscosity inequality similarly implies

$$\min_{\iota}\max_{\tau}[\langle \nabla B(y), g(y,\tau,\iota)\rangle + L(y,\tau,\iota)] \ge 0.$$

These two inequalities together prove the proposition. □

9.3.1.2 Checking the Representation Formula

We consider the points (t,u,v) with $u = u_{\ell}(t)$ and $v < \check{v}(t,u)$, since it is for those points that W has a discontinuity. We also have $W_{\star}(t,u,v) = -q^{-}v$ and $W^{\star} = Q^{+}\check{V} - q^{+}v$.

Call $\delta = (\,\delta_t \;\; \delta_u \;\; \delta_v\,)$ an element of a sub- or superdifferential. We must verify that

$$\forall \delta \in \partial_{-} W_{\star}, \quad \min\{\delta_t + \tau^{\varepsilon}\langle \delta_u u + (\delta_v - 1)v\rangle\,, \delta_v + C^{+}, -\delta_v - C^{-}\} \le 0,$$

$$\forall \delta \in \partial_{+} W^{\star}, \quad \min\{\delta_t + \tau^{\varepsilon}\langle \delta_u u + (\delta_v - 1)v\rangle\,, \delta_v + C^{+}, -\delta_v - C^{-}\} \ge 0.$$

We use Proposition 9.20, with $B(t,u,v) = u - D\exp(-\tau^{+}(T-t))$, hence $\nabla B = (\,-\tau^{+}D\exp(-\tau^{+}(T-t)) \;\; 1 \;\; 0\,)$, and therefore, on $\mathscr{B}$, $\nabla B = (\,-\tau^{+}u \;\; 1 \;\; 0\,)$. This yields

$$\partial_{-} W_{\star} = \begin{cases} (\,\tau^{-}(1+q^{-})v - \alpha\tau^{+}u, \;\alpha, \;-q^{-}\,) & \text{if } t > t_{-}, \\ (\,-\alpha\tau^{+}u, \;\alpha, \;-C^{+}\,) & \text{if } t \le t_{-}, \end{cases}$$

$$\partial_{+} W^{\star} = \begin{cases} (\,Q^{+}\check{V}_t - \alpha\tau^{+}u, \;-\frac{1}{u}(1+q^{+})(\check{v}-v) + Q^{+}\check{V}_u + \alpha, \;-q^{+}\,) & \text{if } u > u_{+}, \\ (\,Q^{+}\check{V}_t - \alpha\tau^{+}u, \;Q^{+}\check{V}_u + \alpha, \;-C^{+}\,) & \text{if } u \le u_{+}, \end{cases}$$

with α an arbitrary positive number.

1. We consider the first viscosity inequality: $\delta \in \partial_{-} W_{\star}$. Two cases arise.

   a. Case $t > t_{-}$: The first term of the min is

$$\begin{aligned} \bar{H}(t,u,v;\delta) &= \tau^{-}(1+q^{-})v - \alpha\tau^{+}u + \tau^{\varepsilon}\langle \alpha u - (1+q^{-})v\rangle \\ &= \begin{cases} \alpha(\tau^{-} - \tau^{+})u < 0 & \text{if } \alpha u - (1+q^{+})v \le 0, \\ (\tau^{-} - \tau^{+})(1+q^{-})v \le 0 & \text{if } \alpha u - (1+q^{+})v \ge 0. \end{cases} \end{aligned}$$

   This is nonpositive in both cases, hence the minimum is nonpositive.

b. Case $t \le t_-$: This concerns only the region $\mathscr{F}$, where, in particular, $-\delta_v - C^- = 0$, hence again the minimum is nonpositive.

2. Consider now the second viscosity inequality: $\delta \in \partial_+ W^\star$. Again, two cases arise.

a. Case $u > u_+$: The first term in the min is

$$\begin{aligned}\bar{H}(t,u,v,\delta) \\ &= Q^+\check{V}_t - \alpha\tau^+u + \tau^\varepsilon\langle -(1+q^+)(\check{v}-v) + Q^+\check{V}_u u + \alpha u - (1+q^+)v\rangle \\ &= Q^+\check{V}_t - \alpha\tau^+u + \tau^\varepsilon\langle -Q^+\mathscr{S}\check{V} + Q^+\check{V}_u u + \alpha u\rangle.\end{aligned}$$

For $\tau^\varepsilon = \tau^+$ this term is null, due to Eq. (8.48). Therefore, the $\max_\tau$ is a-fortiori nonnegative for all α. Moreover, this concerns the region $\mathscr{D}$, where both $\delta_v + C^+ = C^+ - q^+$ and $-\delta_v - C^- = q^+ - C^-$ are positive. Hence the minimum of the three terms is nonnegative.

b. Case $u \le u_+$: The first term in the min is

$$\bar{H}(t,u,v;\delta) = Q^+\check{V}_t - \alpha\tau^+u + \tau^\varepsilon\langle -Q^+\mathscr{S}\check{V} + Q^+\check{V}_u u + \alpha u\rangle.$$

If $\tau^\varepsilon = \tau^+$, then this becomes

$$\bar{H} = Q^+\check{V}_t + \tau^+Q^+\check{V}_u u - Q^+\mathscr{S}\check{V} + (1+C^+)(\check{v}-v) = (1+C^+)(\check{v}-v) \ge 0.$$

The other two terms in the min are never negative. Thus, again, the minimum of the three terms is nonnegative.

The conclusion is that where the representation formula yields a discontinuous function W, it satisfies the viscosity inequalities according to Definition 9.17.

9.3.2 Continuous Regions

9.3.2.1 Regular Regions

We first investigate regions where W and its gradient are continuous.

Trivial Regions

In the so-called trivial regions, we have $W(t,u,v) = M(\mathrm{e}^{-\tau^+(T-t)}u) + q^\varepsilon\langle -v\rangle$, with q^+ as in the vanilla case. Hence, for times such that $q^\varepsilon = C^\varepsilon$,

$$\bar{H}(t,u,v;\mathrm{D}W) = \tau^\varepsilon\langle -(1+q^\varepsilon)v\rangle > 0,$$

and for times such that $|q^\varepsilon| < |C^\varepsilon|$, $\varepsilon = \text{sign}(-v)$

$$\bar{H}(t,u,v;\mathrm{D}W) = \tau^\varepsilon(1+q^\varepsilon)v + \tau^\varepsilon\langle-(1+q^\varepsilon)v\rangle = 0.$$

Hence, in all cases, either $W_v + C^\varepsilon$ or $\bar{H}$ is equal to zero, the other two terms in the DQVI being positive.

Region $\mathscr{F}$

The focal region $t < t_-$, $u < u_+$, proceeds as in the vanilla case (Sect. 8.3.2.1).

Region $\mathscr{E}^-$

In the region $t < t_-$, $u > u_+$, for $v > \check{v}(t,u)$, $\bar{H}(t,u,v;\mathrm{D}W)$ is as previously, hence nonnegative. $W_v + C^+ = C^+ - q^- > 0$ and $-W_v - C^- = 0$. Thus the minimum is zero.

For $v < \check{v}(t,u)$ we get

$$\bar{H}(t,u,v;\mathrm{D}W) = Q^+\check{V}_t + \max_{\tau\in[\tau^-,\tau^+]}[Q^+\check{V}_u u - (1+q^+)(\check{v}-v) - (1+q^+)v].$$

We notice that

$$H(t,u,v;\mathrm{D}W;\tau^+) = Q^+\check{V}_t + \tau^+ Q^+[\check{V}_u u - \mathscr{S}\check{V}] = 0.$$

And since we are on the singular sheet $\{\mathscr{K}\}$, $Q^+\check{V}_t = -n_{\mathscr{K}} = 0 = \sigma$, and therefore $\bar{H} = 0$. Finally, $W_v + C^+ = C^+ - q^+ > 0$ and $-W_v - C^- = q^+ - C^- > 0$. Hence the minimum is zero.

Region $\mathscr{E}^+$

In the region $t > t_-$, $u < u_+$, for $v < \check{v}(t,u)$, $\bar{H}(t,u,v;\mathrm{D}W)$ is as in the focal region, hence nonnegative. For $v > \check{v}$ we get

$$\begin{aligned}\bar{H}(t,u,v;\mathrm{D}W) &= Q^-\check{V}_t + \tau^-[Q^-\check{V}_u u - (1+q^-)v] - \tau^-(1+q^-)(\check{v}-v)\\ &= Q^-\check{V}_t + \tau^- Q^-(\check{V}_u u - \mathscr{S}\check{V}) = 0.\end{aligned}$$

As always, the fact that $Q^-\check{V}_t = -n^- = \tau^-(1+q^-)v \le 0$ guarantees that $\tau = \tau^-$ indeed maximizes the Hamiltonian. We also have $W_v + C^+ = C^+ - q^- > 0$ and $-W_v - C^- = q^- - C^- > 0$. Again the minimum is zero.

Region $\mathscr{D}$

In the region $t > t_-$, $u > u_+$, $\bar{H}(t,u,v)$ is as in the region $\mathscr{E}^-$ if $v < \check{v}(t,u)$ and as in the region $\mathscr{E}^+$ if $v > \check{v}(t,u)$. Hence $\bar{H}(t,u,v;\mathrm{D}W) = 0$. The other two terms are positive. The minimum is zero.

9.3.2.2 Gradient Discontinuities

Proposition 8.49 applies here as in the vanilla case. Hence there is no point in investigating possible discontinuities of $D\check{V}$. Five hypersurfaces of gradient discontinuity thus remain, which we investigate hereafter.

Hyperplane $v = 0$

At $v = 0$, we stressed in Remark 9.15 that W is convex in v, i.e., the subdifferential is nonempty. It is obtained by replacing W_v in DW by a convex combination of the two W_v on both sides of the discontinuity. This makes its v component δ_v smaller than its value in the half space $v > 0$. In the region of interest, it enters $\bar{H}$ with the coefficient $\tau^+ > 0$. Hence the Hamiltonian, which is nonpositive in that region, is a fortiori so. In the trivial regions, the Hamiltonian is $\bar{H} = -q_t^- v + \tau^\varepsilon \langle (\delta_v - 1)v \rangle = 0$.

Hypersurface $v = \check{v}(t,u)$

At $v = \check{v}(t,u)$, the sign ε changes in the representation formula. The analysis here is identical to that of the vanilla case.

Hyperplane $t = t_-$

At $t = t_-$, q_t^- is discontinuous. The analysis here is identical to that of $t = t_\varepsilon$ of the vanilla case, with $\varepsilon = -$.

Hyperplane $u = u_+$

At $u = u_+$, W_u has a discontinuity for $v < \check{v}(t,u)$, going from $W_u = Q^+\check{V}_u$ in the half-space $u < u_+$ to $W_u = Q^+\check{V}_u - (1/u)(1+q^+)(\check{v} - v)$ in the half-space $u > u_+$. This is a decrease in the slope as u increases, hence a place where the superdifferential is nonvoid. It is obtained by replacing W_u in DW by a convex combination of the two gradients; hence, $\lambda \in [0,1]$ and

$$
\begin{aligned}
\bar{H} &= Q^+\check{V}_t + \tau^\varepsilon \langle Q^+\check{V}_u u - \lambda(1+q^+)(\check{v} - v) - (1+q^+)v \rangle \\
&= Q^+\check{V}_t + \tau^+ \langle Q^+[\check{V}_u u - \mathscr{S}\check{V}] + (1-\lambda)(1+q^+)(\check{v} - v) \rangle \\
&= \tau^+(1-\lambda)(1+q^+)(\check{v} - v) \geq 0.
\end{aligned}
$$

The other two terms in the DQVI are always nonnegative, hence so is the min.

Hyperplane $u = u_-$

At $u = u_-$, for $v < \check{v}(t,u)$, W_u goes from $W_u = Q^+\check{V}_u - (1/u)(1+q^+)(\check{v}-v)$ in the half-space $u < u_-$ to $W_u = Q^+\check{V}_u$ in the half-space $u > u_-$. This is therefore a place where the subdifferential is nonempty. One of the three terms in the DQVI should be nonpositive. By construction, we have there $W_v = q^+ = C^-$, insuring that $-W_v - C^- = 0$.

This ends the verification and proves Theorem 9.18.

9.4 Discrete Trading and Algorithms

9.4.1 Algorithms

9.4.1.1 Standard Algorithm

The analysis leading to the standard algorithm of the vanilla case remains valid here, with one notable exception: the Value function generated by the dynamic programming procedure is no longer convex in u since N is not. (Convexity in v is preserved by the same argument.) Hence, the "standard algorithm" (8.78), (8.80) must be slightly modified: in Eq. (8.78) the maximum is to be found for τ ranging over the whole segment $[\tau_h^-, \tau_h^+]$, yielding system (9.13), where the minimization with respect to ξ can still be performed according to the rule (8.79).

While this does complicate the algorithm, if the spread τ^-, τ^+ is not too large, then the search can be done by direct discretization and exploration. We tested both this brute force method and a "golden number" search. The latter was faster, but by a small margin.

Given that the discrete representation formula for this case is based upon a conjecture, the availability of this direct procedure to check it numerically is important.

9.4.1.2 Representation Formula and Fast Algorithm

The discrete trading representation formula of the vanilla case, while very similar to the continuous formula, is based not on an investigation of extremal trajectories but on a detailed analysis of the standard algorithm. However, the corresponding analysis here is intractable [142]. We therefore make the following conjecture.

Conjecture 9.2. *In the domain* $(t,u,v) \in [0,T] \times \mathbb{R}_+ \times \mathbb{R}_+$*, the function* $W^h_{k+\frac{1}{2}}$ *of formula (9.13) is piecewise affine in* v *with two branches.*

It follows from procedure (8.80) that the value function W_k^h itself is piecewise affine. Henceforth, we use the notation $q_\ell^\varepsilon(u)$ without assuming a priori that it coincides with $q^\varepsilon(\ell h, u)$ as defined previously, although this will turn out to be true as a result of the analysis.

Therefore, let

$$\begin{aligned} W^h_{k+\frac{1}{2}}(u,v) \\ &= \max_{\tau\in[\tau_h^-,\tau_h^+]} \left[\check{w}^h_{k+1}((1+\tau)u) + q^\varepsilon_{k+1}((1+\tau)u)\left\langle \check{v}^h_{k+1}((1+\tau)u) - (1+\tau)v\right\rangle - \tau v\right] \\ &= \max_{\tau\in[\tau_h^-,\tau_h^+]} \left\{Q^\varepsilon_{k+1}((1+\tau)u)\check{V}^h_{k+1}((1+\tau)u) - \left[(1+\tau)q^\varepsilon_{k+1}((1+\tau)u)+\tau\right]v\right\}. \end{aligned}$$

For $v=0$, this is just $W^h_{k+\frac{1}{2}}(u,v) = Q^+_{k+1}((1+\tau)u)\check{V}^h_{k+1}((1+\tau)u)$. We will verify that the max is attained at $\tau=\tau_h^+$. For v large enough, in contrast, the dependence on τ is dominated by the term $-\tau v$. The max is therefore obtained at $\tau=\tau_h^-$. If $W_{k+\frac{1}{2}}$ is, as conjectured, piecewise affine in v with two branches, then we can determine it as the intersection of the two branches thus constructed. Hence

$$W^h_{k+\frac{1}{2}}(u,v) = \check{w}^h_k(u) + q^\varepsilon_{k+\frac{1}{2}}\left(\check{v}^h_k(u) - v\right),$$

where

$$q^\varepsilon_{k+\frac{1}{2}}(u) = (1+\tau_h^\varepsilon)q^\varepsilon_{k+1}((1+\tau_h^\varepsilon)u) + \tau_h^\varepsilon,$$

and $(\check{v}_{k+1}(u), \check{w}_{k+1}(u))$ are the coordinates of the intersection of the two branches. The calculations at this point are the same as in the vanilla case. Yet, the above formula has a different consequence for q^+ due to a different initialization.

Let us consider the first step of dynamic programming, from $k=K-1$ to K. If $u_{K-1} < \mathscr{K}/(1+\tau_h^+)$, then only the region $u_K < \mathscr{K}$ can be reached, where $W_K(u,v) = -c^- v$. Similarly, if $u_{K-1} > \mathscr{K}/(1+\tau_h^-)$, then only the region $u_K > \mathscr{K}$ can be reached, where $W_K(u,v) = D - c^- v$. In both cases, we easily see that the maximizing τ is τ^-, and

$$W_{K-\frac{1}{2}}(u,v) = M(u) - [(1+\tau_h^-)c^- + \tau_h^-]v.$$

Thus we set, as in the vanilla case,

$$q^-_{K-\frac{1}{2}} = (1+\tau_h^-)c^- + \tau_h^-.$$

If $\mathscr{K}/(1+\tau_h^+) < u_{K-1} < \mathscr{K}/(1+\tau_h^-)$, then a careful analysis shows that the maximum in τ is reached either for $\tau=\tau_h^-$ or when $u_K=\mathscr{K}$ at $\tau^\star = \mathscr{K}/u - 1$, depending on v, leading to

$$q^+_{K-\frac{1}{2}}(u) = (1+c^-)\frac{\mathscr{K}}{u} - 1$$

and

$$W_{K-\frac{1}{2}}(u,v) = \max\left\{D - q^+_{K-\frac{1}{2}}v,\, -q^-_{K-\frac{1}{2}}v\right\}.$$

Therefore, in this region, we have, as claimed,

$$W_{K-\frac{1}{2}}(u,v) = \check{w}^h_{K-1}(u) + q^\varepsilon_{K-\frac{1}{2}}\left\langle \check{v}^h_{K-1}(u) - v\right\rangle,$$

with

$$\check{V}^h_{K-1} = \begin{pmatrix}\check{v}^h_{K-1}(u)\\ \check{w}^h_{K-1}(u)\end{pmatrix} = \begin{pmatrix}1\\ -q^-_{K-\frac{1}{2}}\end{pmatrix}\frac{D}{q^+_{K-\frac{1}{2}} - q^-_{K-\frac{1}{2}}},$$

but with the $q^+(u)$ of the continuous-time digital case. [As compared to formula (8.19), the term in $Q^-_K\check{V}^h_K((1+\tau^-_h)u)$ is missing because $\check{V}^h_K((1+\tau^-_h)u) = 0$.]

Now we remark that if $q^+_{k+1}(u) = (1+c^-)\mathscr{K}/u - 1$, then the recursion formula yields $q^+_{k+\frac{1}{2}}(u) = q^+_{k+1}(u)$, while the same formula yields the same $q^-_{k+\frac{1}{2}}$ as for the vanilla case since it is initialized in the same way at $k = K$.

Finally, $Q^+_{K-1}\check{V}^h_{K-1}(u) = D$ or $(C^+ - q^-_{K-\frac{1}{2}})D/(q^+_{K-\frac{1}{2}} - q^-_{K-\frac{1}{2}})$, which are both nondecreasing in u, so that the maximum in τ in $Q^+_{K-1}((1+\tau)u)\check{V}^h_{K-1}((1+\tau)u)$ is reached at τ^+, as hypothesized. And this will carry over in the recursion.

Remark 9.21. The singularity of the sheet $\{\mathscr{K}\}$ of the continuous trading case shows up in this discrete trading case also in the fact that as long as $q^+(u) \in (C^-, C^+)$ and $q^- > C^-$, i.e., $q^\varepsilon_{k+\frac{1}{2}} = q^\varepsilon_k$, we find, on the one hand, $Q^+_k(u)\check{V}^h_k(u) = D$, which is the equation of the sheet $\{\mathscr{K}\}$, and on the other hand, the maximization step in the dynamic programming equation reads, for $v \le \check{v}_{k+1}(u)$,

$$\max_{\tau\in[\tau^-_h,\tau^+_h]}\left\{D - \left[(1+\tau)\left(\frac{\mathscr{K}}{(1+\tau)u} - 1\right) + \tau\right]v\right\} = \max_{\tau\in[\tau^-_h,\tau^+_h]}\{D - q^+(u)v\},$$

which is independent of τ.

9.4.2 Convergence

The convergence analysis of the vanilla case holds here, except on two counts that result from the lack of continuity of W.

On the one hand, in the proof of Theorem 8.59, we cannot appeal to Dini's theorem to assert the uniformity of the convergence on any compact. We must therefore be content with the weaker statement of Theorem 9.10. [The convergence is in fact uniform in v over any compact for fixed (t,u), again due to Dini's theorem.]

On the other hand, Lemma 8.62 does not suffice to prove Proposition 8.61 because it applies to continuous functions. In that respect we have the following fact.

Proposition 9.22. *The functions* $(u,v) \to W_k^h(u,v)$ *are u.s.c.*

Proof. Use the notation $(u_k,v_k) = x_k$, $\{\tau_k\}_{k\in\mathbb{K}} = \omega \in \Omega^h$, $\{\xi_k\}_{k\in\mathbb{K}} = \psi \in \Xi^h$, and let Φ^h be the set of nonanticipative strategies from Ω^h to Ξ^h. We may represent W_k^h as

$$W_k^h(x_k) = \inf_{\varphi\in\Phi^h} \sup_{\omega\in\Omega^h} \left[N(x_K) + \sum_{\ell\geq k} (\tau_\ell(v_\ell+\xi_\ell) + C^{\varepsilon_\ell}\xi_\ell) \right].$$

The application $x_k \to x_K$ is continuous for each ω, and since Ω^h is compact in the product topology, the continuity is uniform in ω. The term due to ψ enters the same function additively; it subtracts out in the difference. Hence the continuity is also uniform in ψ. On the other hand, N is u.s.c. Hence the application $x_k \to N(x_K)$ is u.s.c. in x_k, uniformly in (ω,ψ). The integral term on the right-hand side above is also continuous in x_k, uniformly in (ω,ψ) for the same reason. Hence the square bracket in the right-hand side is a function $J(x_k,\omega,\psi)$ u.s.c. in x_k, uniformly in (ω,ψ).

Let $x_k = x$, as well as the strategy $\phi \in \Phi^h$, be fixed. Let $\delta > 0$ be given. There exists a neighborhood $\mathscr{N}$ of x such that

$$\forall x' \in \mathscr{N}, \forall\omega\in\Omega^h, \quad J(x,\omega,\phi(\omega)) \geq J(x',\omega,\phi(\omega)) - \frac{\delta}{2}.$$

For any x' choose an ω' such that

$$J(x',\omega',\phi(\omega')) \geq \sup_{\omega\in\Omega^h} J(x',\omega,\phi(\omega)) - \frac{\delta}{2}.$$

For that ω',

$$J(x,\omega',\phi(\omega')) \geq \sup_{\omega\in\Omega^h} J(x',\omega,\phi(\omega)) - \delta.$$

Hence, a fortiori

$$\forall x' \in \mathscr{N}, \quad \sup_{\omega\in\Omega^h} J(x,\omega,\phi(\omega)) \geq \sup_{\omega\in\Omega^h} J(x',\omega,\phi(\omega)) - \delta.$$

The conclusion is that $x \to \sup_\omega J(x,\omega,\phi(\omega))$ is u.s.c., and this property is conserved upon taking the infimum with respect to $\phi \in \Phi^h$. □

Now we replace Lemma 8.62 of the vanilla case by the following one, strictly more powerful in our case of monotonous convergence.

Lemma 9.23.

1. *Let $\mathscr{N}$ be a compact neighborhood of $\mathbb{R}^n$, $\{\Delta^h\}$ a sequence of u.s.c. functions from $\mathscr{N}$ to $\mathbb{R}$ monotonously decreasing to a pointwise limit Δ as h converges to 0. Let Δ have a strict maximum over $\mathscr{N}$ at $x^\star$. Let x^h be a maximum of Δ^h over $\mathscr{N}$. Then $x^h \to x^\star$ as $h \to 0$.*
2. *Let $\mathscr{N}$ be a compact neighborhood of $\mathbb{R}^n$, $\{\Delta^h\}$ a sequence of functions from $\mathscr{N}$ to $\mathbb{R}$ monotonously decreasing toward a pointwise limit Δ as h converges to 0. Let $\Delta^h_\star$ and $\Delta_\star$ denote their l.s.c. lower envelopes. Let $\Delta_\star$ have a strict minimum over $\mathscr{N}$ at $x_\star$, and let x_h denote a minimum of $\Delta^h_\star$ over $\mathscr{N}$. Then $x_h \to x_\star$ as $h \to 0$.*

Proof. 1. The sequence $\{x^h\}$ lies in the compact $\mathscr{N}$. We may therefore extract a converging subsequence, which we still denote by x^h. Let $\hat{x}$ be its limit. We will show that, necessarily, $\hat{x} = x^\star$, so that it is the whole sequence that converges to $x^\star$.

From the fact that $\Delta^h(x^\star) \downarrow \Delta(x^\star)$ and the definition of x^h, it follows that

$$\Delta(x^\star) \le \Delta^h(x^\star) \le \Delta^h(x^h). \tag{9.22}$$

Assume that $\Delta(x^\star) - \Delta(\hat{x}) =: \delta > 0$. Notice that $\delta = 0$ if and only if $\hat{x} = x^\star$ because the maximum at $x^\star$ is strict by hypothesis. There exists a $\hat{h}$ such that

$$\forall h \le \hat{h}, \quad \Delta^h(\hat{x}) \le \Delta(\hat{x}) + \frac{\delta}{3}.$$

Since $\Delta^{\hat{h}}$ is u.s.c., there exists a neighborhood $\mathscr{N}'$ of $\hat{x}$ such that

$$\forall x \in \mathscr{N}', \quad \Delta^h(x) \le \Delta^h(\hat{x}) + \delta/3 \le \Delta(\hat{x}) + \frac{2\delta}{3}.$$

Since $x^h \to \hat{x}$, there exists a $\tilde{h} \le \hat{h}$ such that for any $h \le \tilde{h}$, x^h lies in $\mathscr{N}'$, hence

$$\Delta^h(x^h) \le \Delta(\hat{x}) + 2\delta/3 = \Delta(x^\star) - \frac{\delta}{3}.$$

This contradicts inequality (9.22). Thus $\delta = 0$, proving assertion 1.

2. The sequence $\{x_h\}$ lies in the compact $\mathscr{N}$. We can extract a subsequence still denoted by $\{x_h\}$ converging to some $\check{x}$. We will show that, necessarily, $\check{x} = x_\star$, so that it is the whole sequence that converges to $x_\star$.

From the definition of $x_\star$, $\Delta_\star(x_\star) \le \Delta_\star(\check{x})$, with equality if and only if $\check{x} = x_\star$. Assume that $\Delta_\star(\check{x}) - \Delta_\star(x_\star) =: \delta > 0$. Since $x_h \to \check{x}$, and $\Delta_\star$ is l.s.c., there exists $\hat{h} > 0$ such that

$$\forall h \le \hat{h}, \quad \Delta_\star(x_h) \ge \Delta_\star(\check{x}) - \frac{\delta}{2} = \Delta_\star(x_\star) + \frac{\delta}{2}.$$

Also, for all x in $\mathcal{N}$, $\Delta^h(x) \geq \Delta(x)$, so that $\Delta^h_\star(x) \geq \Delta_\star(x)$. Hence for x_h we obtain

$$\Delta^h_\star(x_h) \geq \Delta_\star(x_h) \geq \Delta_\star(x_\star) + \frac{\delta}{2}.$$

From the definition of x_h it follows that

$$\forall x \in \mathcal{N}, \quad \Delta^h(x) \geq \Delta^h_\star(x) \geq \Delta^h_\star(x_h) \geq \Delta_\star(x_\star) + \frac{\delta}{2}.$$

Let $h \to 0$ in this last inequality to get

$$\forall x \in \mathcal{N}, \quad \Delta(x) \geq \Delta_\star(x_\star) + \frac{\delta}{2}.$$

From the definition of a lower envelope, there exists a sequence $\{x_n\}$ converging to $x_\star$ such that $\Delta(x_n)$ converges to $\Delta_\star(x_\star)$. And thus

$$\exists x \in \mathcal{N} : \Delta(x) \leq \Delta_\star(x_\star) + \frac{\delta}{3}.$$

But this contradicts the previous inequality. Thus $\delta = 0$, proving assertion 2. This ends the proof of Lemma 9.23 and that of Theorem 9.10 □

Chapter 10
Validation

As a foreword to this chapter, we wish to recall that we have obviously no claim of overall superiority over Black–Scholes. But a careful analysis may reveal some interesting features, and some weaknesses of Black–Scholes that we avoid. Noticeably, in a time of great uncertainty and of crisis, one might be interested in a theory that does not make use of a probabilistic model of the market pricing process.

It should also be noted that the Black–Scholes theory is too strongly entrenched to be challenged as a piece of "natural science" describing what *will* happen or *why* things happen the way they do. Rather, our analysis is always in the spirit of an "engineering" science of decision support, advising one about *what* to do.

10.1 Numerical Results and Comparisons

10.1.1 Numerical Computations

We give here a few numerical results of our theory. Most computations were performed with the"standard" algorithm. Default values of the parameters are, unless otherwise specified, as in Table 10.1. Notice that it follows from (7.1)–(7.3) that if the relative rate of variation of u is τ, then that of S is $\tau + \mu_0$ (Table 10.1).

Figure 10.1 show plots of the functions $\check{v}_k^h(\cdot)$, the optimum hedging portfolio position in the underlying asset, and $W_k^h(\cdot,0)$, the worth of the portfolio, both as functions of the underlying asset's (normalized) market price u, for $t_k = 0, 10, 20, 30, 40$, and $44 = T$. Concerning $\check{v}_k^h$, as t_k increases, it gets closer to the separatrix of Fig. 8.2 and coincides with it for $t_k = T = 44$. In a similar fashion, $W_k^h(u,0)$ decreases as t_k increases to become the graph of M for $t_k = T = 44$. Recall that $\check{v}_0^h(u(0))$ gives the initial portfolio of the hedging strategy and $W_0^h(u(0),0)$ the corresponding premium.

Similarly, Fig. 10.2 shows the same graphs for a digital call. The peak in $\check{v}_k^h$ is more pronounced the closer t_k is to T. But because of transaction costs, it remains

P. Bernhard et al., *The Interval Market Model in Mathematical Finance*, Static & Dynamic Game Theory: Foundations & Applications, DOI 10.1007/978-0-8176-8388-7_10,

Table 10.1 Default parameters for numerical computations

μ_0	$\mathscr{K}$	D	T	$\tau^- + \mu_0$	$\tau^+ + \mu_0$	C^-	C^+	c^-	c^+
0.0123%	1	1	44	−5%	3%	−0.6986%	0.7014%	$0.5 \times C^-$	$0.5 \times C^+$

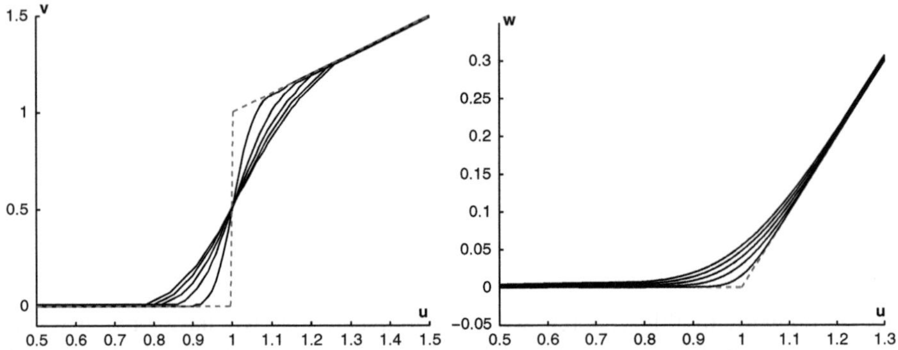

Fig. 10.1 Graphs of $\check{v}_k^h(u)$ and $W_k^h(u,0)$ for $t_k = 0, 10, 20, 30, 40, 44$ for a vanilla call

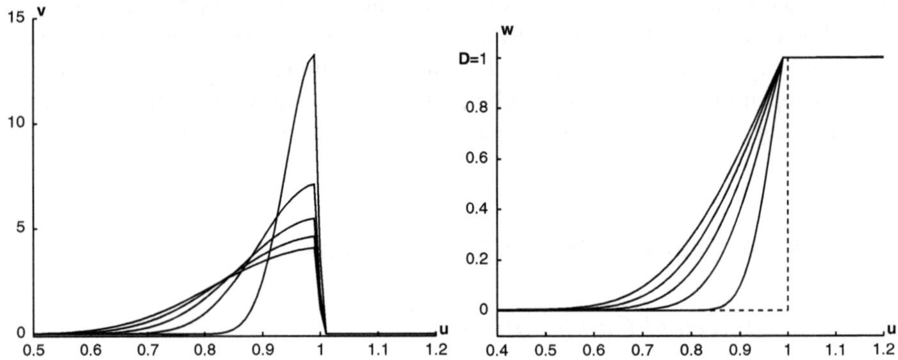

Fig. 10.2 Graphs of $\check{v}_k^h(u)$ and $W_k^h(u,0)$ for $t_k = 0, 10, 20, 30, 40, 44$ for a digital call

bounded by D/C^+. Concerning $W_k^h(u,0)$, the same remark applies that was made previously, but now with the step function M of a digital call.

10.1.2 Numerical Comparison with Black–Scholes

The qualitative aspects of the graphs of W in Fig. 10.1 bear a notable similarity to that of the Black–Scholes theory. To see that more clearly, we provide in Fig. 10.3 a

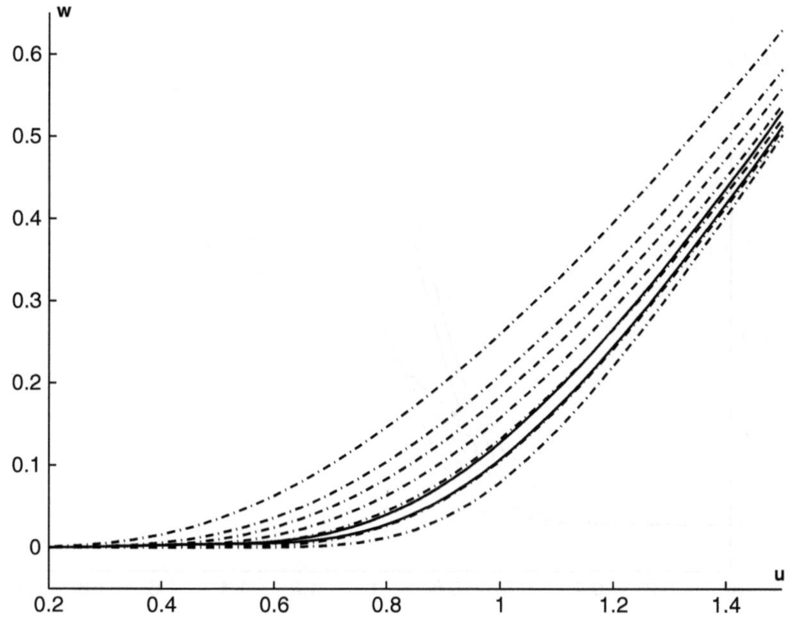

Fig. 10.3 Comparison of Black–Scholes premium for $\sigma = 0.03, 0.04, 0.05, 0.08, 0.1, 0.2, 0.3$ (*dotted lines*) with our premium with and without transaction costs (*solid lines*)

superposition of the graphs of $W(0,u,0)$, both with and without transaction costs[1] (the two solid lines) and the graphs of the Black–Scholes premiums, for volatilities $\sigma = 0.03, 0.04, 0.05, 0.08, 0.1, 0.2, 0.3$ (dotted lines). For this computation, in the case with transaction costs, we have used the partial differential equation (PDE) (8.12) and representation formula (8.13) of our continuous-time theory, integrated with an order-2 Runge Kutta method in time, and a centered (order 2) finite difference scheme in u. This is also a way to verify the convergence of the discrete algorithm to the continuous-time solution.

We notice that our premium in the absence of transaction costs is very close to the Black–Scholes premium for $\sigma = 0.04$. This is not surprising in view of the fact that for a convex terminal payment, and without transaction costs, our theory coincides with that of Cox, Ross, and Rubinstein, which, for a small enough step size, is close to Black–Scholes.

Remark 10.1. This convergence argument must be used with care. Indeed, in the convergence of the theory of Cox, Ross, and Rubinstein toward Black–Scholes, the interval $[d-1, u-1]$ must decrease toward zero as $\sqrt{h}$. In our theory, the underlying

[1] Our continuous-time theory trivializes to the stop-loss strategy and premium in the absence of transaction costs. But this is not so for the discrete-time theory. Here it is used with a step size equal to 1 (day).

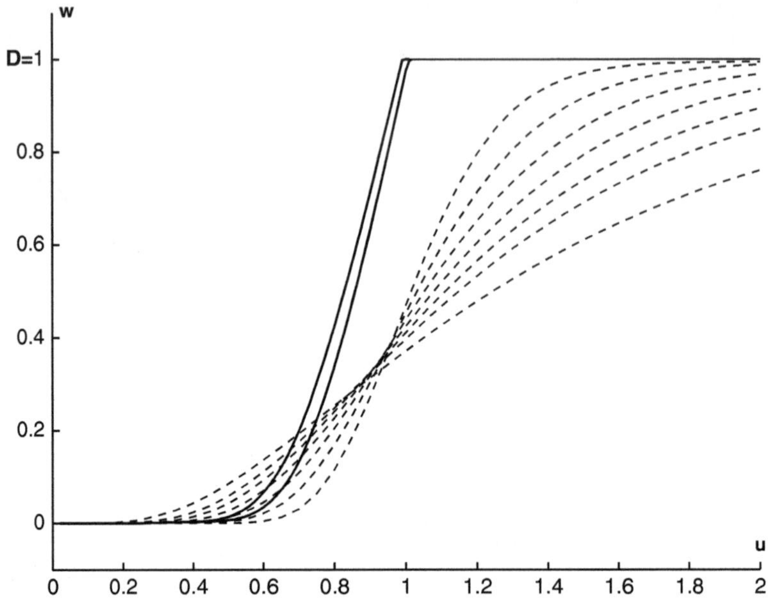

Fig. 10.4 Comparison of premium for a digital call as derived from the Black–Scholes theory with $\sigma = 0.03, 0.04, 0.05, 0.08, 0.1, 0.2, 0.3$ (*dotted lines*) with ours with and without transaction costs (*solid lines*)

market model is continuous and fixed, which makes the interval $[\tau_h^-, \tau_h^+]$ decrease as h. This is why in the limit and without transaction costs, our theory converges to the stop-loss strategy and not to a Black–Scholes equivalent.

The convergence argument does not apply to a nonconvex terminal payment, so that we should not expect to see the digital options premium coincide with that of formula (2.16). Moreover, the worst-case analysis is clearly more different from the classical theory in the case of digital options, where our theory gives a premium equal to D as soon as $u(0) \geq u_- \simeq \mathscr{K}$. Indeed, this shows up in the graphs we provide in Fig. 10.4. Here, due to the complex partitioning of the (t, u) space exhibited by Fig. 9.1, using the fundamental PDE would be exceedingly difficult. Hence the premium of our theory was computed with the discrete-time fast algorithm, with a step size of 0.1.

10.1.3 Explaining the Volatility Smile

A question raised by a reviewer is: does the new theory say anything about the volatility smile?

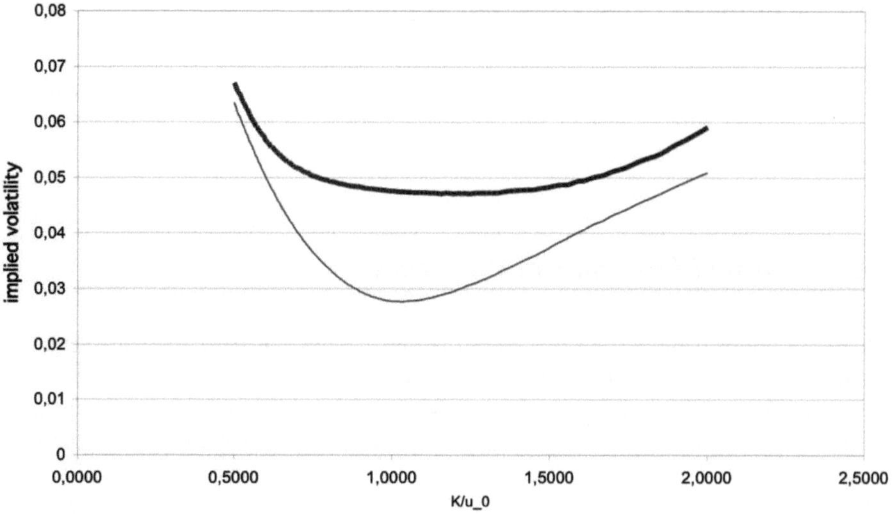

Fig. 10.5 Black–Scholes implied volatility for premiums of our theory as a function of $\mathscr{K}/u_0$. *Thick line*: default parameters, *thin line*: $\tau^+ = -\tau^- = 0.02$

Note first that the interval model does not have a concept of volatility. The nearest thing might be the width $\tau^+ - \tau^-$ of the interval of admissible relative variations. But as stressed in Part II, this is a measure of *risk* and not of variability.

Yet, we may offer one remark.

Remark 10.2. The premiums computed by the new theory, analyzed with the tools of the Black–Scholes theory, display a volatility smile (at least for realistic parameter values).

To show this, we took the premiums computed with our theory for a fixed maturity, a fixed initial stock price u_0, and exercise prices $\mathscr{K}$ ranging from $.5u_0$ to $2u_0$. Then, for each premium we computed the implied volatility of the Black–Scholes theory. We systematically found that the implied volatility was minimal for $\mathscr{K}$ slightly above u_0 and maximum at the extremes, drawing a rather typical "smile." Hence, if the premiums on the market follow our theory, one will indeed observe a smile. Figure 10.5 shows such curves for two sets of parameter values.

We do not give too much credit to this verification, though, as there is no reason why the actual premiums should obey our theory. On the other hand, that theory is not known by the agents on the market, and on the other hand, is not known engineering experience teaches us that Milton Friedman's "as if" argument [76, Chap. 1] is more credible for the long-term adaptive processes of evolution – as suggested by Friedman himself but contrary to his billard expert example – or economics than for such complex technical matters. We consider our theory more as a normative one, a decision help, than as a positive theory.

Yet, if one believes that real-life premiums obey a rational rule, this favors our theory.

10.2 Compared Strengths and Weaknesses

To go further in a comparison, one must distinguish between what the mathematics strictly say and what is practically possible beyond the mathematically grounded facts, thanks to some robustness in the theory.

10.2.1 Strict Mathematical Properties

Clearly, a major weakness of our model is that it rules out from the start very fast price variations in the market. If we try to take τ^- and τ^+ so large that the model is (essentially) always satisfied, then we will end up with an unrealistically large premium. This relates to the classic fact that because our market model is incomplete we must resort to superreplication, potentially ending up with that unrealistically large premium. Hence, only by tolerating some violations of the market model will we get a reasonable premium.

Now, the Black–Scholes theory has its own theoretical shortcomings. On the one hand, it fundamentally assumes that trading is continuous and with no delay. It is impossible, within Samuelson's model, to achieve hedging if the trading is not done continuously, except with the trivial – and too expensive – "buy and hold" strategy $v = u$. On the other hand, within Samuelson's model, *there is no nontrivial hedging portfolio for option pricing with transaction costs* [138].

The first problem derives from the fact that Samuelson's model may display arbitrarily large price variations in any finite time, while the second problem stems from the closely related fact that it has almost surely trajectories of unbounded total variation. (This in itself could be considered as not very realistic.)

Let us concentrate on the continuous versus discrete trading issue. Real trading must be discrete, forcing a discrepancy between real trading and the Black–Scholes theory. This is of little consequence as long as the price of the underlying stock does not change too quickly. But when it does, that discrepancy becomes potentially fatal.

Hence *both theories fail under the same circumstances* – when there are unusually fast-moving variations in the price of the underlying stock. In our theory, the market model is violated; in Black–Scholes, it is the portfolio model that fails.

Mathematically, it is impossible to reconcile a model that allows for arbitrarily large stock price variations within one time step with discrete-time hedging. Hence a mathematical theory must give up one of the two features. Black–Scholes gives up the (theoretical) ability to do discrete trading. We wanted to develop a theory of discrete trading, the discrete-time market model being consistent with (i.e., the time sampling of) a continuous-time underlying market model, kept fixed as the time step is decreased. Thus we had to give up a model that would allow for arbitrarily large price variations in one time step. Yet we wanted a model less idealized than that of Cox, Ross, and Rubinstein – and not dependent on the time step. Thus we were forced to invent the interval model,[2] at the price of relinquishing market completeness. And it is no surprise that other authors came up with the same model.

Turning now to the transaction-cost issue, such costs are a natural ingredient of our theory. Indeed we were forced to introduce them to avoid the naive stop-loss strategy, which is the only solution of the continuous hedging problem in the absence of transaction costs. While we view our new theory's ability to deal exactly with transaction costs, even large ones and with a moderate computational load, as a strength of the theory, the fact that in their absence the only solution is the naive one may be viewed as a limitation of any model with bounded variation trajectories. In contrast, one needs the difficult theory of diffusion limits to deal approximately with small transaction costs in the Black–Scholes theory, giving up exact replication, and at the price of very large computational loads [1].

Finally, once transaction costs are introduced, it is only natural to assume that there are closing costs as well. The introduction of those costs creates a difference between closure in kind or closure in cash. This is a fact of life. Yet, if this difference is deemed annoying, closing costs may be removed, just by setting c^- and c^+ to zero, this time with no detrimental effect on the theory.

10.2.2 Robustness

Now, it is well known that in practice, the Black–Scholes theory, and the derived δ-hedging strategy, can be used with discrete transactions, provided that they are frequent enough. Also, small transaction costs can be tolerated in practice. These are features of robustness of that theory to small violations of the hypotheses used to derive it.

The new theory also seems to exhibit a fair degree of robustness, as illustrated by the following numerical experiments, reported here for a vanilla call only. (Similar experiments with a digital call are reported in [142].)

10.2.2.1 Stochastic Spillover and Expected Cost

In a first series of experiments, we assumed that the relative one-step variation $\tau_k = (u(t_{k+1}) - u(t_k))/u(t_k)$ obeyed a simple independent identically distributed stochastic model, with compact support $[\sigma^-, \sigma^+]$ possibly larger than $[\tau^-, \tau^+]$, and we computed the expected cost to the seller of applying the strategy $\varphi^\star$ derived from our theory for $\tau \in [\tau^-, \tau^+]$. Specifically, we chose a *relative spillover* $\Delta > 0$,

$$\sigma^\varepsilon = (1+\Delta)\tau^\varepsilon, \quad \varepsilon \in \{-,+\},$$

and we used two different probability laws: uniform and "hat," i.e., a law with piecewise affine density, equal to zero at both end points σ^ε and maximum at zero.

[2]But not the name, which was coined by Engwerda et al. [132].

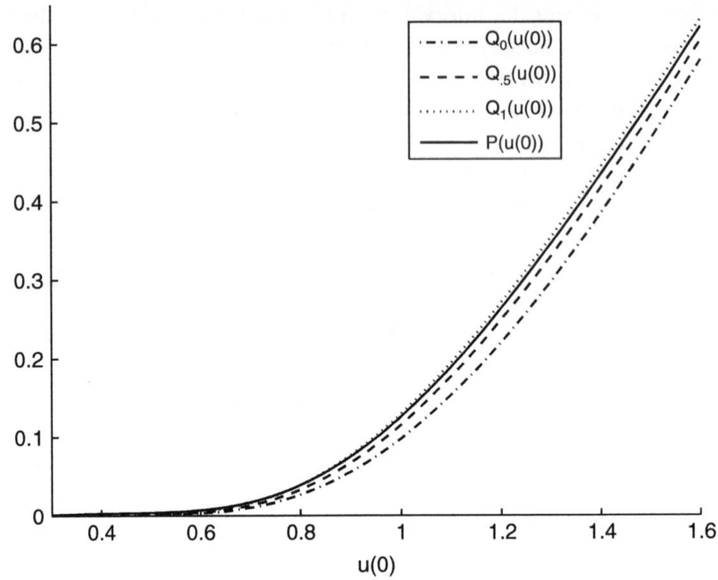

Fig. 10.6 Theoretical premium P and expected cost Q_Δ for a vanilla call, a uniformly distributed τ, and a relative spillover $\Delta = 0, 0.5, 1$

The computation of the expected cost was done using Kolmogorov's equation,

$$W_k^h(u,v) = \mathbb{E}[W_{k+1}^h((1+\tau)u,(1+\tau)(v+\xi_k^\star)) - \tau(v+\xi_k^\star)] + C^\varepsilon\langle\xi_k^\star\rangle,$$
$$W_K^h(u,v) = N(u,v).$$

(This requires only a small modification of the computer code for the standard algorithm.) The expectation at each time step was computed using the trapezoid formula.

Vanilla call

The results for a vanilla call and the uniform law are drawn in Fig. 10.6. We call $P(u(0))$ the premium obtained from our theory and Q_Δ the expected cost computed as described. We plotted P and Q_Δ for $\Delta = 0, 0.5, 1$. The curve for $\Delta = 0.85$ would be indistinguishable from the curve of P. The results for the hat law are given in Fig. 10.7 for $\Delta = 0.5, 1, 2$. The curve for $\Delta = 1.62$ would be indistinguishable from the curve of P. Finally, a better representation of the role of Δ can be obtained from Fig. 10.8, where we plotted $P(\mathscr{K}) - Q_\Delta(\mathscr{K})$ as a function of Δ.

Obviously, according to our previous remarks, the strategy $\varphi^\star$ remains safe in expectation up to a relative spillover of 85% for a uniform law and 162% for the hat law.

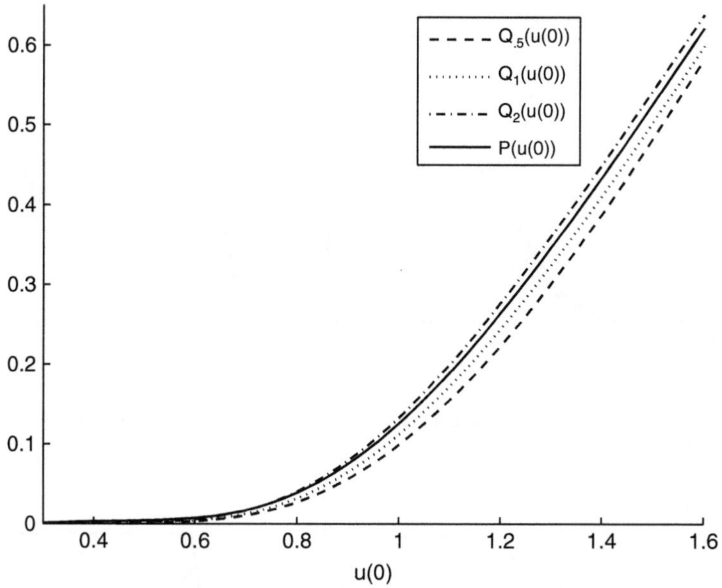

Fig. 10.7 Theoretical premium P and expected cost Q_Δ for a vanilla call, a "hat" distributed τ, and a relative spillover $\Delta = 0.5, 1, 2$

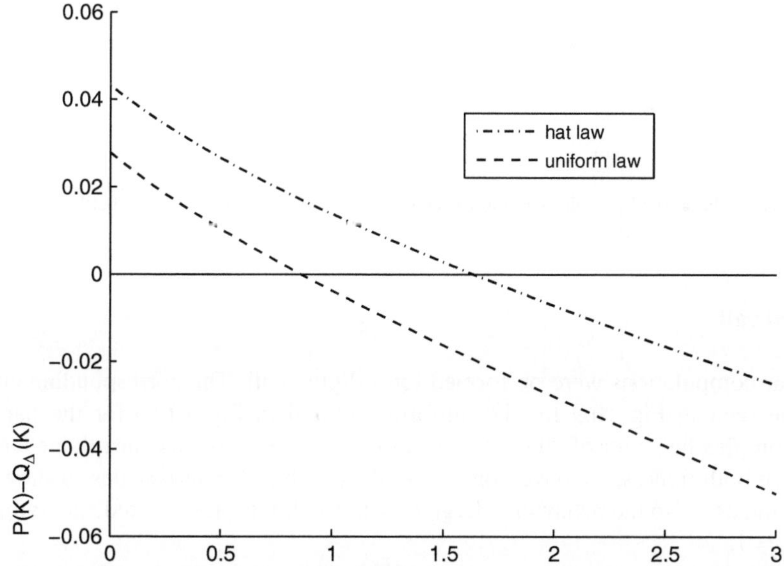

Fig. 10.8 Vanilla call: $P - Q_\Delta$ for $u(0) = \mathscr{K}$ as a function of Δ

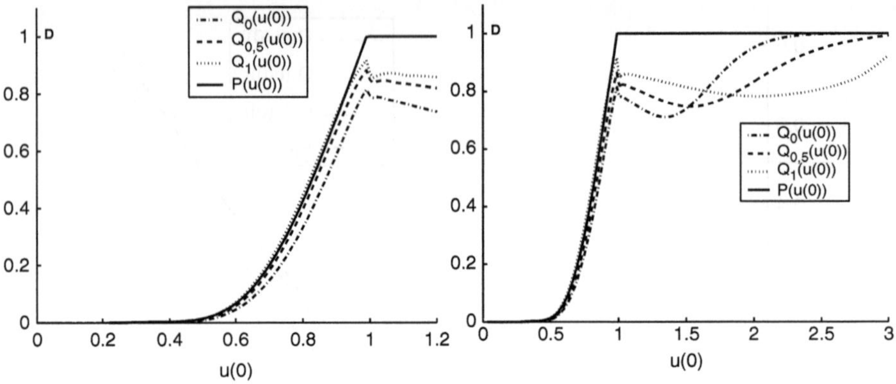

Fig. 10.9 Theoretical premium P and expected cost Q_Δ for a digital call, a uniformly distributed τ, and a relative spillover $\Delta = 0, 0.5, 1$

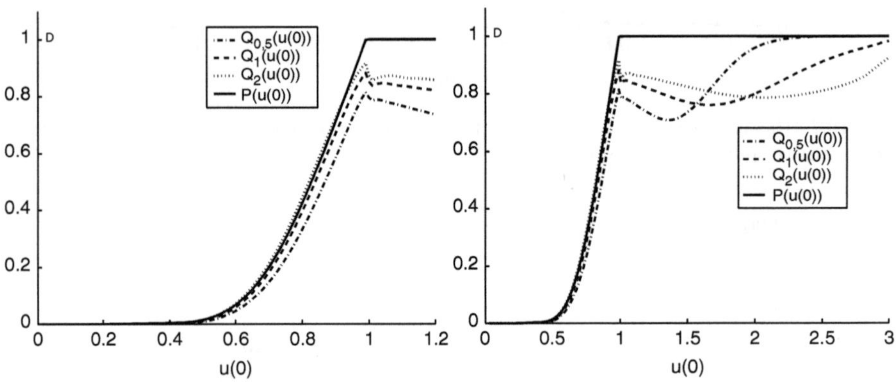

Fig. 10.10 Theoretical premium P and expected cost Q_Δ for a digital call, a "hat" distributed τ, and a relative spillover $\Delta = 0.5, 1, 2$

Digital call

Similar computations were performed for a digital call. The corresponding curves can be seen in Fig. 10.9 for the uniform law and in Fig. 10.10 for the hat law. The complex behavior of these curves close to $u = \mathscr{K}$ remains unexplained at this time. In both figures, we have drawn two plots with different u scales to show that behavior, and also the behavior at large u, which exhibits, as expected, an asymptote at $Q_\Delta = 1$.

In Fig. 10.11, we give the equivalent of Fig. 10.8, but for the digital call, and with P and Q_Δ evaluated at $u(0) = 0.8\mathscr{K}$ to stay away from the region close to $\mathscr{K}$ where the "worst-case" effect is strong. The implication is that, here again, a spillover of 80% for the uniform law and one of 150% for the hat law still provide a hedge in expectation.

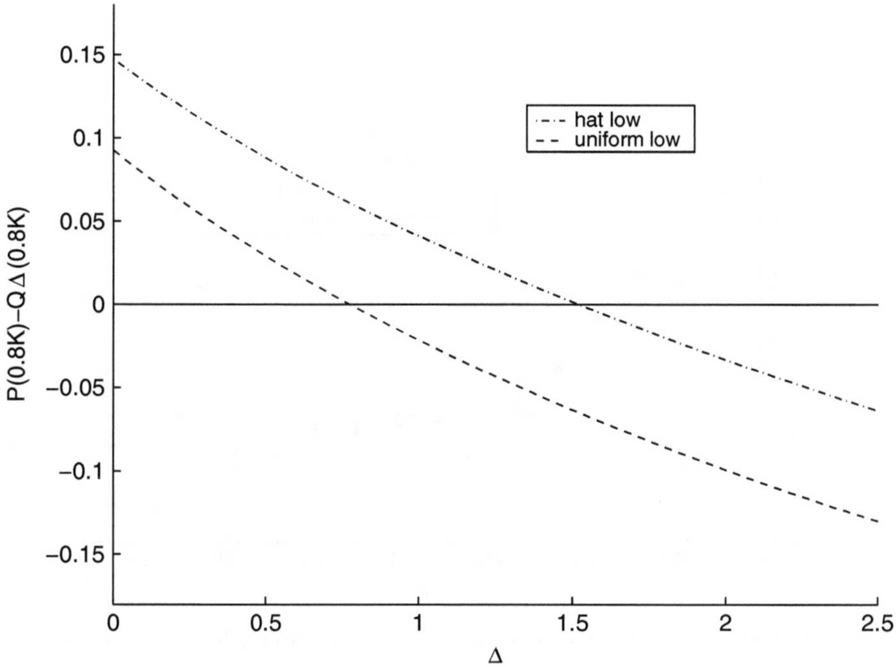

Fig. 10.11 Digital call: $P - Q_\Delta$ for $u(0) = 0.8\mathscr{K}$ as a function of Δ

10.2.2.2 Deterministic Spillover and Simulated Cost

In a last set of experiments, we used time series of underlying prices, either real or simulated as a lognormal sequence à la Samuelson. On each time series, we narrowed the interval $[\tau^-, \tau^+]$ by rejecting a symmetric and increasing proportion of the samples τ_k. The proportion p of samples *outside* the interval used is therefore a measure of the relative spillover. Next, we computed the hedging strategy derived from our theory and simulated its application to the given time series, and we plotted the cost incurred by the seller together with the premium computed according to our theory. In the following figures, we plotted both against the relative spillover p. To have a better comparison with Black–Scholes, we also plotted the cost incurred without the transaction costs, and we have estimated the empirical volatility of the time series to show the Black–Scholes premium.

Figures 10.12–10.15 show the results of these experiments for two real-time series, Airfrance shares from August 18 to October 20, 1998 (estimated volatility 0.03), Paris' stock exchange CAC40 index from March 6 to May 8, 1998 (estimated volatility 0.01), and two simulated lognormal time series with zero drift and volatilities 0.1 and 0.01.

We remark that the results are more satisfactory for the simulated lognormal series, where a relative spillover of 30% seems acceptable, than for the real ones,

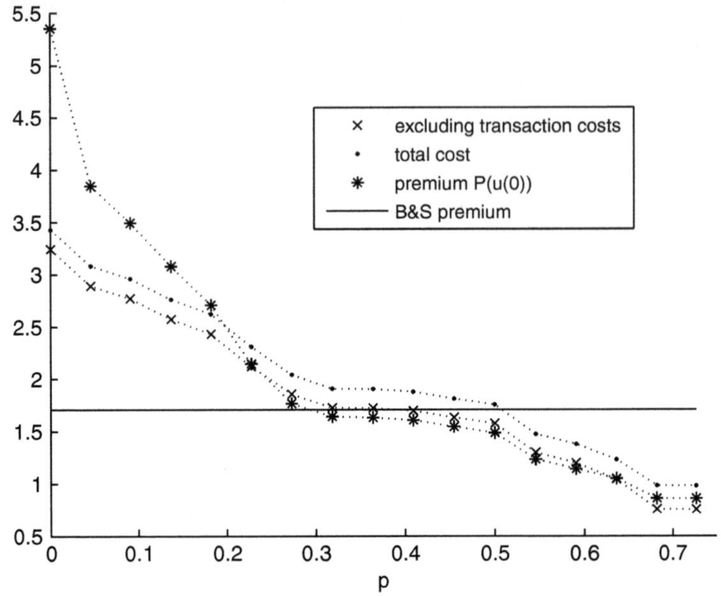

Fig. 10.12 Theoretical premium and cost incurred as a function of spillover, Airfrance series

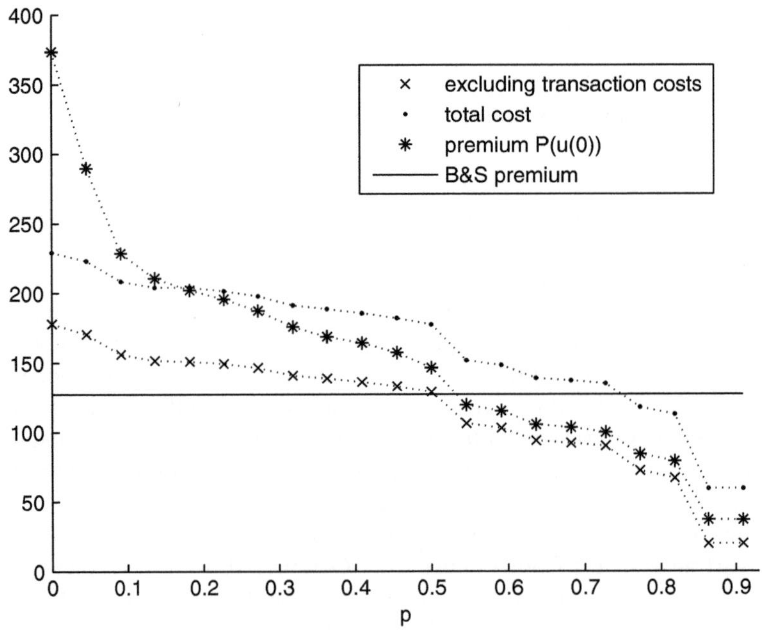

Fig. 10.13 Theoretical premium and cost incurred as a function of spillover, CAC40 series

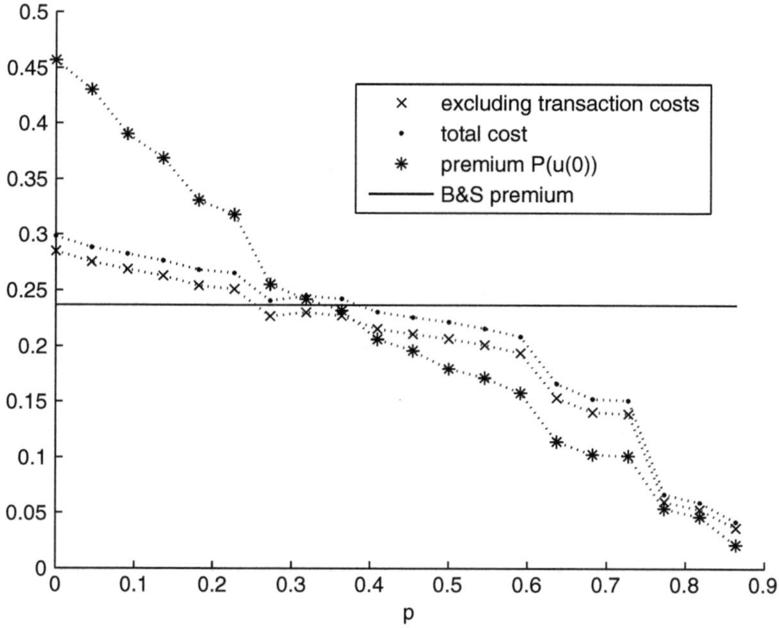

Fig. 10.14 Theoretical premium and cost incurred as a function of spillover, lognormal series, $\sigma = 0.09$

where only a 20% relative spillover yields an effective hedge, but with a higher premium than Black–Scholes. Also, the gap with a Black–Scholes premium is larger for low volatilities. Indeed, for low volatilities, the worst-case analysis leads to too high a premium.

10.3 Conclusion

A careful analysis shows that it is rather natural to resort to such "interval models." Furthermore, for the strict problem of hedging a contingent claim, the robust control approach used in this volume allows us to proceed without endowing the set of admissible stock price trajectories with a probability law. This is so since what is sought is a hedge for *every* possible trajectory. (And this remark carries over to the Black–Scholes theory, as we showed in Chap. 2.)

The resulting theory exhibits a strong mathematical structure that can be exploited to get semiexplicit formulas leading to a fast algorithm, whether in discrete or continuous trading. The latter is the limit of the former with vanishing step size, this, we stress, keeping *the same* continuous-time model for the underlying price trajectories. Thus the discrete trading strategy, which is very simple to implement, is a good and safe approximation of the theoretical continuous strategy.

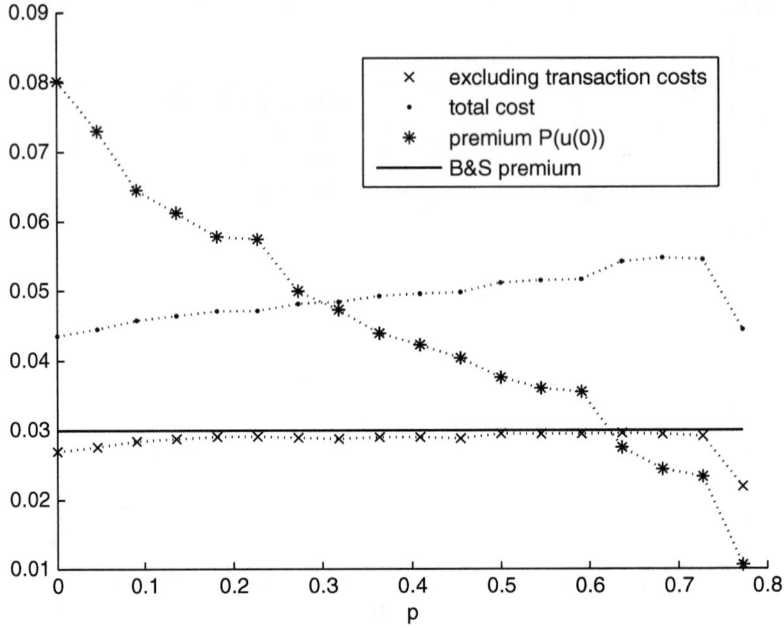

Fig. 10.15 Theoretical premium and cost incurred as a function of spillover, lognormal series, $\sigma = 0.01$

The premiums computed qualitatively and quantitatively resemble Black–Scholes premiums, although the presence of transaction costs obviously makes them larger. Moreover, they display (explain?) the volatility smile observed in real-life premiums.

The hedging strategies derived from the theory exhibit a fair degree of robustness to violations of the market model. Our simulations show that a relative spillover of 30% of the relative one-step price variations, as compared to the interval model used, often leads to a premium comparable to the Black–Scholes premium and an effective hedge. Yet, these simulations were run using a posteriori "statistical" information on the price series. If an approach based on the interval model is to be routinely used, then one must also develop new statistical tools to inform it.

This robustness analysis is carried out in a *normative* perspective to show that this theory can be used on the actual market as a *decision aid*. In spite of the remark concerning the volatility smile, we have at this point no claim to a *positive* theory that would *explain* premiums actually used by operators, the less so because the current market is overwhelmingly dominated by the Black–Scholes theory, which is, in that respect, more or less self-enforcing.

While we obviously claim no overall superiority over the Black–Scholes theory, we feel that the results provided so far point to the conclusion that our theory might be useful in some situations, for instance, when transaction costs are too high to be neglected or when time discretization is critical. And in any case, it provides a sensible alternative to the analysis of the problem of option pricing.

Yet this is by all means a young theory. There remains much work to be done in sensitivity analysis, simulations, and validation, and many theoretical questions remain open. We hope to have proved that such work is worthwhile pursuing.

Part IV
Game-Theoretic Analysis of Rainbow Options in Incomplete Markets

Author: Vassili N. Kolokoltsov
Department of Statistics,
University of Warwick,
UK

In this part, the author expands the ideas in the papers [95, 96] to develop a pure game-theoretic approach to option pricing in a multidimensional market (rainbow options), where risk-neutral probabilities emerge automatically from the robust-control evaluation. The process of investment is considered as a zero-sum game between an investor and Nature. We also extend the analysis to credit derivatives such as credit default swaps.

Notation

- T: Exercise time
- $t \in [0, T]$: Current time
- τ: Time step
- m: Current time step
- n: Last time step; $T = n\tau$
- $S^j, j \in \{1, \ldots, J\}$: Market price of asset j
- d^j, u^j: Minimum and maximum ratios S^j_{m+1}/S^j_m
- ξ^j: Actual ratio S^j_{m+1}/S^j_m
- r: Riskless interest rate
- $\rho = 1 + r\tau$: Yield of riskless asset
- γ^j_m: Number of shares of asset j in portfolio at time step m
- X_m: Portfolio worth at time step m
- K: Exercise price, or "strike"
- f: Terminal payment faced by seller
- β: Proportional transaction cost rate

Chapter 11
Introduction

11.1 Introduction to Game-Theoretic Pricing

Expanding the ideas of the author's papers [95,96] we develop a pure game-theoretic approach to option pricing in a multidimensional market (rainbow options), where risk-neutral probabilities emerge automatically from robust control evaluation. The process of investment is considered as a zero-sum game of an investor with Nature.

For basic examples of complete markets, like binomial models or geometric Brownian motion, our approach yields the same results as the classic (by now) risk-neutral evaluation developed by Cox–Ross–Rubinstein or Black–Scholes. However, for incomplete markets, such as those for rainbow options in multidimensional binomial or interval models, the coexistence of an infinite number of risk-neutral measures precludes unified pricing of derivative securities by the usual methods. Several competing methods have been proposed for pricing options under these circumstances (see, e.g., a review in Bingham and Kiesel [45]), most of them using certain subjective criteria, say a utility function for payoff or a certain risk measure. In our game-theoretic approach, no subjectivity enters the game. We define and calculate a hedge price, which is the minimal capital needed to meet the obligation for all performances of the markets, within the rules specified by the model (dominated hedging).

Though our price satisfies the so-called no strictly acceptable opportunities (NSAO) condition suggested in Carr et al. [53], one still could argue, of course, that this is not a completely fair price, because the absence of an exogenously specified initial probability distribution does not allow us to speak about a.s. performance and implies necessarily the possibility of an additional surplus. However, together with the hedging price for buying a security, which may be called an upper price, one can equally reasonably define a lower price that can be seen as a hedge for selling the security. The difference between these two values can be considered as a precise measure of the intrinsic risk that underlies incomplete markets. An alternative way

P. Bernhard et al., *The Interval Market Model in Mathematical Finance*, Static & Dynamic Game Theory: Foundations & Applications, DOI 10.1007/978-0-8176-8388-7_11,

to deal with possible unpredictable surplus, as suggested, for example, in Lyons [108] for models with unknown volatility, consists in specifying a possible cash back that the holder of an option should receive when price moves (unpredictable initially) turn out to be favorable.

Our method is robust enough to be able to accommodate various market rules and settings including path-dependent payoffs, American options, real options, and transaction costs. On the other hand, it leads to rather simple numerical algorithms. The continuous-time limit is described by nonlinear or fractional Black–Scholes type equations.

For completeness, we also demonstrate how the standard stochastic models of financial dynamics come very naturally into play in our game-theoretic setting.

A more detailed description is available at the beginning of each chapter. The main results are given in Chaps. 12 and 13.

11.2 Related Works

Some bibliographical comments seem to be in order. A game-theoretic (or robust control) approach to options was used in McEneaney [112], though in that paper the main point was to prove that the option prices of standard models could be characterized as viscosity solutions of the corresponding Hamilton–Jacobi equation. As a byproduct it was confirmed (similarly to analogous results in Avellaneda et al. [18] and Lyons [108]) that one could hedge prices in stochastic volatility models by the Black–Scholes strategies specified by the maximal volatility. A related paper is Olsder [123], where only a basic one-dimensional model was analyzed, though with some transaction costs included. An ideologically related paper is De Meyer and Moussa Saley [66] devoted to the strategic origins of Brownian motion. A completely different approach to bringing game theory to option pricing is developed in Ziegler [148, 149].

The reasonability of the extension of the binomial model allowing for price jumps inside an interval (interval model) was realized by several authors; see Kolokoltsov [95], Bernhard [37], Aubin et al. [17], and Roorda et al. [132]. In the last-named paper the term *interval model* was coined. The series of papers by Bernhard et al. [36, 37, 39] deals with one-dimensional models with very general strategies and transaction costs including both continuous and jump-type trading. The corresponding Hamilton–Jacobi–Bellman equations have peculiar degeneracies that require subtle techniques to handle.

Hedging by domination (superreplication), rather than replication, is well established in the literature, especially in connection with models incorporating transaction costs; see, for example, [23]. Problems with transaction costs in standard models are well known, as the title ("There is no nontrivial hedging portfolio for option pricing with transaction costs") of Soner et al.'s [138] paper indicates.

This problem, similar to the story with incomplete markets, leads to the development of optimizations based on a subjectively chosen utility function; see, for example, Davis and Norman [65] and Barles and Soner [23].

Upper and lower values for prices have been discussed in many works; see, for example, El Karoui and Quenez [71] and Roorda et al. [132]. An abstract definition of lower and upper prices can be given in the general game-theoretic approach to probability and finance advocated in the monograph by Shafer and Vovk [136].

The well-known fact that the existing (however complicated) stochastic models are far from being precise reflections of the real dynamics of market prices leads naturally to attempts to relax the assumed stochastic restrictions of models. For instance, Avellaneda et al. [18] and Lyons [108] work with unknown volatilities leading to nonlinear Black–Scholes type equations (though still nondegenerate, unlike those obtained here subsequently). On the other hand, Hobson [84] (see also [83, 85] and references therein) suggests model-independent estimates based on the observed prices of traded securities, the main technique being the Skorohod embedding problem (SEP). These approaches still build the theory on some basic underlying stochastic model (e.g., geometric Brownian motion), unlike our method, which starts up front with robust control. Similarly, hedging with respect to several (or all) equivalent martingale measures, based on optional decomposition (see Föllmer and Kramkov [74] and Kramkov [106]), depend on some initial probability law (with respect to which equivalence is considered). The risk-neutral or martingale measures that arise from our approach are not linked to any initial law. They are not equivalent but represent extreme points of risk-neutral measures on all possible realizations of a stock price process.

"Fractional everything" has become a popular topic in modern literature; see, for example, the recent monograph by Tarasov [141]. For the study of financial markets, this is, of course, a natural step in moving from a discussion of power laws in economics (see, e.g., the various perspectives in Uchaikin and Zolotarev [144], Maslov [111], and references therein) to the applicability of fractional dynamics in financial markets (see, e.g., Meerschaert and Scala [114], Meerschaert et al. [113], Jumarie [92], Wang [145], and references therein. Our game-theoretic analysis leads to degenerate or nonlinear versions of fractional Black–Scholes type equations.

The main ideas of this exposition were presented in brief in [102].

Chapter 12
Emergence of Risk-Neutral Probabilities from a Game-Theoretic Origin

12.1 Geometric Risk-Neutral Probabilities and Their Extreme Points

In this section we will introduce risk-neutral probabilities in an abstract geometric setting revealing their basic properties and describing their extreme points. This discussion belongs to convex analysis and is independent of either games or finance, but it will provide us with basic tools to use later.

For a compact metric space E we denote by $\mathscr{P}(E)$ the set of probability laws on E and by $C(E)$ the Banach space of bounded continuous functions on E. For our purpose here we will mostly need finite subsets $E = \{\xi_1, \ldots, \xi_k\}$ of $\mathbb{R}^d$ (in which case probability laws are given by the sets of positive numbers $\{p_1, \ldots, p_k\}$ summing up to one). We will work in a bit more general setting anticipating further applications; more general compact sets pop in very naturally sometimes, especially when dealing with mixed strategies.

For $f \in C(E)$, $\mu \in \mathscr{P}(E)$, the standard pairing is given by the integration or, in more probabilistic notation, by the expectation (where we will denote, with some abuse of notation, random and integration variables by the same letter):

$$(f,\mu) = \int_E f(x)\mu(\mathrm{d}x) = \mathbf{E}_\mu f(x) = \mathbf{E}_\mu f.$$

This pairing also extends to vector-valued functions f.

The set $\mathscr{P}(E)$ is known to be a compact set in its weak topology [where $\mu_n \to \mu$, as $n \to \infty$, whenever $(f, \mu_n) \to (f, \mu)$ for any $f \in C(E)$, as $n \to \infty$].

Definition 12.1. Let us say that a probability law $\mu \in \mathscr{P}(E)$ on $E \subset \mathbb{R}^d$ is *risk neutral with respect to the origin* or, more concisely, *risk neutral* if the origin is its barycenter, that is, $\int_E \xi \mu(d\xi) = 0$. The set of all risk-neutral laws on E will be denoted by $\mathscr{P}_{rn}(E)$.

P. Bernhard et al., *The Interval Market Model in Mathematical Finance*, Static & Dynamic Game Theory: Foundations & Applications, DOI 10.1007/978-0-8176-8388-7_12,

It will be convenient to work in a bit more general setting of probability laws with linear constraints. That is, for a compact subset $E \subset \mathbb{R}^n$ and a continuous mapping $F: E \to \mathbb{R}^d$ let

$$\mathscr{P}(E;F) = \{\mu \in \mathscr{P}(E) : (F,\mu) = \int F(x)\mu(\mathrm{d}x) = 0\}.$$

This is clearly a convex closed subset of $\mathscr{P}(E)$ (which may be empty of course), and $\mathscr{P}_{rn}(E) = \mathscr{P}(E;Id)$, where Id is the identical mapping of E.

The key role for the analysis of risk-neutral probabilities belongs to the following two conditions for a subset of $\mathbb{R}^d$.

Definition 12.2. A set $E \subset \mathbb{R}^d$ is called *weakly* (resp. *strongly*) *positively complete* if there exists no $\omega \in \mathbb{R}^d$ such that $(\omega,\xi) > 0$ [resp. $(\omega,\xi) \geq 0$] for all $\xi \in E$. Geometrically, this means that E does not belong to any open (resp. closed) half-space of $\mathbb{R}^d$.

Clearly E is strongly positively complete if for any $\omega \in \mathbb{R}^d$ there exist vectors $\xi_1, \xi_2 \in E$ such that $(\omega,\xi_1) > 0$ and $(\omega,\xi_2) < 0$. If $E \subset \mathbb{R}^d$ is a compact convex set, then E is weakly positively complete if and only if it contains the origin. It is, moreover, strongly positively complete whenever the origin is not its boundary point.

Remark 12.3. Roorda et al. [131] call a finite set E positively complete if it is weakly positively complete in our sense.

The importance of these notions is revealed in the following statement.

Proposition 12.4. *Let $E \subset \mathbb{R}^n$ be compact and a mapping $F: E \to \mathbb{R}^d$ continuous.*

(1) The set $\mathscr{P}(E;F)$ is not empty if and only if $F(E)$ is weakly positively complete in $\mathbb{R}^d$.
(2) Let E' be the support of a measure $\mu \in \mathscr{P}(E;F)$. If $F(E')$ does not coincide with the origin, then it is strongly positively complete in the subspace $\mathbb{R}^m \subset \mathbb{R}^d$ generated by $F(E')$.
(3) Let $F(E)$ be weakly positively complete in $\mathbb{R}^d$. Then either the support of any $\mu \in \mathscr{P}(E;F)$ is contained in $F^{-1}(0)$ or there exists a subspace $\mathbb{R}^m \subset \mathbb{R}^d$ such that $F(E) \cap \mathbb{R}^m$ is strongly positively complete in $\mathbb{R}^m$ and any $\mu \in \mathscr{P}(E;F)$ has a support in $F^{-1}(\mathbb{R}^m)$.

Proof. (1) Let $F(E)$ be not weakly positively complete, so that there exists $\omega \in \mathbb{R}^d$ such that $(\omega, F(x)) > 0$ for all $x \in E$. Then

$$(\omega, \int F(x)\mu(\mathrm{d}x)) > 0 \tag{12.1}$$

for any $\mu \in \mathscr{P}(E)$, which cannot hold for $\mu \in \mathscr{P}(E;F)$. Conversely, suppose the set $\mathscr{P}(E;F)$ is empty. Then the image of $\mathscr{P}(E)$ under the linear mapping

$$\mu \to \int F(x)\mu(\mathrm{d}x)$$

does not contain the origin. As it is a compact convex set, it follows (by the Banach separation theorem) that there exists an $\omega \in \mathbb{R}^d$ such that

$$(\omega, \int F(x)\mu(\mathrm{d}x)) > 0$$

for all $\mu \in \mathscr{P}(E)$. In particular, using point measures we conclude that $(\omega, F(x)) > 0$ for all $x \in E$.

(2) Suppose $F(E') \neq \{0\}$, so that it generates a subspace $\mathbb{R}^m \subset \mathbb{R}^d$ with $m > 0$. Then for any $\omega \in \mathbb{R}^m$ there exists $x \in E'$ such that $(\omega, F(x)) \neq 0$. But since

$$(\omega, \int F(x)\mu(\mathrm{d}x)) = 0,$$

there must exist another $y \in E'$ such that the signs of $(\omega, F(y))$ and $(\omega, F(x))$ are different.

(3) Suppose $F(E)$ is weakly positively complete. Then either $F(E)$ is strongly positively complete (and we are done) or there exists $\omega \in \mathbb{R}^d$ such that $(\omega, F(x)) \geq 0$ for all $x \in E$ and the set $E_1 = \{x \in E : (\omega, F(x)) = 0\}$ is nonempty. Let Π be the subspace generated by $F(E_1)$. As it belongs to the subspace orthogonal to ω, its dimension is strictly less than d. Moreover, the support of any $\mu \in \mathscr{P}(E;F)$ is contained in E_1, so that $\mathscr{P}(E;F) = \mathscr{P}(E_1;F)$. As this set is not empty, $F(E_1)$ is weakly positively complete in $\mathbb{R}^m$. The proof is now completed by induction in the dimension d.

□

Let us turn to the extreme points of sets $\mathscr{P}(E;F)$. It is a well-known (and a simple) fact that the set of extreme points of $\mathscr{P}(E)$ coincides with the set of Dirac measures (or atoms) δ_x, $x \in E$, that assign the total mass to a single point x. The next statement is a straightforward extension of this fact.

Proposition 12.5. *Let $E \subset \mathbb{R}^n$ be compact, a mapping $F = (F^1, \ldots, F^d) : E \to \mathbb{R}^d$ be continuous, and μ be an extreme point of the set $\mathscr{P}(E;F)$. Then μ is a linear combination of not more than $d+1$ Dirac measures.*

Proof. Assuming our claim does not hold, one can find a partition of E in $d+2$ Borel subsets $E = E_1 \cup \cdots \cup E_{d+2}$ that are pairwise disjoint with $\mu(E_i) > 0$ for all i. As $\mu \in \mathscr{P}(E;F)$,

$$(F^i, \mu) = \sum_{j=1}^{d+2} \int_{E_j} F^i(x)\mu(\mathrm{d}x) = 0, \quad i = 1, \ldots, d.$$

For any collection of $(d+1)$ numbers $\varepsilon_i \in (-1,1)$ let

$$\mu_+ = \sum_{j=1}^{d+2} (1+\varepsilon_j)\mu|_{E_j}, \quad \mu_- = \sum_{j=1}^{d+2} (1-\varepsilon_j)\mu|_{E_j}.$$

Clearly, $\mu_\pm$ are positive measures such that $\mu = (\mu_+ + \mu_-)/2$. To show that μ is not an extreme point of $\mathscr{P}(E;F)$ (and hence to complete the proof), it is sufficient to show that there exists a family of ε_j such that $\mu_\pm \in \mathscr{P}(E;F)$, that is,

$$\sum_{j=1}^{d+2} \varepsilon_j \mu(E_j) = 0,$$

$$\sum_{j=1}^{d+2} \varepsilon_j \int_{E_j} F^i(x)\mu(\mathrm{d}x) = 0, \quad i = 1,\ldots,d.$$

But this is a homogeneous system of $d+1$ linear equations for $d+2$ variables, so that the solution space has a positive dimension. □

Let us now give more specific results for risk-neutral laws.

From Proposition 12.4 it follows that the set of risk-neutral laws $\mathscr{P}_{rn}(E)$ on a compact subset $E \subset \mathbb{R}^d$ is not empty if and only if E is weakly positively complete and the support of any risk-neutral law is strongly positively complete in the subspace it generates.

Let us start with the most important case of finite subsets $E = \{\xi_1,\ldots,\xi_k\}$ of $\mathbb{R}^d$. If such E is strongly positively complete, then $k > d$ (by the Banach separation theorem). Hence the minimal families of vectors in $\mathbb{R}^d$ satisfying strong positive completeness contain precisely $d+1$ vectors.

Let us say that a finite family of vectors $E = \{\xi_1,\ldots,\xi_k\}$ in $\mathbb{R}^d$ is in *general position* if the vectors of any subset of $\{\xi_1,\ldots,\xi_k\}$ of size d are linearly independent (in particular, all vectors in E are nonvanishing).

Proposition 12.6. *Let a finite set $E = \{\xi_1,\ldots,\xi_{d+1}\}$ be strongly positively complete in $\mathbb{R}^d$. Then*

(1) The family E is in general position;
(2) The origin belongs to the interior of the simplex $\Pi[\xi_1,\ldots,\xi_{d+1}]$, defined as the convex hull of the family $\{\xi_1,\ldots,\xi_{d+1}\}$;
(3) There exists a unique risk-neutral probability law $\{p_1,\ldots,p_{d+1}\}$ on the set $\{\xi_1,\ldots,\xi_{d+1}\}$, with

$$p_i = C^{-1}(-1)^{i-1}\det\begin{pmatrix} \xi_1^1 & \cdots & \xi_{i-1}^1 & \xi_{i+1}^1 & \cdots & \xi_{d+1}^1 \\ \xi_1^2 & \cdots & \xi_{i-1}^2 & \xi_{i+1}^2 & \cdots & \xi_{d+1}^2 \\ & & \cdots & & & \\ \xi_1^d & \cdots & \xi_{i-1}^d & \xi_{i+1}^d & \cdots & \xi_{d+1}^d \end{pmatrix}, \quad i = 1,\ldots,d, \tag{12.2}$$

where

$$C = \det \begin{pmatrix} 1 & 1 & \cdots & 1 \\ \xi_1^1 & \xi_2^1 & \cdots & \xi_{d+1}^1 \\ & & \cdots & \\ \xi_1^d & \xi_2^d & \cdots & \xi_{d+1}^d \end{pmatrix}. \tag{12.3}$$

Proof. (1) If the vectors $\xi_1, \dots \xi_d$, say, are dependent, then they generate a proper subspace Π of $\mathbb{R}^d$. Then, for any $\omega \in \mathbb{R}^d$ that is orthogonal to Π we will have either $(\omega, \xi_i) \geq 0$ for all i or $(\omega, \xi_i) \leq 0$ for all i (depending on the position of the vector ξ_{d+1}), contradicting the strong positive completeness.

(2) If the origin does not belong to the interior of the convex compact set $\Pi[\xi_1, \dots, \xi_{d+1}]$, then by the Banach separation theorem there exists a closed half-space of $\mathbb{R}^d$ containing all $\xi_1, \dots, \xi_{d+1}$. This again contradicts the strong positive completeness.

(3) The existence and uniqueness of a risk-neutral probability law on the set $\{\xi_1, \dots, \xi_{d+1}\}$ follow directly from (2) and (1), respectively. To prove (12.2), let us consider the following vector-valued determinant:

$$D = \det \begin{pmatrix} \xi_1 & \xi_2 & \cdots & \xi_{d+1} \\ \xi_1^1 & \xi_2^1 & \cdots & \xi_{d+1}^1 \\ \xi_1^2 & \xi_2^2 & \cdots & \xi_{d+1}^2 \\ & & \cdots & \\ \xi_1^d & \xi_2^d & \cdots & \xi_{d+1}^d \end{pmatrix},$$

defined via its expansion with respect to the first row. Clearly $D = 0$, as each of its coordinates is given by a determinant with two coinciding rows. Hence, expanding it with respect to the first row yields

$$\sum_{i=1}^{d+1} p_i \xi_i = 0,$$

with p_i given by (12.2). By (1) and (3), a collection of real numbers $\{p_i\}$, $i = 1, \dots, d+1$ (not even necessarily positive) is defined uniquely (up to a constant multiplier) by the condition $\sum_{i=1}^{d+1} p_i \xi_i = 0$. Hence it remains to observe that the probabilities $\{p_i\}$ sum up to one, which one sees by expanding the determinant defining the normalizing constant (with respect to the first row). □

A geometric interpretation of the probabilities $\{p_i\}$ will be revealed in the next section, where they will arise naturally from certain game-theoretic (or robust-control) evaluation.

For arbitrary k we have the following proposition.

Proposition 12.7. *Let a family $E = \{\xi_1, \ldots, \xi_k\}$ be strongly positively complete and in general position. Then the extreme points of the convex set of risk-neutral probabilities on $\{\xi_1, \ldots, \xi_k\}$ are risk-neutral probabilities with supports on strongly positively complete subsets of E of size precisely $d+1$.*

Proof. By Proposition 12.5, any extreme risk-neutral law is supported on no more than $d+1$ points.

On the other hand, it is clear that risk-neutral probabilities with supports on strongly positively complete subsets of size precisely $d+1$ are extreme points. In fact, if this were not the case for such a probability law, then it could be presented as a convex combination of other risk-neutral laws. But these risk-neutral laws would necessarily have the same support as the initial law, which would contradict the uniqueness of the risk-neutral law supported by $d+1$ points in general position.

Finally, by our nondegeneracy assumption, any set of $k < d+1$ points forms a basis in the subspace $\mathbb{R}^k$ it generates and hence is not even weakly positively complete (and hence cannot support a risk-neutral measure). □

Remark 12.8. The preceding statement implies that there exist strongly positively complete subsets of size precisely $d+1$ because the set of risk-neutral probabilities is not empty and hence contains extreme points (as any convex compact set). It is also easy to prove this fact explicitly using Carathéodory's theorem: if a point x in $\mathbb{R}^d$ lies in the convex hull of a set P, then there is a subset P' of P consisting of $d+1$ or fewer points such that x lies in the convex hull of P'.

The most general result is as follows.

Proposition 12.9. *Let a compact set $E \subset \mathbb{R}^d$ be strongly positively complete. Then the extreme points of the set of risk-neutral probabilities on E are the Dirac mass at zero (only when E contains the origin) and the risk-neutral measures with support on families of size $m+1$, $0 < m \le d$, that generate a subspace of dimension m and are strongly positively complete in this subspace.*

Proof. It is clear that the laws specified in the proposition are extreme points. Conversely, assume that the support E' of an extreme risk-neutral measure contains $m+1 > 0$ points with $m > 0$. By Proposition 12.5, $m \le d$. If E' generates a subspace of dimension less than m, then, again by Proposition 12.5, it cannot support an extreme risk-neutral measure. Finally, by Proposition 12.4 (2), E' is strongly positively complete in the subspace it generates. □

12.2 Game-Theoretic Origin of Risk-Neutral Laws: Preliminaries

We will now develop a general technique for the evaluation of minimax expressions depending linearly on one of the controls showing how naturally the extreme risk-neutral probabilities arise in such an evaluation. That is, let E be a compact subset of

$\mathbb{R}^d$ and f a continuous function on E. We aim at the evaluation of the game-theoretic (or minimax) expression

$$\Pi[E](f) = \inf_{\gamma \in \mathbb{R}^d} \sup_{\xi \in E} [f(\xi) - (\xi, \gamma)] \tag{12.4}$$

and at finding the corresponding minimizing γ (whenever the minimum exists). More generally, for a compact metric space E and continuous functions $f : E \to \mathbb{R}$, $g : E \to \mathbb{R}^d$, we are interested in the expression

$$\Pi[E](f;g) = \inf_{\gamma \in \mathbb{R}^d} \sup_{\xi \in E} [f(\xi) - (g(\xi), \gamma)]. \tag{12.5}$$

However, this problem can be reduced to the previous one because one has

$$\Pi[E](f;g) = \inf_{\gamma \in \mathbb{R}^d} \sup_{\eta \in g(E)} [\max_{\xi \in g^{-1}(\eta)} f(\xi) - (\eta, \gamma)]. \tag{12.6}$$

Similar problems with γ, not from the whole $\mathbb{R}^d$ but restricted to its bounded subset, are also of interest.

In this section we initiate this analysis by looking at Problem (12.4) with a finite set E having precisely $d+1$ points. Two remarkable facts that we will reveal are that in this case expression (12.4) depends linearly on f and the minimizing γ is unique and also depends linearly on f.

To introduce the main ideas in a simple setting, let us start with a three-point set E in $\mathbb{R}^2$, that is, with the problem of calculating

$$\Pi[\xi_1, \xi_2, \xi_3](f) = \min_{\gamma \in \mathbb{R}^2} \max_{\xi_1, \xi_2, \xi_3} [f(\xi_i) - (\xi_i, \gamma)], \tag{12.7}$$

assuming that the set $E = \{\xi_1, \xi_2, \xi_3\}$ is strongly positively complete in $\mathbb{R}^2$.

Suppose the min in (12.7) is attained on a vector γ_0 and the corresponding max on a certain ξ_i. Suppose this max is unique, so that

$$f(\xi_i) - (\xi_i, \gamma_0) > f(\xi_j) - (\xi_j, \gamma_0) \tag{12.8}$$

for all $j \neq i$. As $\xi_i \neq 0$, changing γ_0 by a small amount we can reduce the left-hand side of (12.8) by preserving inequality (12.8). This possibility contradicts the assumption that γ_0 is a minimal point. Hence, if γ_0 is a minimal point, the corresponding maximum must be attained on at least two vectors. Suppose it is attained on precisely two vectors, that is,

$$f(\xi_i) - (\xi_i, \gamma_0) = f(\xi_j) - (\xi_j, \gamma_0) > f(\xi_m) - (\xi_m, \gamma_0) \tag{12.9}$$

for some different i, j, m. Since the angle between ξ_i, ξ_j is strictly less than π [by Proposition 12.6 (1)], adding a vector

$$\varepsilon(\xi_i/|\xi_j| + \xi_j/|\xi_i|)$$

to γ_0 will reduce simultaneously the first two expressions from the left-hand side of (12.9) but preserve (for small enough ε) the inequality on the right-hand side of (12.9). This again contradicts the assumption that γ_0 is a minimal point. Hence, if γ_0 is a minimal point, it must satisfy the equation

$$f(\xi_1) - (\xi_1, \gamma) = f(\xi_2) - (\xi_2, \gamma) = f(\xi_3) - (\xi_3, \gamma), \tag{12.10}$$

which is equivalent to the system

$$\begin{cases} (\xi_2 - \xi_1, \gamma_0) = f(\xi_2) - f(\xi_1), \\ (\xi_3 - \xi_1, \gamma_0) = f(\xi_3) - f(\xi_1). \end{cases} \tag{12.11}$$

Again by Proposition 12.6 (1), the vectors $\xi_2 - \xi_1, \xi_3 - \xi_1$ are independent. Hence system (12.11) has a unique solution γ_0.

For a pair of vectors $u, v \in \mathbb{R}^2$, let $D(u,v)$ denote the oriented area of the parallelogram built on u, v and $R(u)$ the result of the rotation of u on $90°$ anticlockwise. That is, for $u = (u^1, u^2)$, $v = (v^1, v^2)$,

$$D(u,v) = u^1 v^2 - u^2 v^1, \quad R(u) = (u^2, -u^1).$$

Notice that the determinant of system (12.11) is

$$D(\xi_2 - \xi_1, \xi_3 - \xi_1) = D(\xi_2, \xi_3) + D(\xi_3, \xi_1) + D(\xi_1, \xi_2),$$

and by the standard formulas of linear algebra, the unique solution γ_0 is

$$\gamma_0 = \frac{f(\xi_1)R(\xi_2 - \xi_3) + f(\xi_2)R(\xi_3 - \xi_1) + f(\xi_3)R(\xi_1 - \xi_2)}{D(\xi_2, \xi_3) + D(\xi_3, \xi_1) + D(\xi_1, \xi_2)} \tag{12.12}$$

and the corresponding optimal value

$$\Pi[\xi_1, \xi_2, \xi_3](f) = \frac{f(\xi_1)D(\xi_2, \xi_3) + f(\xi_2)D(\xi_3, \xi_1) + f(\xi_3)D(\xi_1, \xi_2)}{D(\xi_2, \xi_3) + D(\xi_3, \xi_1) + D(\xi_1, \xi_2)}. \tag{12.13}$$

Hence we arrive at the following proposition.

Proposition 12.10. *Let $E = \{\xi_1, \xi_2, \xi_3\}$ be strongly positively complete in $\mathbb{R}^2$, and let $f(\xi_1), f(\xi_2), f(\xi_3)$ be arbitrary numbers. Then expression* (12.7) *is given by* (12.13)*, and the minimum is attained on the single γ_0 given by* (12.12)*.*

Proof. The preceding discussion shows that if γ_0 is a minimum point, then it is unique and given by (12.12). It remains to observe that a minimal point does exist because

$$\max_{\xi_1, \xi_2, \xi_3} [f(\xi_i) - (\xi_i, \gamma)] \to \infty$$

as $\gamma \to \infty$. □

Corollary 12.11. *Expression* (12.13) *can be written equivalently as*

$$\Pi[\xi_1,\xi_2,\xi_3](f) = \mathbf{E}f(\xi),$$

where the expectation is defined with respect to the unique risk-neutral probability law $\{p_1,p_2,p_3\}$ *on* ξ_1,ξ_2,ξ_3*:*

$$p_i = \frac{D(\xi_j,\xi_m)}{D(\xi_2,\xi_3)+D(\xi_3,\xi_1)+D(\xi_1,\xi_2)}$$

[(i,j,k) is either (1,2,3) or (2,3,1) or (3,1,2)].

The main objective of this section is to extend the result obtained to strongly positively complete sets $E = \{\xi_1,\dots,\xi_{d+1}\}$ of $d+1$ points in $\mathbb{R}^d$. Methodologically, it is quite straightforward, but technically it requires certain lengthy manipulations with determinants. The result is given subsequently by Proposition 12.12, which is the natural extension of Proposition 12.10 and its corollary (so the reader may wish to skip the calculations leading to it).

Thus we are interested in evaluating the expression

$$\Pi[\xi_1,\dots,\xi_{d+1}](f) = \min_{\gamma\in\mathbb{R}^d}\max_{i=1,\dots,d+1}[f(\xi_i)-(\xi_i,\gamma)]. \qquad (12.14)$$

Assume that $\mathbb{R}^d$ is equipped with the standard basis $e_1,\dots,e_d$ fixing the orientation. Without loss of generality we will assume now that the vectors $\xi_1,\dots,\xi_{d+1}$ are ordered so that the vectors $\{\xi_2,\xi_3,\dots,\xi_{d+1}\}$ form an oriented basis of $\mathbb{R}^d$. The fact that the vector ξ_1 lies outside any half-space containing this basis allows one to identify the orientation of other subsets of $\xi_1,\dots,\xi_{d+1}$ of size d. That is, let $\{\hat{\xi}_i\}$ denote the ordered subset of $\xi_1,\dots,\xi_{d+1}$ obtained by removing ξ_i. The basis $\{\hat{\xi}_i\}$ is oriented if and only if i is odd. For instance, if $d=3$, then the oriented bases form the triples $\{\xi_2,\xi_3,\xi_4\}$, $\{\xi_1,\xi_2,\xi_4\}$, $\{\xi_1,\xi_4,\xi_3\}$, and $\{\xi_1,\xi_3,\xi_2\}$.

The same argument as for $d=2$ leads us to the conclusion that a minimal point γ_0 must satisfy the equation

$$f(\xi_1)-(\xi_1,\gamma) = \cdots = f(\xi_{d+1})-(\xi_{d+1},\gamma), \qquad (12.15)$$

which is equivalent to the system

$$(\xi_i-\xi_1,\gamma_0) = f(\xi_i)-f(\xi_1), \quad i=2,\dots,d+1. \qquad (12.16)$$

By Proposition 12.6, this system has a unique solution, say γ_0.

To write it down explicitly, we will use the natural extensions of the notations used previously for $d=2$. For a collection of d vectors $u_1,\dots,u_d\in\mathbb{R}^d$, let $D(u_1,\dots,u_d)$ denote the oriented volume of the parallelepiped built on $u_1,\dots,u_d$

and $R(u_1,\dots,u_{d-1})$ the rotor of the family $(u_1,\dots,u_{d-1})$. That is, denoting by upper scripts the coordinates of vectors,

$$D(u_1,\dots,u_d)=\det\begin{pmatrix} u_1^1 & \cdots & u_1^d \\ u_2^1 & \cdots & u_2^d \\ & \cdots & \\ u_d^1 & \cdots & u_d^d \end{pmatrix}, \quad R(u_1,\dots,u_{d-1})=\det\begin{pmatrix} e_1 & \cdots & e_d \\ u_1^1 & \cdots & u_1^d \\ & \cdots & \\ u_{d-1}^1 & \cdots & u_{d-1}^d \end{pmatrix}$$

$$=e_1\det\begin{pmatrix} u_1^2 & \cdots & u_1^d \\ & \cdots & \\ u_{d-1}^2 & \cdots & u_{d-1}^d \end{pmatrix}-e_2\det\begin{pmatrix} u_1^1 & u_1^3 & \cdots & u_1^d \\ & & \cdots & \\ u_{d-1}^1 & u_{d-1}^3 & \cdots & u_{d-1}^d \end{pmatrix}+\cdots.$$

Finally, let us define a multilinear operator $\tilde{R}$ from an ordered set $\{u_1,\dots,u_d\}$ of d vectors in $\mathbb{R}^d$ to $\mathbb{R}^d$:

$$\begin{aligned}\tilde{R}(u_1,\dots,u_d)&=R(u_2-u_1,u_3-u_1,\dots,u_d-u_1)\\ &=R(u_2,\dots,u_d)-R(u_1,u_3,\dots,u_d)+\cdots+(-1)^{d-1}R(u_1,\dots,u_{d-1}).\end{aligned}$$

Returning to system (12.16) observe that its determinant, which we denote by D, equals

$$D=D(\xi_2-\xi_1,\dots,\xi_{d+1}-\xi_1)=\det\begin{pmatrix} \xi_2^1-\xi_1^1 & \xi_2^2-\xi_1^2 & \cdots & \xi_2^d-\xi_1^d \\ & \cdots & & \\ \xi_{d+1}^1-\xi_1^1 & \xi_{d+1}^2-\xi_1^2 & \cdots & \xi_{d+1}^d-\xi_1^d \end{pmatrix}.$$

Using the linear dependence of a determinant on columns, this is rewritten as

$$D(\xi_2,\dots,\xi_{d+1})-\xi_1^1\det\begin{pmatrix} 1 & \xi_2^2 & \cdots & \xi_2^d \\ & \cdots & & \\ 1 & \xi_{d+1}^2 & \cdots & \xi_{d+1}^d \end{pmatrix}-\xi_1^2\det\begin{pmatrix} \xi_2^1 & 1 & \xi_2^3 & \cdots & \xi_2^d \\ & & \cdots & & \\ \xi_{d+1}^1 & 1 & \xi_{d+1}^3 & \cdots & \xi_{d+1}^d \end{pmatrix}-\cdots,$$

implying that

$$D=D(\xi_2-\xi_1,\dots,\xi_{d+1}-\xi_1)=\sum_{i=1}^{d+1}(-1)^{i-1}D(\{\hat{\xi}_i\})=\det\begin{pmatrix} 1 & \cdots & 1 \\ \xi_1^1 & \cdots & \xi_{d+1}^1 \\ & \cdots & \\ \xi_1^d & \cdots & \xi_{d+1}^d \end{pmatrix}. \tag{12.17}$$

Notice that according to the orientation specified above, $D(\{\hat{\xi}_i\})$ are positive (resp. negative) for odd (resp. even) i, implying that all terms in (12.17) are positive, so that the collection of numbers

$$p_i = \frac{1}{D}(-1)^{i-1}D(\{\hat{\xi}_i\}) = \frac{(-1)^{i-1}D(\{\hat{\xi}_i\})}{D(\xi_2-\xi_1,\dots,\xi_d-\xi_1)}, \quad i=1,\dots,d+1, \tag{12.18}$$

defines a probability law on the set $\xi_1,\dots,\xi_{d+1}$ with a full support. As one sees directly, this law is precisely the unique risk-neutral law on $\{\xi_1,\dots,\xi_{d+1}\}$ constructed previously in Sect. 12.1.

By linear algebra, the unique solution γ_0 to system (12.16) is given by the formulas

$$\gamma_0^1 = \frac{1}{D}\det\begin{pmatrix} f(\xi_2)-f(\xi_1) & \xi_2^2-\xi_1^2 & \cdots & \xi_2^d-\xi_1^d \\ & \cdots & & \\ f(\xi_{d+1})-f(\xi_1) & \xi_{d+1}^2-\xi_1^2 & \cdots & \xi_{d+1}^d-\xi_1^d \end{pmatrix}, \tag{12.19}$$

$$\gamma_0^2 = \frac{1}{D}\det\begin{pmatrix} \xi_2^1-\xi_1^1 & f(\xi_2)-f(\xi_1) & \cdots & \xi_2^d-\xi_1^d \\ & \cdots & & \\ \xi_{d+1}^1-\xi_1^1 & f(\xi_{d+1})-f(\xi_1) & \cdots & \xi_{d+1}^d-\xi_1^d \end{pmatrix}, \tag{12.20}$$

and similarly for other γ_0^i. One sees by inspection that for any i

$$f(\xi_i)-(\gamma_0,\xi_i) = \frac{1}{D}\sum_{i=1}^{d+1}[f(\xi_i)(-1)^{i+1}D(\{\hat{\xi}_i\})] \tag{12.21}$$

and

$$\begin{aligned}\gamma_0 =& \frac{1}{D}(f(\xi_2)-f(\xi_1))R(\xi_3-\xi_1,\dots,\xi_{d+1}-\xi_1)\\ &-\frac{1}{D}(f(\xi_3)-f(\xi_1))R(\xi_2-\xi_1,\xi_4-\xi_1,\dots,\xi_{d+1}-\xi_1)\\ &+\cdots+\frac{1}{D}(-1)^{d+1}(f(\xi_{d+1})-f(\xi_1)R(\xi_2-\xi_1,\dots,\xi_d-\xi_1),\end{aligned}$$

which is rewritten as

$$\gamma_0 = -\frac{1}{D}\left[f(\xi_1)\tilde{R}(\{\hat{\xi}_1\})-f(\xi_2)\tilde{R}(\{\hat{\xi}_2\})+\cdots+(-1)^d f(\xi_{d+1})\tilde{R}(\{\hat{\xi}_{d+1}\})\right]. \tag{12.22}$$

For example, if $d=3$, then we have

$$\Pi[\xi_1,\dots,\xi_4](f)=\frac{f(\xi_1)D_{234}+f(\xi_2)D_{143}+f(\xi_3)D_{124}+f(\xi_4)D_{132}}{D_{234}+D_{143}+D_{124}+D_{132}},$$

$$\gamma_0=-\frac{f(\xi_1)R_{234}+f(\xi_2)R_{143}+f(\xi_3)R_{124}+f(\xi_4)R_{132}}{D_{234}+D_{143}+D_{124}+D_{132}},$$

where $D_{ijm}=D(\xi_i,\xi_j,\xi_m)$ and

$$R_{ijm}=R(\xi_i,\xi_j)+R(\xi_j,\xi_m)+R(\xi_m,\xi_i).$$

As in the case where $d=2$, we arrive at the following proposition.

Proposition 12.12. *Let a family* $\{\xi_1,\dots,\xi_{d+1}\}$ *in* $\mathbb{R}^d$ *be strongly positively complete, and let* $f(\xi_1),\dots,f(\xi_{d+1})$ *be arbitrary numbers. Then*

$$\Pi[\xi_1,\dots,\xi_{d+1}](f)=\frac{1}{D}\sum_{i=1}^{d+1}[f(\xi_i)(-1)^{i+1}D(\{\hat{\xi}_i\})]=\mathbf{E}f(\xi), \tag{12.23}$$

and the minimum in (12.14) *is attained on the single* γ_0 *given by* (12.22) *or, equivalently, by*

$$\gamma_0=\mathbf{E}\left[f(\xi)\frac{\tilde{R}(\{\hat{\xi}\})}{D(\{\hat{\xi}\})}\right], \tag{12.24}$$

where the expectation is with respect to the probability law (12.18)*, which is the unique risk-neutral probability law on* $\{\xi_1,\dots,\xi_{d+1}\}$.

To better visualize the preceding formulas, it would be helpful to delve a bit into their geometric meaning. Each term $(-1)^{i-1}D(\{\hat{\xi}_i\})$ in (12.17) equals $d!$ times the volume of the pyramid (polyhedron) with vertices $\{0\cup\{\hat{\xi}_i\}\}$. The determinant D, being the volume of the parallelepiped built on $\xi_2-\xi_1,\dots,\xi_{d+1}-\xi_1$, equals $d!$ times the volume of the pyramid $\Pi[\xi_1,\dots,\xi_{d+1}]$ in the affine space $\mathbb{R}^d$ with vertices being the end points of the vectors ξ_i, $i=1,\dots,d+1$. Consequently, formula (12.17) expresses the decomposition of the volume of the pyramid $\Pi[\xi_1,\dots,\xi_{d+1}]$ into $d+1$ parts, the volumes of the pyramids $\Pi[\{0\cup\{\hat{\xi}_i\}\}]$ obtained by sectioning from the origin, and the weights of distribution (12.18) are the ratios of these parts to the entire volume. Furthermore, the magnitude of the rotor $R(u_1,\dots,u_{d-1})$ is known to equal the volume of the parallelepiped built on $u_1,\dots,u_{d-1}$. Hence $\|\tilde{R}(\{\xi_i\})\|$ is equal to $(d-1)!$ times the volume (in the affine space $\mathbb{R}^d$) of the $(d-1)$-dimensional face of the pyramid $\Pi[\xi_1,\dots,\xi_{d+1}]$ with vertices $\{\hat{\xi}_i\}$. Hence the magnitudes of the ratios $\tilde{R}(\{\hat{\xi}_i\})/D(\{\hat{\xi}_i\})$, playing the roles of weights in (12.24), are the ratios of the $(d-1)!$ times $(d-1)$-dimensional volumes of the bases of the pyramids $\Pi[\{0\cup\{\hat{\xi}_i\}\}]$ to $d!$ times their full d-dimensional volumes. Consequently,

$$\frac{\|\tilde{R}(\{\hat{\xi}_i\})\|}{D(\{\hat{\xi}_i\})} = \frac{1}{h_i}, \tag{12.25}$$

where h_i is the length of the perpendicular from the origin to the affine hyperspace generated by the end points of the vectors $\{\hat{\xi}_i\}$. Hence

$$|\gamma_0| \leq \|f\| \max_{i=1,\ldots,d+1} h_i^{-1}. \tag{12.26}$$

On the other hand,

$$|\gamma_0| \leq \|f\| \sum_{i=1}^{d+1} p_i h_i^{-1},$$

and by the foregoing geometric interpretation of $\{p_i\}$,

$$|\gamma_0| \leq \|f\| \frac{1}{d} \frac{S(\xi_1,\ldots,\xi_{d+1})}{V(\xi_1,\ldots,\xi_{d+1})}, \tag{12.27}$$

where $S(\xi_1,\ldots,\xi_{d+1})$ is the surface volume of the pyramid $\Pi[\xi_1,\ldots,\xi_{d+1}]$ [the sum of $(d-1)$-dimensional volumes of all its $d+1$ faces] and $V(\xi_1,\ldots,\xi_{d+1})$ is its volume. Thus we obtained the following corollary.

Corollary 12.13. *The minimizing value* $\gamma_0 = \gamma_0(\xi_1,\ldots,\xi_{d+1})$ *from* (12.24) *satisfies estimates* (12.26) *and* (12.27).

These estimates are important for numerical calculations of γ_0 (yielding some kind of stability estimates with respect to the natural parameters). On the other hand, we will need them subsequently for nonlinear extensions of Proposition 12.12.

12.3 Game-Theoretic Origin of Risk-Neutral Laws: Main Result

Let us now move to general Problem (12.4) starting with the case of a finite E, that is, with the expression

$$\Pi[\xi_1,\ldots,\xi_k](f) = \inf_{\gamma \in \mathbb{R}^d} \max_i [f(\xi_i) - (\xi_i, \gamma)] \tag{12.28}$$

for a finite set $E = \{\xi_1,\ldots,\xi_k\}$ in $\mathbb{R}^d$.

Theorem 12.14. *Let a family of vectors* $\xi_1,\ldots,\xi_k$ *in* $\mathbb{R}^d$ *be strongly positively complete. Then*

$$\Pi[\xi_1,\ldots,\xi_k](f) = \max_{\mu} \mathbf{E}_\mu f(\xi), \tag{12.29}$$

where max *is taken over all extreme points* μ *of risk-neutral laws on* $\{\xi_1,\ldots,\xi_k\}$, *given by Proposition 12.9, and* inf *in* (12.28) *is attained on the corresponding* γ *given by Proposition 12.12.*

Proof. Clearly

$$\Pi[\xi_1,\ldots,\xi_k](f) = \inf_{\gamma\in\mathbb{R}^d}\max_{p\in\mathscr{P}(E)}\sum_{i=1}^k p_i[f(\xi_i)-(\xi_i,\gamma)]$$

$$\geq \inf_{\gamma\in\mathbb{R}^d}\max_{p\in\mathscr{P}_{rn}(E)}\sum_{i=1}^k p_i[f(\xi_i)-(\xi_i,\gamma)] = \max_{\mu}\mathbf{E}_\mu f(\xi),$$

where the last max is taken over the extreme points μ of the (nonempty) set $\mathscr{P}_{rn}(E)$ of risk-neutral laws on E. Consequently the left-hand side is bounded from below. Moreover, as for the support E' of an extreme risk-neutral law

$$\max_{\xi\in E'}[f(\xi)-(\xi,\gamma)]\to\infty,$$

as $\gamma\to\infty$, and hence also

$$\max_{i=1,\ldots,k}[f(\xi_i)-(\xi_i,\gamma)]\to\infty,$$

the infinum in (12.28) is attained on some finite γ. Assuming that γ_0 is such a minimum point, let E' denote the subset of all ξ, where the maximum is attained in the expression $f(\xi)-(\xi,\gamma_0)$. We can now conclude, as in the proof of Proposition 12.10 above, that E' is weakly positively complete in the subspace it generates. Suppose E' is not strongly positively complete in the subspace it generates. Then there exists ω such that $(\omega,\xi)\geq 0$ for all $\xi\in E'$ and $(\omega,\xi)=0$ for some proper subset E'' of E'. Then for sufficiently small ε, $\gamma_0'=\gamma_0+\varepsilon\omega$ is still a minimum point for (12.28), but the maximum $f(\xi)-(\xi,\gamma_0')$ is attained only on E''. Following the same procedure with E'' and so on, we will be able to find eventually a subset $\tilde{E}$ of E' and a minimal point $\tilde{\gamma}$ such that $\tilde{E}$ is the set of maximum points of $f(\xi)-(\xi,\tilde{\gamma})$ and either $\tilde{E}$ coincides with the origin or is strongly positively complete in the subspace it generates. □

Remark 12.15. If $d=2$, then the number of allowed triples $\{i,j,m\}$ supporting μ on the right-hand side of (12.29) is two for $k=4$, can be 3, 4, or 5 (depending on the position of the origin inside Π) for $k=5$, and can be 4, 6, or 8 for $k=6$. This number seems to increase exponentially as $k\to\infty$.

Remark 12.16. As is easy to see, the max in (12.29) is attained on a family $\{\xi_i\}_{i\in I}$ if and only if

$$f(\xi_i)-(\gamma_I,\xi_i)\geq f(\xi_r)-(\gamma_I,\xi_r) \tag{12.30}$$

for any $i\in I$ and any r, where γ_I is the corresponding optimal value. Consequently, on the convex set of functions f satisfying inequalities (12.30) for all r, the mapping $\Pi[\xi_1,\ldots,\xi_k](f)$ is linear:

$$\Pi[\xi_1,\ldots,\xi_k](f)=\mathbf{E}_I f(\xi).$$

Remark 12.17. Notice that min and max in (12.28) are not interchangeable, as

$$\max_i \min_{\gamma\in\mathbb{R}^d} [f(\xi_i)-(\xi_i,\gamma)] = -\infty.$$

It is often important to know the size of a region that the minimizing γ in expression (12.28) belongs to. For $d+1$-point sets E such an estimate is given by (12.26). Let us supply an appropriate estimate for an arbitrary E. Let

$$\varkappa_0 = \varkappa_0(E) = \min_{\omega\in\mathbb{R}^d,|\omega|=1} \max_{\xi\in E}(-\xi,\omega). \quad (12.31)$$

It is easy to see that $\varkappa_0 > 0$ for any strongly positively complete set E.

Proposition 12.18. *Under the assumptions of Theorem 12.14, all minimizing γ in expression* (12.28) *(which exist by Theorem 12.14) satisfy the estimate*

$$|\gamma| \le \operatorname{osc}(f)/\varkappa_0, \quad (12.32)$$

where

$$\operatorname{osc}(f) = \max_{\xi\in E} f(\xi) - \min_{\xi\in E} f(\xi).$$

In particular,

$$\Pi[\xi_1,\dots,\xi_k](f) = \min_{|\gamma|\le osc(f)/\varkappa_0} \max_i [f(\xi_i)-(\xi_i,\gamma)]. \quad (12.33)$$

Proof. For any γ

$$\max_i [f(\xi_i)-(\xi_i,\gamma)] \ge \min_i f(\xi_i) + |\gamma|\varkappa_0.$$

On the other hand,

$$\Pi[\xi_1,\dots,\xi_k](f) \le \max_i f(\xi_i).$$

Therefore, if the infinum in (12.28) is attained on a γ, then

$$\min_i f(\xi_i) + |\gamma|\varkappa_0 \le \max_i f(\xi_i),$$

implying (12.32). □

Let us now formulate a mirror image of Theorem 12.14, where min and max are reversed. Its proof is almost literally the same as the proof of Theorem 12.14.

Theorem 12.19. *Under the assumptions of Theorem 12.14 the expression*

$$\underline{\Pi}[\xi_1,\dots,\xi_k](f) = sup_{\gamma\in\mathbb{R}^d} \min_{\xi_1,\dots,\xi_k} [f(\xi_i)+(\xi_i,\gamma)] \quad (12.34)$$

can be evaluated by the formula

$$\underline{\Pi}[\xi_1,\dots,\xi_k](f) = \min_{\mu} \mathbf{E}_\mu f(\xi), \tag{12.35}$$

where min *is taken over all extreme points μ of risk-neutral laws on $\{\xi_1,\dots,\xi_k\}$. Moreover, sup in* (12.34) *is attained on some γ and all maximizing γ satisfy* (12.32)*.*

Remark 12.20. The right-hand side of (12.35) is similar to the formula for a coherent acceptability measure [2,131]. However, in the theory of acceptability measures, the collection of measures with respect to which minimization is performed is subjectively specified. In our model, this collection is the collection of all extreme points that arises objectively as an evaluation tool for our game-theoretic problem.

It is worth mentioning that the coherent acceptability measures ϕ introduced in Artzner et al. [2] represent particular cases of nonlinear averages in the sense of Kolmogorov [111]. The distinguishing feature that leads to the representation of ϕ as an infimum over probability measures is its superadditivity. Postulating subadditivity instead of superadditivity would lead similarly to the representation as a supremum over probability measures and, hence, to the analog of (12.29).

Let us turn to infinite sets E. The crucial argument used at the beginning of Sect. 12.2 for the identification of optimal γ does not work here. We will substitute it with a limiting procedure from finite subsets.

Theorem 12.21. *Let a compact set $E \subset \mathbb{R}^d$ be strongly positively complete. Then*

$$\Pi[E](f) = \max_{\mu} \mathbf{E}_\mu f(\xi), \tag{12.36}$$

where max *is taken over all extreme points μ of risk-neutral laws on E given by Proposition 12.9,* inf *in* (12.28) *is attained on some γ, and all minimizing γ satisfy* (12.32)*.*

Proof. As in the proof of Theorem 12.14 we show that the left-hand side of (12.29) is bounded from below by its right-hand side, which is finite and equal to $\mathbf{E}_{\mu_0} f(\xi)$, where μ_0 is the corresponding maximizing extreme risk-neutral law. By Proposition 12.5, the support of μ_0 is a finite set $E' = \{\xi_1,\dots,\xi_k\}$, with $k \le d+1$. Let $\{E_n\}$ be an increasing sequence of $1/n$-nets in E containing E' and such that the corresponding $\varkappa_0(E_n)$ converge to $\varkappa_0(E)$ as $n \to \infty$. By Theorem 12.21,

$$\Pi[E_n](f) = \max_{\mu} \mathbf{E}_\mu f(\xi) = \mathbf{E}_{\mu_0} f(\xi)$$

for all n. It remains to show that $\Pi[E](f)$ coincides with the limit of $\Pi[E_n](f)$ as $n \to \infty$. But this follows from the fact (which we show literally in the same way as for a finite set) that inf in the expression for $\Pi[E]$ is attained, and all minimizing γ satisfy (12.32). In fact, if an increasing sequence of bounded continuous functions on a compact set converges to a continuous function, then their minima converge to the minimum of the limiting function. □

12.4 Nonlinear Extension

A distinguishing feature of the expression under minimax in (12.4) or (12.5) is its linearity with respect to γ. What can be said in the case of a nonlinear dependence on γ? One of the standard approaches to nonlinear problems is by attempting to represent them as small perturbations of a linear one. Using this idea in our setting (and reducing our attention to finite sets of strategies of a maximizer), we will look at the problem of evaluating the minimax expression

$$\Pi[\xi_1,\ldots,\xi_k](f) = \min_{\gamma\in\mathbb{R}^d}\max_{\xi_1,\ldots,\xi_k}[f(\xi_i,\gamma)-(\xi_i,\gamma)], \tag{12.37}$$

where f is a continuous function. Its "smallness" will be measured in terms of its Lipschitz constant. The results obtained here will be used for the analysis of transaction costs.

We will also assume here that the set $E=\{\xi_1,\ldots,\xi_k\}\subset\mathbb{R}^d$ is strongly positively complete and is in general position (recall the latter means that the vectors of any subset of E of size d are linearly independent). Let us introduce two characteristics of such a set E that measure numerically a spread of the elements of this system around the origin.

Let $\varkappa_1=\varkappa_1(\xi_1,\ldots,\xi_k)$ be the minimum among the numbers $\varkappa$ such that for any subfamily ξ_i, $i\in I\subset\{\xi_1,\ldots,\xi_k\}$, that is not itself strongly positively complete in $\mathbb{R}^d$ one can choose a vector $\omega_I\in R^d$ of unit norm such that

$$(\xi_i,\omega_I)\geq\varkappa,\quad i\in I. \tag{12.38}$$

Due to our assumptions on E, this $\varkappa_1$ is strictly positive. Let $\varkappa_2=\varkappa_2(\xi_1,\ldots,\xi_k)$ be the minimum of the lengths of all perpendiculars from the origin to the affine hypersubspaces generated by the end points of any subfamily containing d vectors.

Theorem 12.22. *Let $E=\{\xi_1,\ldots,\xi_k\}$, $k>d$, be a family of vectors in $\mathbb{R}^d$, which is strongly positively complete and in general position.*

Let a function f be bounded below and Lipschitz continuous in γ, i.e.,

$$|f(\xi_i,\gamma_1)-f(\xi_i,\gamma_2)|\leq\varkappa|\gamma_1-\gamma_2| \tag{12.39}$$

for all i, and assume

$$\varkappa<\min(\varkappa_1,\varkappa_2). \tag{12.40}$$

Then the minimum in (12.37) *is finite and is attained on some γ_0 and*

$$\Pi[\xi_1,\ldots,\xi_k](f)=\max_I[\mathbf{E}_I f(\xi,\gamma_I)], \tag{12.41}$$

where max *is taken over all strongly positively complete families $\{\xi_i\}_{i\in I}$, $I\subset\{1,\ldots,k\}$, of size $|I|=d+1$, $\mathbf{E}_I$ denotes the expectation with respect to the unique risk-neutral probability on $\{\xi_i\}_{i\in I}$ (given by Proposition 12.12), and γ_I is the corresponding (unique) optimal value, constructed below.*

In particular, if $k = d+1$, then γ_0 is the unique solution of Eq. (12.44) *below.*

Proof. Arguing now as at the beginning of Sect. 12.2, suppose the min in (12.37) is attained on a vector γ_0 and the corresponding max is attained precisely on a subfamily ξ_i, $i \in I \subset \{\xi_1, \dots, \xi_k\}$, so that

$$f(\xi_i, \gamma_0) - (\gamma_0, \xi_i)$$

coincide for all $i \in I$ and

$$f(\xi_i, \gamma) - (\gamma, \xi_i) > f(\xi_j, \gamma) - (\gamma, \xi_j) \tag{12.42}$$

for $j \notin I$ and $\gamma = \gamma_0$, but this family is not strongly positively complete. (This is of course always the case for subfamilies of size $|I| < d+1$.) Let us pick up a unit vector ω_I satisfying (12.38). As for $\gamma = \gamma_0 + \varepsilon\omega_I$,

$$f(\xi_i, \gamma) - (\xi_i, \gamma) = f(\xi_i, \gamma_0) - (\xi_i, \gamma_0) - \varepsilon[(\xi_i, \omega_I) + \beta],$$

with some $|\beta| \leq \varkappa$, and $(\xi_i, \omega_I) > \varkappa_1$, this expression is less than

$$f(\xi_i, \gamma_0) - (\xi_i, \gamma_0)$$

for all $\varepsilon > 0$ and all $i \in I$. But at the same time (12.42) is preserved for small ε, contradicting the minimality of γ_0. Hence, if γ_0 is a minimal point, then the corresponding max must be attained on a strongly positively complete family. But, by item (2) of Proposition 12.7 (see remark following it), any such family contains a subfamily with $d+1$ elements only. Consequently, if γ_0 is a minimal point, then the corresponding max must be attained on a strongly positively complete family containing $d+1$ elements.

To proceed, let us assume first that $k = d+1$. Then a possible value of γ_0 satisfies the system

$$(\xi_i - \xi_1, \gamma_0) = f(\xi_i, \gamma_0) - f(\xi_1, \gamma_0), \quad i = 2, \dots, d+1, \tag{12.43}$$

which by (12.24) is rewritten as

$$\gamma_0 = \mathbf{E}\left[f(\xi, \gamma_0) \frac{\tilde{R}(\{\hat{\xi}\})}{D(\{\hat{\xi}\})} \right], \tag{12.44}$$

where the expectation is with respect to the probability law (12.18). This is a fixed-point equation. Condition (12.39), (12.40), the definition of $\varkappa_2$, and estimate (12.26) imply that the mapping on the right-hand side is a contraction, and hence (12.44) has a unique solution γ_0. Moreover, the minimum in (12.37) exists and is attained on some finite γ because

$$\max_{\xi_1, \dots, \xi_k} [f(\xi_i, \gamma) - (\xi_i, \gamma)] \to \infty, \tag{12.45}$$

as $\gamma \to \infty$ (as this holds already for vanishing f). And consequently it is attained on the single possible candidate γ_0.

Now let $k > d+1$ be arbitrary. Using the case $k = d+1$ we can conclude that

$$\Pi[\xi_1,\ldots,\xi_k](f) \geq \max_I \mathbf{E}_I f(\xi,\gamma_I),$$

and hence the left-hand side is bounded below and (12.45) holds. Thus the minimum in (12.37) is attained on some γ_0, which implies (12.41) due to the characterization of optimal γ given previously. □

In applications to options we need to use Theorem 12.22 recursively under expanding systems of vectors ξ. To this end, we require some estimates indicating the change in the basic coefficients of spread under linear scaling of all coordinates.

For a vector $z \in \mathbb{R}^d_+$ with positive coordinates, let

$$\delta(z) = \max_i z^i / \min_i z^i.$$

We will use the *Hadamard product* on vectors defined as $(\eta \circ \xi)^i = \eta^i \xi^i$.

Proposition 12.23. *Let a system* $\{\xi_1,\ldots,\xi_k\}$ *of vectors in* $\mathbb{R}^d$ *be strongly positively complete and in general position. Let* $\varkappa_1,\varkappa_2$ *be the characteristics of the system* $\{\xi_1,\ldots,\xi_k\}$ *introduced previously, and for a vector* $z \in \mathbb{R}^d$ *with positive coordinates let* $\varkappa_1(z),\varkappa_2(z)$ *denote the characteristics of the system* $\{z\circ\xi_1,\ldots,z\circ\xi_k\}$. *Then*

$$\varkappa_1(z) \geq |z|\varkappa_1(d\delta(z))^{-1}, \quad \varkappa_2(z) \geq |z|\varkappa_2(\sqrt{d}\delta(z))^{-1}. \tag{12.46}$$

Proof. Let us denote by z^{-1}, just for this proof, the vector in $\mathbb{R}^d$ with coordinates $1/z^i$.

For a unit vector $\phi = |z^{-1}|^{-1} z^{-1} \circ \omega$ we get, using $(\xi_i,\omega) \geq \varkappa_1$, that

$$(z\circ\xi_i,\phi) = |z^{-1}|(\xi_i,\omega) \geq |z|\varkappa_1 \frac{1}{|z|\,|z^{-1}|}.$$

Hence, to get the first inequality in (12.46), it remains to observe that

$$|z|\,|z^{-1}| = (\sum_{i=1}^d (z^i)^2 \sum_{i=1}^d (z^i)^{-2})^{1/2} \leq d\delta(z).$$

Turning to the proof of the second inequality in (12.46) let us recall that for any subsystem of d elements that we denote by $u_1,\ldots,u_d$ the length of the perpendicular h from the origin to the affine hyperspace generated by the end points of vectors $\{u_1,\ldots,u_d\}$ is expressed, by (12.25), as

$$h = \frac{D(u_1,\ldots,u_d)}{\|\tilde{R}(u_1,\ldots,u_d)\|}. \tag{12.47}$$

From the definition of D as a determinant it follows that

$$D(\{z \circ u_i\}) = \prod_{l=1}^{d} z^l D(\{u_i\}).$$

Next, for the jth coordinate of the rotor $R(\{z \circ u_i\})$ we have

$$R^j(\{z \circ u_i\}) = \frac{1}{z^j} \prod_{l=1}^{d} z^l R^j(\{u_i\}),$$

so that

$$\begin{aligned} \|R(\{z \circ u_i\})\| &= \prod_{l=1}^{d} z^l \left(\sum_j \frac{1}{(z^j)^2} (R^j(\{u_i\}))^2 \right)^{1/2} \\ &\leq \prod_{l=1}^{d} z^l \frac{1}{\min_j z^j} \|R(\{u_i\})\| \leq \frac{1}{|z|} \prod_{l=1}^{d} z^l \sqrt{d} \delta(z) \|R(\{u_i\})\|. \end{aligned}$$

Hence

$$h(z) = \frac{D(\{z \circ u_i\})}{\|\tilde{R}(\{z \circ u_i\})\|} \geq |z| (\sqrt{d} \delta(z))^{-1},$$

implying the second inequality in (12.46). □

12.5 Infinite-Dimensional Setting and Finite-Dimensional Projections

We saw previously that probability laws emerge naturally from the robust-control (or minimax) evaluations of payoffs on multidimensional geometric objects. However, in practice one often must deal with real-valued random variables, describing, for instance, possible stock price jumps. In our setting, only two-valued real random variables appear naturally from a risk-neutral evaluation (as $d+1$ equals 2 for $d=1$), which in finance corresponds to binomial models of option pricing. Problems with the extensions of binomial models are well known, as already trinomial models (in usual setting) lead to incomplete markets. However, if stock price jumps come from various factors (explicitly or implicitly), then various discrete laws on the sizes of these jumps can be naturally obtained as projections from multidimensional objects.

To obtain (via projections) random variables with a countable range, it is helpful to first extend our construction of risk-neutral probabilities to the infinite-dimensional setting, which we will carry out for the simplest geometry of orthogonal basis vectors and related pyramids.

Let $\{e_1, e_2, \ldots\}$ be an orthonormal basis in a Hilbert space H. Let $\omega^1, \omega^2, \ldots$ be an arbitrary sequence of nonvanishing real numbers and $\alpha^1, \alpha^2, \ldots$ a sequence of positive numbers with converging $\sum_{j=1}^{\infty}(\alpha^j \omega^j)^2$. Let us consider a collection of vectors $\{\xi_0, \xi_1, \xi_2, \ldots\}$, where $\xi_j = \omega^j e_j$, $j = 1, 2, \ldots$, and

$$\xi_0 = -\sum_{j=1}^{\infty} \alpha^j \omega^j e_j,$$

and the corresponding minimax problem

$$\Pi[\xi_0, \xi_1, \ldots](f) = \min_{\gamma \in H} \max_{i=0,1,\ldots} [f(\xi_i) - (\xi_i, \gamma)]. \tag{12.48}$$

Arguing as in a finite-dimensional setting we conclude that if γ_0 is a minimum point, then all expressions $f(\xi_i) - (\xi_i, \gamma)$, $i = 0, 1, \ldots$, should be equal, leading to the system of equations

$$f(\xi_j) - f(\xi_i) = (\xi_j - \xi_i, \gamma_0), \quad i, j = 0, 1, \ldots. \tag{12.49}$$

For $i, j > 0$, this is rewritten as

$$\gamma_0^j \omega^j = \gamma_0^i \omega^i + f(\xi_j) - f(\xi_i). \tag{12.50}$$

Substituting this into (12.49) with $j = 0$ yields

$$\gamma_0^i \omega^i + \sum_{k=1}^{\infty} \gamma_0^k \alpha^k \omega^k = f(\xi_i) - f(\xi_0),$$

or, using (12.50),

$$\gamma_0^i \omega^i (1 + \sum_{k=1}^{\infty} \alpha^k) + \sum_{k=1}^{\infty} \alpha^k (f(\xi_k) - f(\xi_i)) = f(\xi_i) - f(\xi_0)$$

(whenever the sum converges), which implies

$$\gamma_0^i \omega^i = f(\xi_i) - \left(1 + \sum_{k=1}^{\infty} \alpha^k\right)^{-1} \left(f(\xi_0) + \sum_{k=1}^{\infty} \alpha^k f(\xi_k)\right) \tag{12.51}$$

and

$$f(\xi_i) - (\gamma_0, \xi_i) = \left(1 + \sum_{k=1}^{\infty} \alpha^k\right)^{-1} \left(f(\xi_0) + \sum_{k=1}^{\infty} \alpha^k f(\xi_k)\right). \tag{12.52}$$

Hence we obtain the following theorem.

Theorem 12.24. *Let $\omega^1, \omega^2, \ldots$ be an arbitrary sequence of nonvanishing real numbers and $\alpha^1, \alpha^2, \ldots$ a sequence of positive numbers such that the sums*

$$\sum_{j=1}^{\infty} (\alpha^j \omega^j)^2, \quad \sum_{j=1}^{\infty} \alpha^j$$

converge. Then there exists a unique risk-neutral probability law $\{p_0, p_1, \ldots\}$ on $\{\xi_0, \xi_1, \xi_2, \ldots\}$ (that is, satisfying $\sum p_i \xi_i = 0$), where

$$p_0 = \left(1 + \sum_{k=1}^{\infty} \alpha^k\right)^{-1}, \quad p_i = \alpha^i \left(1 + \sum_{k=1}^{\infty} \alpha^k\right)^{-1}, \quad i = 1, \ldots. \tag{12.53}$$

Moreover, if the expectation $\mathbf{E}f(\xi_.)$ with respect to this risk-neutral probability law exists, then expression (12.48) *is well defined with*

$$\Pi[\xi_0, \xi_1, \ldots](f) = \mathbf{E}f(\xi_.),$$

and the minimum is attained on a single γ_0 given by (12.51) *or, equivalently, by*

$$\gamma_0^i = (\omega^i)^{-1}[f(\xi_i) - \mathbf{E}f(\xi_.)].$$

Projecting the vectors $\{\xi_i\}$ on $-\xi_0$ and shifting transfers the probability law from the Hilbert space to the discrete law on real numbers $\beta_j = -(\xi_j, \xi_0) + q$, $j = 0, 1, \ldots$, $q \in \mathbb{R}$, with

$$\beta_0 = -\sum_{j=1}^{\infty} (\alpha^j \omega^j)^2 + q, \quad \beta_j = \alpha^j (\omega^j)^2 + q, \quad j = 1, 2, \ldots.$$

For instance, to obtain a discrete approximation to a power (or a stable) law with the probability of tails (values higher than y for large y) of order $y^{-\alpha}$, $\alpha > 1$, we need p_j of order $j^{-r\alpha}$ on points $\beta_j = j^r$ for some $r > 0$ with $r\alpha > 1$. In terms of the foregoing α^j and ω^j, this gives α^j of order $j^{-r\alpha}$ and $(\alpha^j \omega^j)^2$ of order $j^{-r(\alpha-1)}$.

Similarly, distributions on finite subsets of $\mathbb{R}$ or $\mathbb{R}^d$ can be obtained by projections from geometric risk-neutral probabilities on vectors from Euclidean spaces of higher dimension.

In financial interpretation, we can consider the jumps of size β_j as coming from a large number of factors ξ_j, which can be explicit (as for indices based on the average of several stocks) or hidden (as for individual stocks depending on several factors of market performance).

12.6 Extension to a Random Geometry

Here we extend the results of Sect. 12.3 to the case of random vectors, reducing again our attention to finite sets E. To begin with, let us consider the minimax expression

$$\Pi[\xi_1,\ldots,\xi_k](f) = \min_{\gamma\in\mathbb{R}^d} \max_i \widetilde{\mathbf{E}}[f(\xi_i) - (\xi_i,\gamma)], \tag{12.54}$$

where $\{\xi_1,\ldots,\xi_k\}$ is a given family of $\mathbb{R}^d$-valued random variables and $\widetilde{\mathbf{E}}$ denotes the corresponding expectation or, equivalently,

$$\Pi[\xi_1,\ldots,\xi_k](f) = \min_{\gamma\in\mathbb{R}^d} \max_i [\widetilde{\mathbf{E}}f(\xi_i) - (\widetilde{\mathbf{E}}\xi_i,\gamma)]. \tag{12.55}$$

This expression can be evaluated directly by Theorem 12.14 applied to vectors $\widetilde{\mathbf{E}}\xi_i$ instead of ξ_i. In particular, if $k = d+1$ and the set $\{\widetilde{\mathbf{E}}\xi_1,\ldots,\widetilde{\mathbf{E}}\xi_k\}$ is strongly positively complete, then that implies the existence of a probability law $\{p_1,\ldots,p_{d+1}\}$ such that

$$\sum_{i=1}^{d+1} p_i\widetilde{\mathbf{E}}\xi_i = 0, \quad \Pi[\xi_1,\ldots,\xi_{d+1}](f) = \sum_{i=1}^{d+1} p_i\widetilde{\mathbf{E}}f(\xi_i).$$

Alternatively, there exists a probability space and two random variables on it, an $\mathbb{R}^d$-valued ξ, and a $\{1,\ldots,k\}$-valued η such that $\mathbf{E}(\xi|\eta = i)$ is distributed like ξ_i, $i \in \{1,\ldots,k\}$, and

$$\mathbf{E}\xi = \mathbf{E}(\mathbf{E}(\xi|\eta)) = 0, \quad \Pi[\xi_1,\ldots,\xi_{d+1}](f) = \mathbf{E}f(\xi).$$

From a game-theoretic point of view, expression (12.54) describes the payoff of an investor in the game, where first he chooses γ, then Nature chooses i, and finally a random element is chosen according to the distribution of ξ_i.

On the other hand, one can imagine another scenario, where first a random event is chosen, then the investor chooses γ, and finally Nature makes a choice. Then the investor's payoff turns out to be

$$\begin{aligned}&\Pi[\xi_1,\ldots,\xi_k](f)\\ &= \widetilde{\mathbf{E}} \min_{\gamma\in\mathbb{R}^d} \max_i [f(\xi_i) - (\xi_i,\gamma)] \sum_{j=1}^{m} p_j \left(\min_{\gamma\in\mathbb{R}^d} \max_i [f(\xi_i(j)) - (\xi_i(j),\gamma)] \right),\end{aligned} \tag{12.56}$$

where $\widetilde{\mathbf{E}}$ denotes the expectation with respect to a given distribution $\{p_1,\ldots,p_m\}$ on m possible families $\{\xi_1,\ldots,\xi_k\}$. If $k = d+1$ and the sets $\{\widetilde{\mathbf{E}}\xi_1,\ldots,\widetilde{\mathbf{E}}\xi_k\}$ are strongly positively complete almost surely, then by Theorem 12.14 there exist, for each $j = 1,\ldots,m$, the laws on the sets $\{\widetilde{\mathbf{E}}\xi_1(j),\ldots,\widetilde{\mathbf{E}}\xi_k(j)\}$ such that

$$\mathbf{E}_j\xi_.(j) = 0, \quad \Pi[\xi_1,\ldots,\xi_k](f) = \sum_{j=1}^{m} p_j\mathbf{E}_jf(\xi_.(j)).$$

Let us discuss two examples related to stochastic volatility and correlated stock models.

The first example is one-dimensional. Suppose $\xi_1(\sigma) = a\sigma$, $\xi_2(\sigma) = -b\sigma$, where a, b are positive constants and the "size scale" σ is a positive random variable (volatility) with a given distribution. Then the minimax expression

$$\Pi[\xi_1,\xi_2](f) = \widetilde{\mathbf{E}} \min_{\gamma\in\mathbb{R}} \max[f(\xi_1(\sigma)) - (\xi_1(\sigma),\gamma), f(\xi_2(\sigma)) - (\xi_2(\sigma),\gamma)], \quad (12.57)$$

where $\widetilde{\mathbf{E}}$ denotes the expectation with respect to σ, can be evaluated as

$$\Pi[\xi_1,\xi_2](f) = \widetilde{\mathbf{E}} \left[\frac{b}{a+b} f(a\sigma) + \frac{a}{a+b} f(-b\sigma) \right]. \quad (12.58)$$

For the second example, let $\phi_1,\ldots,\phi_d$ be independent vectors in $\mathbb{R}^d$, and our family of vectors $\{\xi\}$ consists of $2d$ vectors

$$\xi_j^{\pm} = \pm\phi_j, \quad j = 1,\ldots,d,$$

and $d(d-1)/2$ vectors

$$\xi_{jk}^{\pm} = \pm(\phi_j + \phi_k), \quad j < k.$$

If the vectors $\phi_1,\ldots,\phi_d$ are interpreted as (normalized) stock price jumps, including only vectors $\xi_i^{\pm}$ in the model corresponds to the independence of jumps (independent jumps do not occur simultaneously). Allowing for vectors $\xi_{jk}^{\pm}$ includes possible positive correlations.

Remark 12.25. Negative correlations would yield the vectors

$$\tilde{\xi}_{jk}^{\pm} = \pm(\phi_j - \phi_k), \quad j \neq k,$$

which could be analyzed analogously.

Introducing an exogeneous probability law on correlations, that is, the numbers p_i, $i = 1,\cdots,d$, and p_{jk}, $j,k \in \{1,\cdots,d\}$, $j < k$, summing up to one, leads to the following expression of type (12.56):

$$\sum_{i=1}^{d} p_i \min_{\gamma\in\mathbb{R}^d} \max[f(\phi_i) - (\phi_i,\gamma), f(-\phi_i) + (\phi_i,\gamma)]$$
$$+ \sum_{j<k} p_{jk} \min_{\gamma\in\mathbb{R}^d} \max[f(\phi_j + \varphi_k) - (\phi_j + \phi_k,\gamma), f(-\phi_j - \phi_k) + (\phi_j + \phi_k,\gamma)], \quad (12.59)$$

which by (trivial application of) Theorem 12.14 is equal to

$$\frac{1}{2}\sum_{i=1}^{d} p_i[f(\phi_i) + f(-\phi_i)] + \frac{1}{2}\sum_{j<k} p_{jk}[f(\phi_j + \varphi_k) + f(-\phi_j - \phi_k)]. \quad (12.60)$$

12.7 Mixed Strategies with Linear Constraints

In this section we extend the results of Sect. 12.3 to risk-neutral probabilities with linear constraints. Firstly, a trivial extension allows a player choosing ξ in (12.4) to use mixed strategies, i.e., probability laws on E. Secondly, instead of allowing all mixed strategies, we restrict them to those compatible with certain linear constraints, which practically can arise from given market prices (say, of vanilla options). These restrictions may lead to a completion (at least partial) of a market. That is, if without restrictions one can have an immense set of risk-neutral laws, algebraic restrictions can essentially reduce the set, sometimes even to a single element.

Notice first that (12.36) of Theorem 12.21 can be equivalently rewritten in the following way:

$$\Pi[E](f) = \inf_{\gamma\in\mathbb{R}^d} \max_{\mu\in\mathscr{P}(E)} \mathbf{E}_\mu[f(\xi)-(\gamma,\xi)] = \max_{\mu\in\mathscr{P}_{rn}(E)} \mathbf{E}_\mu f(\xi) \tag{12.61}$$

because for a linear function its maximum on a convex compact set is always attained on an extreme point of this set.

Let $E\subset\mathbb{R}^d$ be a compact set and $\widetilde{\mathscr{P}}(E)$ a closed convex subset of $\mathscr{P}(E)$ [the main example is a set of type $\mathscr{P}(E;F)$ from Sect. 12.1]. We are interested in the expression

$$\begin{aligned}\widetilde{\Pi}[E](f) &= \inf_{\gamma\in\mathbb{R}^d} \max_{\mu\in\widetilde{\mathscr{P}}(E)} \mathbf{E}_\mu[f(\xi)-(\gamma,\xi)]\\ &= \inf_{\gamma\in\mathbb{R}^d} \max_{\mu\in\widetilde{\mathscr{P}}(E)} \left[\int f(\xi)\mu(\mathrm{d}\xi)-\left(\gamma,\int\xi\mu(\mathrm{d}\xi)\right)\right].\end{aligned} \tag{12.62}$$

Let B denote the linear mapping $\widetilde{\mathscr{P}}(E)\to\mathbb{R}^d$ given by

$$B\mu = \mathbf{E}_\mu\xi = \int \xi\mu(\mathrm{d}\xi)$$

(the barycenter or center of mass). Its image $B(\widetilde{\mathscr{P}}(E))$ is a compact convex set in $\mathbb{R}^d$.

The following main result extends Theorem 12.21 to the case of mixed strategies with constraints.

Theorem 12.26. *The set $\widetilde{\mathscr{P}}(E)\cap\mathscr{P}_{rn}(E)$ is empty if and only if the set $B(\widetilde{\mathscr{P}}(E))$ is not weakly positively complete, in which case $\widetilde{\Pi}[E](f)=-\infty$. Otherwise,*

$$\widetilde{\Pi}[E](f) = \max_{\mu\in\widetilde{\mathscr{P}}(E)\cap\mathscr{P}_{rn}(E)} \mathbf{E}_\mu f(\xi). \tag{12.63}$$

Proof. A convex compact set $B(\widetilde{\mathscr{P}}(E))$ is not weakly positively complete if and only if it does not contain the origin, which is equivalent to $\widetilde{\mathscr{P}}(E)\cap\mathscr{P}_{rn}(E)=\emptyset$.

If this is the case, then there exists a unit vector ω in $\mathbb{R}^d$ such that

$$(\omega, \mathbf{E}_\mu(\xi)) \geq \delta$$

for some $\delta > 0$ and all $\mu \in \widetilde{\mathscr{P}}(E)$. Hence

$$\lim_{r\to\infty} \max_{\mu\in\widetilde{\mathscr{P}}(E)} \mathbf{E}_\mu[f(\xi) - r(\omega,\xi)] = -\infty,$$

implying $\widetilde{\Pi}[E](f) = -\infty$.

Next, by literally the same argument as at the beginning of the proof of Theorem 12.14, we show that the left-hand side of (12.63) is bounded from below by its right-hand side. The main point is to show that the equality holds.

Notice that the expression for $\widetilde{\Pi}[E](f)$ is of type (12.5) with $\widetilde{\mathscr{P}}(E)$ instead of E and with linear functions $\mu \mapsto \int f(\xi)\mu(\mathrm{d}\xi)$ and $\mu \mapsto B\mu$ instead of f and g. Consequently, applying (12.6) we get

$$\widetilde{\Pi}[E](f) = \inf_{\gamma\in\mathbb{R}^d} \sup_{z\in B(\widetilde{\mathscr{P}}(E))} \left[\max_{\eta\in B^{-1}(z)} \int f(\xi)\eta(\mathrm{d}\xi) - (\gamma, z) \right]. \tag{12.64}$$

This expression is already of the form (12.4) and can be calculated via Theorem 12.21 yielding

$$\widetilde{\Pi}[E](f) = \max_{p\in\mathscr{P}_{rn}(B(\widetilde{\mathscr{P}}(E)))} \int_{B(\widetilde{\mathscr{P}}(E))} \left[\max_{\eta\in B^{-1}(z)} \int_E f(\xi)\eta(\mathrm{d}\xi) \right] p(\mathrm{d}z) \tag{12.65}$$

whenever $B(\widetilde{\mathscr{P}}(E))$ is at least weakly positively complete. We must find out whether this expression equals the right-hand side of (12.63).

To this end, let us consider the multivalued mapping Ω on $B(\widetilde{\mathscr{P}}(E))$ that maps $z \in B(\widetilde{\mathscr{P}}(E))$ to the (closed and convex) set $\Omega(z)$ of maximum points of the linear function $\eta \mapsto \int f(\xi)\eta(\mathrm{d}\xi)$ defined on the (closed and convex) set $B^{-1}(z)$. This map is clearly upper semicontinuous with nonempty (also closed and convex) set values. Hence by the standard theory of measurable selections (see, e.g., Jayne and Rogers [90], Chap. 6) there exists a measurable selector $\omega(z)$ of this multivalued mapping, that is, a measurable function $z \mapsto \omega(z) \in \widetilde{\mathscr{P}}(E)$ such that $\omega(z) \in \Omega(z)$ [in particular, $B\omega(z) = z$] for all z. (This ω can be chosen to be of the first Baire class, i.e., to be represented as a pointwise limit of continuous functions [90].)

Consequently, denoting by p_0 a maximum point for the right-hand side of (12.65), we have

$$\widetilde{\Pi}[E](f) = \int_{B(\widetilde{\mathscr{P}}(E))} \int_E f(\xi)\omega(z)(\mathrm{d}\xi)p_0(\mathrm{d}z) = \int_E f(\xi)\mu_0(\mathrm{d}\xi), \tag{12.66}$$

where

$$\mu_0(d\xi) = \int_{B(\widetilde{\mathscr{P}}(E))} \omega(z)(d\xi)p_0(dz).$$

Clearly $\mu_0 \in \mathscr{P}_{rn}(E)$ since

$$\int_E \xi\mu_0(d\xi) = \int_{B(\widetilde{\mathscr{P}}(E))} B\omega(z)p_0(dz) = \int_{B(\widetilde{\mathscr{P}}(E))} zp_0(dz) = 0$$

and $\mu_0 \in \widetilde{\mathscr{P}}(E)$, as a limit of convex combinations of measures from $\widetilde{\mathscr{P}}(E)$. Thus we found a representation for $\widetilde{\Pi}[E](f)$ as an integral of f with respect to a measure from $\widetilde{\mathscr{P}}(E) \cap \mathscr{P}_{rn}(E)$, completing the proof of (12.63) and, hence, of the theorem. □

Remark 12.27. To any probability law p on $B(\widetilde{\mathscr{P}}(E))$ there corresponds its image measure $p^* = \omega^*(p)$ on $\widetilde{\mathscr{P}}(E)$ with respect to the mapping ω, i.e., $p^* \in \mathscr{P}(\widetilde{\mathscr{P}}(E))$ and

$$\int_{\widetilde{\mathscr{P}}(E)} \phi(\eta)p^*(d\eta) = \int_{B(\widetilde{\mathscr{P}}(E))} \phi(\omega(z))p(dz)$$

for any continuous ϕ on $\widetilde{\mathscr{P}}(E)$, so that in particular

$$\int_{\widetilde{\mathscr{P}}(E)} B\eta\, p^*(d\eta) = \int_{B(\widetilde{\mathscr{P}}(E))} zp(dz).$$

Due to the natural projection $\mathscr{P}(\widetilde{\mathscr{P}}(E)) \to \widetilde{\mathscr{P}}(E)$ that maps a measure P on $\widetilde{\mathscr{P}}(E)$ to a measure $\pi(P) \in \widetilde{\mathscr{P}}(E)$ given by

$$\int f(\xi)\pi(P)(d\xi) = \int_{\widetilde{\mathscr{P}}(E)} \int_E f(\xi)d\eta(\xi)P(d\eta),$$

the chain $p \mapsto \omega^*(p) = p^* \mapsto \pi(p^*)$ defines a mapping $\mathscr{P}(B(\widetilde{\mathscr{P}}(E))) \to \widetilde{\mathscr{P}}(E)$ preserving risk neutrality. Thus what we proved above was that the maximum on the right-hand side of (12.63) is always attained on an image measure of a mapping $p \mapsto \pi(p^*)$.

Chapter 13
Rainbow Options in Discrete Time, I

13.1 Colored European Options as a Game Against Nature

Recall that a European *option* is a contract between two parties where one party has the right, but not the obligation, to complete a transaction in the future (at a previously agreed amount, date, and price). More precisely, consider a financial market dealing with several securities: risk-free bonds (or bank accounts) and J common stocks, $J = 1, 2 \ldots$. If $J > 1$, then the corresponding options are called *colored* or *rainbow options* (J-color option for a given J). Suppose the prices of the units of these securities, B_m and S^i_m, $i \in \{1, 2, \ldots, J\}$, change in discrete moments of time $m = 1, 2, \ldots$ according to the recurrent equations $B_{m+1} = \rho B_m$, where the $\rho \geq 1$ is an interest rate that remains unchanged over time, and $S^i_{m+1} = \xi^i_{m+1} S^i_m$, where $\xi^i_m, i \in \{1, 2, \ldots, J\}$, are unknown sequences taking values in some fixed intervals $M_i = [d_i, u_i] \subset \mathbb{R}$. This model generalizes the colored version of the classic Cox–Ross–Rubinstein (CRR) model in a natural way. In the latter, a sequence ξ^i_m is confined to taking values only among two boundary points d_i, u_i, and it is supposed to be random with some given distribution. In our model any value in the interval $[d_i, u_i]$ is allowed, and no probabilistic assumptions are made. Hence it is often referred to as an *interval model*.

An option's type is specified by a given premium function f of J variables. The following examples are standard.

Option delivering the best of J risky assets and cash:

$$f(S^1, S^2, \ldots, S^J) = \max(S^1, S^2, \ldots, S^J, K); \tag{13.1}$$

Calls on the maximum of J risky assets:

$$f(S^1, S^2, \ldots, S^J) = \max(0, \max(S^1, S^2, \ldots, S^J) - K); \tag{13.2}$$

Multiple-strike options:

$$f(S^1, S^2, \ldots, S^J) = \max(0, S^1 - K_1, S^2 - K_2, \ldots, S^J - K_J); \tag{13.3}$$

P. Bernhard et al., *The Interval Market Model in Mathematical Finance*, Static & Dynamic Game Theory: Foundations & Applications, DOI 10.1007/978-0-8176-8388-7_13,

Portfolio options:

$$f(S^1,S^2,\ldots,S^J) = \max(0, n_1S^1 + n_2S^2 + \cdots + n_JS^J - K); \tag{13.4}$$

And *spread* options:

$$f(S^1,S^2) = \max(0,(S^2 - S^1) - K). \tag{13.5}$$

Here, the $S^1,S^2,\ldots,S^J$ represent the expiration date values of the underlying assets (in principle unknown at the start), and $K,K_1,\ldots,K_J$ represent the (agreed from the beginning) strike prices. The presence of max in all these formulas reflects the basic assumption that the buyer is not obligated to exercise her right and would do it only in case of a positive gain.

The investor is supposed to control the growth of her capital in the following way. Let X_m denote the investor's capital at the time $m = 1,2,\ldots$. At each time $m-1$ the investor determines her portfolio by choosing the number γ_m^j of common stock of each kind to be held, so that the structure of the capital is represented by the formula

$$X_{m-1} = \sum_{j=1}^{J} \gamma_m^j S_{m-1}^j + \left[X_{m-1} - \sum_{j=1}^{J} \gamma_m^j S_{m-1}^j\right],$$

where the expression in brackets corresponds to the part of the investor's capital held in a bank account. The control parameters γ_m^j can take all real values, i.e., short selling and borrowing are allowed. The value ξ_m becomes known at the moment m, and thus the capital at the moment m becomes

$$X_m = \sum_{j=1}^{J} \gamma_m^j \xi_m^j S_{m-1}^j + \rho\left[X_{m-1} - \sum_{j=1}^{J} \gamma_m^j S_{m-1}^j\right] \tag{13.6}$$

if transaction costs are not taken into account.

If n is the prescribed *maturity date*, then this procedure repeats n times starting from some initial capital $X = X_0$ (selling price of an option), and at the end the investor is obliged to pay the premium f to the buyer. Thus the (final) income of the investor equals

$$G(X_n,S_n^1,S_n^2,\ldots,S_n^J) = X_n - f(S_n^1,S_n^2,\ldots,S_n^J). \tag{13.7}$$

The evolution of the capital can thus be described by the n-step game of the investor with Nature, the behavior of the latter being characterized by unknown parameters ξ_m^j. The strategy of the investor is by definition any sequence of vectors $(\gamma_1,\ldots,\gamma_n)$ such that each γ_m could be chosen using all previous information: the sequences $X_0,\ldots,X_{m-1}$ and $S_0^i,\ldots,S_{m-1}^j$ (for every stock $j = 1,2,\ldots,J$). The control parameters γ_m^j can take all real values, i.e., short selling and borrowing are allowed. A position of the game at any time m is characterized by $J+1$ nonnegative numbers $X_m,S_m^1,\ldots,S_m^J$, with the final income specified by the function

$$G(X,S^1,\ldots,S^J) = X - f(S^1,\ldots,S^J). \tag{13.8}$$

The main definition of the theory is as follows. A strategy $\gamma_1,\dots,\gamma_n$ of an investor is called a *hedge* if for any sequence $(\xi_1,\dots,\xi_n)$ the investor is able to meet her obligations, i.e.,

$$G(X_n, S_n^1,\dots,S_n^J) \geq 0.$$

The minimal value of the capital X_0 for which the hedge exists is called the *hedging price* H of an option.

Looking for guaranteed payoffs means looking for the worst-case scenario (the so-called *robust-control approach*), i.e., for the minimax strategies. Thus if the final income is specified by a function G, then the investor's guaranteed income (discounted to the initial time) in a one-step game with initial conditions $X, S^1,\dots,S^J$ is given by the *Bellman* (or *Shapley*) operator

$$\begin{aligned}\mathbf{B}G(X,S^1,\dots,S^J)\\ = \frac{1}{\rho}\max_{\gamma}\min_{\{\xi^j\in[d_j,u_j]\}} G\left(\rho X + \sum_{i=1}^J \gamma^i\xi^i S^i - \rho\sum_{i=1}^J \gamma^i S^i, \xi^1 S^1,\dots,\xi^J S^J\right),\end{aligned} \tag{13.9}$$

and (as follows from the standard backward induction argument; see, e.g., Bellman [25] or Kolokoltsov and Malafeyev [104]) the investor's guaranteed income in the n-step game with initial conditions $X_0, S_0^1,\dots,S_0^J$ is given by the formula

$$\mathbf{B}^n G(X_0, S_0^1,\dots,S_0^J).$$

In our model, G is given by (13.8). For G of this form,

$$\mathbf{B}G(X,S^1,\dots,S^J) = X - \frac{1}{\rho}\min_{\gamma}\max_{\xi}[f(\xi^1S^1,\xi^2S^2,\dots,\xi^JS^J) - \sum_{j=1}^J \gamma^j S^j(\xi^j-\rho)],$$

and hence

$$\mathbf{B}^n G(X,S^1,\dots,S^J) = X - (\mathcal{B}^n f)(S^1,\dots,S^J),$$

where the *reduced Bellman operator* of the European option contract specified by the payoff f is defined as

$$(\mathcal{B}f)(z^1,\dots,z^J) = \frac{1}{\rho}\min_{\gamma}\max_{\{\xi^j\in[d_j,u_j]\}}\left[f(\xi^1z^1,\xi^2z^2,\dots,\xi^Jz^J) - \sum_{j=1}^J \gamma^j z^j(\xi^j-\rho)\right], \tag{13.10}$$

or, more concisely, using the Hadamard product notation introduced in Sect. 12.4,

$$(\mathcal{B}f)(z) = \frac{1}{\rho}\min_{\gamma}\max_{\{\xi^j\in[d_j,u_j]\}}[f(\xi\circ z) - (\gamma, \xi\circ z - \rho z)]. \tag{13.11}$$

This leads to the following result from Kolokoltsov [95].

Theorem 13.1. *The minimal value of X_0 for which the income of an investor is nonnegative (and which by definition is the hedge price H^n in an n-step game) is given by*

$$H^n = (\mathscr{B}^n f)(S_0^1, \dots, S_0^J). \tag{13.12}$$

Changing variables $\xi = (\xi^1, \dots, \xi^J)$ to $\eta = \xi \circ z$ yields

$$(\mathscr{B}f)(z^1, \dots, z^J) = \frac{1}{\rho} \min_{\gamma} \max_{\{\eta \in [z^i d_i, z^i u_i]\}} \left[f(\eta) - \sum_{i=1}^{J} \gamma^i (\eta^i - \rho z^i) \right], \tag{13.13}$$

or, by shifting,

$$(\mathscr{B}f)(z^1, \dots, z^J) = \frac{1}{\rho} \min_{\gamma} \max_{\{\eta \in [z^i(d_i - \rho), z^i(u_i - \rho)]\}} [f(\eta + \rho z) - (\gamma, \eta)]. \tag{13.14}$$

Thus we are in the setting of Sect. 12.3. Moreover, assuming f is convex (which is often the case for option payoffs), we can apply Theorem 12.14, where the max is taken over the set of vectors

$$\eta_I = \xi_I \circ z - \rho z,$$

which are the vertices of the rectangular parallelepiped

$$\Pi_{z,\rho} = \times_{i=1}^{J} [z^i(d_i - \rho), z^i(u_i - \rho)],$$

where

$$\xi_I = \{d_i|_{i \in I}, u_j|_{j \notin I}\}$$

are the vertices of

$$\Pi = \times_{i=1}^{J} [d_i, u_i] \tag{13.15}$$

parameterized by all subsets (including the empty one) $I \subset \{1, \dots, J\}$.

Since the origin is an internal point of Π (because $d_i < \rho < u_i$), the family $\{\eta_I\}$ is strongly positively complete. The condition of the general position is rough (or generic) in the sense that it is fulfilled for an open dense subset of pairs (d_i, u_i). Applying Theorem 12.14 (and Remark 12.16) to (13.14) and returning to ξ yields the following theorem.

Theorem 13.2. *If the vertices ξ_I of the parallelepiped Π are in a general position in the sense that for any J subsets $I_1, \dots, I_J$ the vectors $\{\xi_{I_k} - \rho \mathbf{1}\}_{k=1}^J$ are independent in $\mathbb{R}^J$, then*

$$(\mathscr{B}f)(z) = \frac{1}{\rho} \max_{\{\Omega\}} \mathbf{E}_\Omega f(\xi \circ z), \quad z = (z^1, \dots, z^J), \tag{13.16}$$

where $\{\Omega\}$ is the collection of all subsets $\Omega = \xi_{I_1}, \dots, \xi_{I_{J+1}}$ of the set of vertices of Π, of size $J+1$, such that their convex hull contains $\rho \mathbf{1}$ as an interior point ($\mathbf{1}$ is the vector with all coordinates 1), and where $\mathbf{E}_\Omega$ denotes the expectation with respect *to the unique probability law* $\{p_I\}$, $\xi_I \in \Omega$, *on the set of vertices of* Π, *which is supported on* Ω *and is risk neutral with respect to* $\rho\mathbf{1}$, *that is,*

$$\sum_{I \subset \{1,\dots,J\}} p_I \xi_I = \rho \mathbf{1}. \tag{13.17}$$

Moreover, if

$$f(\xi \circ z) - (\gamma_{I_1,\dots,I_{J+1}}, (\xi - \rho\mathbf{1}) \circ z) \geq f(\zeta \circ z) - (\gamma_{I_1,\dots,I_{J+1}}, (\zeta - \rho\mathbf{1}) \circ z)$$

for all vertices ξ, ζ *such that* $\xi \in \Omega$ *and* $\zeta \notin \Omega$, *where* $\gamma_{I_1,\dots,I_{J+1}}$ *is the corresponding optimal value for the polyhedron* $\Pi[\xi_{I_1}, \dots, \xi_{I_{J+1}}]$, *then*

$$(\mathscr{B}f)(z^1, \dots, z^J) = \frac{1}{\rho} \mathbf{E}_\Omega f(\xi \circ z). \tag{13.18}$$

Risk neutrality now corresponds to its usual meaning in finance, i.e., (13.17) means that all discounted stock prices are martingales.

Notice that the max in (13.16) is over a finite number of explicit expressions, which is of course a great achievement as compared with the initial minimax over an infinite set. In particular, it reduces the calculation of the iterations $\mathscr{B}^n f$ to the calculation on a controlled Markov chain. Let us also stress that the number of eligible Ω in (13.16) is the number of different pyramids (convex polyhedrons with $J+1$ vertices) with vertices taken from the vertices of Π and containing $\rho\mathbf{1}$ as an interior point. Hence this number can be effectively calculated.

13.2 Nonexpansion and Homogeneity of Solutions

Let us point out some properties of the operator $\mathscr{B}$ given by (13.16) that are obvious but important for practical calculations. $\rho\mathscr{B}$ is *nonexpansive*:

$$\|\mathscr{B}(f_1) - \mathscr{B}(f_2)\| \leq \frac{1}{\rho} \|f_1 - f_2\|$$

and *homogeneous* (both with respect to addition and multiplication):

$$\rho\mathscr{B}(\lambda + f) = \lambda + \rho\mathscr{B}(f), \quad \mathscr{B}(\lambda f) = \lambda \mathscr{B}(f)$$

for any function f and $\lambda \in \mathbb{R}$ (resp. $\lambda > 0$) for the first (resp. second) equation.

Next, if f_p is a power function, that is,

$$f_p(z) = (z^1)^{i_1} \cdots (z^J)^{i_J},$$

then $f_p(\xi \circ z) = f_p(\xi) f_p(z)$, implying

$$(\mathscr{B}^n f_p)(z) = ((\mathscr{B} f_p)(\mathbf{1}))^n f_p(z). \tag{13.19}$$

Therefore, power functions are invariant under $\mathscr{B}$ (up to multiplication by a constant). Consequently, if for a payoff f one can find a reasonable approximation by a power function, that is, there exists a power function f_p such that $\|f - f_p\| \leq \varepsilon$, then

$$\|\mathscr{B}^n f - \lambda^n f_p\| \leq \frac{1}{\rho^n} \|f - f_p\| \leq \frac{\varepsilon}{\rho^n}, \quad \lambda = (\mathscr{B} f_p)(\mathbf{1}), \tag{13.20}$$

so that an approximate calculation of $\mathscr{B}^n f$ is reduced to the calculation of one number λ. This implies the following scheme for an approximate evaluation of $\mathscr{B}$. First, find the best fit to f in terms of the functions $\alpha + f_p$ (where f_p is a power function and α a constant), and then use (13.20).

13.3 Submodular Payoffs: Two Colors

One can obtain an essential reduction in the combinatorics of Theorem 13.2 (i.e., in the number of eligible Ω) under additional assumptions on the payoff f. The most natural one in the context of options turns out to be the notion of submodularity. A function $f : \mathbb{R}_+^2 \to \mathbb{R}_+$ is called *submodular* if the inequality

$$f(x_1, y_2) + f(x_2, y_1) \geq f(x_1, y_1) + f(x_2, y_2)$$

holds whenever $x_1 \leq x_2$ and $y_1 \leq y_2$. Similarly, a function $f : \mathbb{R}_+^d \to \mathbb{R}_+$ is called *submodular* if

$$f\left(x \bigvee y\right) + f\left(x \bigwedge y\right) \leq f(x) + f(y),$$

where $\bigvee$ (resp. $\bigwedge$) denotes the Pareto (coordinatewise) maximum (resp. minimum).

Remark 13.3. If f is twice continuously differentiable, then it is submodular if and only if $\frac{\partial^2 f}{\partial z_i \partial z_j} \leq 0$ for all $i \neq j$.

As one easily sees, the payoffs of the first three examples of rainbow options (Sect. 13.1), i.e., those defined by (13.1)–(13.3), are submodular. Let us explain how the assumptions of submodularity can simplify Theorem 13.2.

First, let $J = 2$. The polyhedron Π from (13.15) is then a rectangle. From the submodularity of f it follows that if Ω from Theorem 13.2 is either

$$\Omega_{12} = \{(d_1, d_2), (d_1, u_2), (u_1, u_2)\}$$

or

$$\Omega_{21} = \{(d_1, d_2), (u_1, d_2), (u_1, u_2)\},$$

then $(f,\xi)-(\gamma_0,\xi)$ coincide for all vertices ξ of Π. Hence Ω_{12} and Ω_{21} can be discarded in Theorem 13.2, i.e., the maximum is always achieved on either

$$\Omega_d = \{(d_1,d_2),(d_1,u_2),(u_1,d_2)\}$$

or

$$\Omega_u = \{(d_1,u_2),(u_1,d_2),(u_1,u_2)\}.$$

But the interiors of the triangles formed by Ω_u and Ω_d do not intersect, so that each point of Π (in a general position) lies only in one of them (and this position no longer depends on f). Hence, depending on the position of $\rho\mathbf{1}$ in Π, expression (13.16) reduces to either $\mathbf{E}_{\Omega_u}$ or $\mathbf{E}_{\Omega_d}$. This yields the following result (obtained in Kolokoltsov [95]).

Theorem 13.4. *Let $J=2$ and f be convex submodular. Denote*

$$\kappa = \frac{(u_1u_2-d_1d_2)-\rho(u_1-d_1+u_2-d_2)}{(u_1-d_1)(u_2-d_2)} = 1-\frac{\rho-d_1}{u_1-d_1}-\frac{\rho-d_2}{u_2-d_2}. \tag{13.21}$$

If $\kappa \geq 0$, then $\rho(\mathcal{B}f)(z_1,z_2)$ equals

$$\frac{\rho-d_1}{u_1-d_1}f(u_1z_1,d_2z_2)+\frac{\rho-d_2}{u_2-d_2}f(d_1z_1,u_2z_2)+\kappa f(d_1z_1,d_2z_2), \tag{13.22}$$

and the corresponding optimal strategies are

$$\gamma^1 = \frac{f(u_1z_1,d_2z_2)-f(d_1z_1,d_2z_2)}{z_1(u_1-d_1)}, \quad \gamma^2 = \frac{f(d_1z_1,u_2z_2)-f(d_1z_1,d_2z_2)}{z_2(u_2-d_2)}.$$

If $\kappa \leq 0$, then $\rho(\mathcal{B}f)(z_1,z_2)$ equals

$$\frac{u_1-\rho}{u_1-d_1}f(d_1z_1,u_2z_2)+\frac{u_2-\rho}{u_2-d_2}f(u_1z_1,d_2z_2)+|\kappa| f(u_1z_1,u_2z_2), \tag{13.23}$$

and

$$\gamma^1 = \frac{f(u_1z_1,u_2z_2)-f(d_1z_1,u_2z_2)}{z_1(u_1-d_1)}, \quad \gamma^2 = \frac{f(u_1z_1,u_2z_2)-f(u_1z_1,d_2z_2)}{z_2(u_2-d_2)}.$$

Clearly the linear operator $\mathcal{B}$ preserves the set of convex submodular functions. Hence one can use this formula recursively to obtain all powers of $\mathcal{B}$ in a closed form. For instance, if $\kappa=0$, then one obtains for the hedge price the following *two-color extension of the classic CRR formula* (for the latter see, e.g., Bingham and Kiesel [45]):

$$\mathcal{B}^n f(S_0^1,S_0^2) = \rho^{-n}\sum_{k=0}^{n}\binom{n}{k}\left(\frac{\rho-d_1}{u_1-d_1}\right)^k\left(\frac{\rho-d_2}{u_2-d_2}\right)^{n-k} f\left(u_1^k d_1^{n-k}S_0^1, d_2^k u_2^{n-k}S_0^2\right). \tag{13.24}$$

Similar formulas hold for nonvanishing κ.

13.4 Submodular Payoffs: Three or More Colors

Here we continue the discussion of submodular payoffs looking at options with more than two colors. Trying to argue as in the case $J = 2$, suppose first that Ω from Theorem 13.2 contains the vertex $(d_1, \ldots, d_J)$. Next, if it then contains any other vertex

$$\xi_I = \{d_i|_{i \in I}, u_j|_{j \notin I}\},$$

then, due to submodularity, the set where the max is attained in (13.11) must contain all $\xi(j)$, $j \notin I$, which have only one u_j on the jth place and d_i on other places. On the other hand, the whole set Ω has to be strongly positively complete, so that there can be no i such that all vectors from Ω have d_i on this place. From these two facts it follows that if Ω contains $(d_1, \ldots, d_J)$, then the set Ω_d consisting of $(d_1, \ldots, d_J)$ and J other vertices with only one coordinate u_j, $j = 1, \ldots, J$, is also maximizing in (13.16). Consequently, if $\rho\mathbf{1} \in \Omega_d$, then the max in (13.16) is reduced to only one term arising from Ω_d. Next, $\rho\mathbf{1} \in \Omega_d$ means that the vector v with coordinates $\{(\rho - d_i)/(u_i - d_i)\}$ belongs to the simplex

$$\Sigma = \{x_1, \cdots x_J \geq 0 : x_1 + \cdots + x_J < 1\}$$

or, equivalently,

$$\sum_{i=1}^{J} \frac{\rho - d_i}{u_i - d_i} < 1, \tag{13.25}$$

and the risk-neutral probabilities $p_0, p_1, \ldots, p_J$ on the vertices of $\Sigma - v$ for a $v \in \Sigma$ are clearly $1 - \sum_{j=1}^{J} v^j, v^1, \ldots, v^J$.

Similarly, if $\rho\mathbf{1} \in \Omega_u$, that is,

$$\sum_{i=1}^{J} \frac{u_i - \rho}{u_i - d_i} < 1 \tag{13.26}$$

holds, then the max in (13.16) is reduced to only one term arising from Ω_u, where the set Ω_u consists of $(u_1, \ldots, u_J)$ and J other vertices with only one coordinate d_j, $j = 1, \ldots, J$.

This yields the following result, where we use the following notation: for a set $I \subset \{1, 2, \ldots, J\}$, $f_I(z)$ [resp. $\tilde{f}_I(z)$] is $f(\xi^1 z_1, \ldots, \xi^J z_J)$ with $\xi^i = d_i$ for $i \in I$ and $\xi_i = u_i$ for $i \notin I$ (resp. $\xi^i = u_i$ for $i \in I$ and $\xi_i = d_i$ for $i \notin I$).

Theorem 13.5. *Let f be convex and submodular.*

(1) If (13.25) *holds, then*

$$(\mathcal{B}f)(z) = \frac{1}{\rho}\left[\tilde{f}_{\emptyset}(z) + \sum_{j=1}^{J} \frac{\rho - d_j}{u_j - d_j}(\tilde{f}_j(z) - \tilde{f}_{\emptyset})\right]. \tag{13.27}$$

(2) If (13.26) *holds, then*

$$(\mathscr{B}f)(z) = \frac{1}{\rho}\left[f_{\emptyset}(z) + \sum_{j=1}^{J}\frac{u_j-\rho}{u_j-d_j}(f_j(z)-f_{\emptyset})\right]. \tag{13.28}$$

Hence, in these cases our $\mathscr{B}$ again reduces to a linear form, allowing for a straightforward calculation of its iterations, i.e., for a *multicolor extension of the CRR formula*, as in the previous case $J=2$.

When condition (13.25) or (13.26) does not hold, the reduced Bellman operator does not turn to a linear form, even though essential simplifications are still possible in case of submodular payoffs.

Let us sort our this combinatorics only for three colors, $J=3$. Suppose that $\rho\mathbf{1}$ lies neither in the tetrahedron Ω_d nor in Ω_u (i.e., neither of the conditions of Theorem 13.5 holds). From the above reductions of possible Ω it follows that in that case one can discard all Ω containing either (d_1,d_2,d_3) or (u_1,u_2,u_3). Hence only six vertices are left for eligible Ω. From the consideration of a general position we further deduce that altogether only six Ω are possible, namely, the three tetrahedra containing the vertices (d_1,d_2,u_3), (d_1,u_2,d_3), (u_1,d_2,d_3) and one vertex from (d_1,u_2,u_3), (u_1,u_2,d_3), (u_1,d_2,u_3), and symmetrically the three tetrahedra containing the vertices (d_1,u_2,u_3), (u_1,u_2,d_3), (u_1,d_2,u_3) and one vertex from (d_1,d_2,u_3), (d_1,u_2,d_3), (u_1,d_2,d_3). However, any particular point in a general position belongs to only three out of these six, leaving in formula (13.16) the max over three possibilities only. The particular choice of these three tetrahedra depends on the coefficients

$$\alpha_{12} = \left(1-\frac{u_1-r}{u_1-d_1}-\frac{u_2-r}{u_2-d_2}\right),$$
$$\alpha_{13} = \left(1-\frac{u_1-r}{u_1-d_1}-\frac{u_3-r}{u_3-d_3}\right),$$
$$\alpha_{23} = \left(1-\frac{u_2-r}{u_2-d_2}-\frac{u_3-r}{u_3-d_3}\right), \tag{13.29}$$

and leads to the following result obtained in Hucki and Kolokoltsov [86] (though with a much more elaborate proof than here).

Theorem 13.6. *Let $J=3$ and f be convex and submodular, and neither of the conditions of Theorem 13.5 holds.*

(1) If $\alpha_{12}\geq 0$, $\alpha_{13}\geq 0$ and $\alpha_{23}\geq 0$, then

$$(\mathscr{B}f)(\mathbf{z}) = \frac{1}{\rho}\max\left\{\begin{array}{l}(-\alpha_{123})f_{\{1,2\}}(\mathbf{z})+\alpha_{13}f_{\{2\}}(\mathbf{z})+\alpha_{23}f_{\{1\}}(\mathbf{z})+\frac{u_3-r}{u_3-d_3}f_{\{3\}}(\mathbf{z})\\ (-\alpha_{123})f_{\{1,3\}}(\mathbf{z})+\alpha_{12}f_{\{3\}}(\mathbf{z})+\alpha_{23}f_{\{1\}}(\mathbf{z})+\frac{u_2-r}{u_2-d_2}f_{\{2\}}(\mathbf{z})\\ (-\alpha_{123})f_{\{2,3\}}(\mathbf{z})+\alpha_{12}f_{\{3\}}(\mathbf{z})+\alpha_{13}f_{\{2\}}(\mathbf{z})+\frac{u_1-r}{u_1-d_1}f_{\{1\}}(\mathbf{z})\end{array}\right\}.$$

(2) If $\alpha_{ij} \le 0$, $\alpha_{jk} \ge 0$ and $\alpha_{ik} \ge 0$, where $\{i,j,k\}$ is an arbitrary permutation of the set $\{1,2,3\}$, then

$$(\mathscr{B}f)(\mathbf{z}) = \frac{1}{\rho}\max\left\{\begin{array}{l} (-\alpha_{ijk})\, f_{\{i,j\}}(\mathbf{z}) + \alpha_{ik} f_{\{j\}}(\mathbf{z}) + \alpha_{jk} f_{\{i\}}(\mathbf{z}) + \frac{u_k - r}{u_k - d_k} f_{\{k\}}(\mathbf{z}) \\ \alpha_{jk} f_{\{i\}}(\mathbf{z}) + (-\alpha_{ij}) f_{\{i,j\}}(\mathbf{z}) + \frac{u_k - r}{u_k - d_k} f_{\{i,k\}}(\mathbf{z}) - \frac{d_i - r}{u_i - d_i} f_{\{j\}}(\mathbf{z}) \\ \alpha_{ik} f_{\{j\}}(\mathbf{z}) + (-\alpha_{ij}) f_{\{i,j\}}(\mathbf{z}) + \frac{u_k - r}{u_k - d_k} f_{\{j,k\}}(\mathbf{z}) - \frac{d_j - r}{u_j - d_j} f_{\{i\}}(\mathbf{z}) \end{array}\right\}.$$

(3) If $\alpha_{ij} \ge 0$, $\alpha_{jk} \le 0$ and $\alpha_{ik} \le 0$, where $\{i,j,k\}$ is an arbitrary permutation of the set $\{1,2,3\}$, then

$$(\mathscr{B}f)(\mathbf{z}) = \frac{1}{\rho}\max\left\{\begin{array}{l} \alpha_{ij} f_{\{k\}}(\mathbf{z}) + (-\alpha_{jk}) f_{\{j,k\}}(\mathbf{z}) + \frac{u_i - r}{u_i - d_i} f_{\{i,k\}}(\mathbf{z}) - \frac{d_k - r}{u_k - d_k} f_{\{j\}}(\mathbf{z}) \\ \alpha_{ij} f_{\{k\}}(\mathbf{z}) + (-\alpha_{ik}) f_{\{i,k\}}(\mathbf{z}) + \frac{u_j - r}{u_j - d_j} f_{\{j,k\}}(\mathbf{z}) - \frac{d_k - r}{u_k - d_k} f_{\{i\}}(\mathbf{z}) \\ (\alpha_{123} + 1) f_{\{k\}}(\mathbf{z}) - \alpha_{jk} f_{\{j,k\}}(\mathbf{z}) - \alpha_{ik} f_{\{i,k\}}(\mathbf{z}) - \frac{d_k - r}{u_k - d_k} f_{\{i,j\}}(\mathbf{z}) \end{array}\right\}.$$

One must stress here that the application of Theorem 13.6 is rather limited: since $\mathscr{B}$ is not reduced to a linear form, it is not clear how to use it for the iterations of $\mathscr{B}$ because the submodularity does not seem to be preserved under such $\mathscr{B}$.

13.5 Transaction Costs

Let us now extend the model of Sect. 13.1 to include possible transaction costs. They can depend on transactions in various ways. The simplest for the analysis are the so-called *fixed transaction costs*, which are equal to a fixed fraction $(1-\beta)$ (with β a small constant) of the entire portfolio. Hence for fixed costs, Eq. (13.6) changes to

$$X_m = \beta \sum_{j=1}^{J} \gamma_m^j \xi_m^j S_{m-1}^j + \rho \left(X_{m-1} - \sum_{j=1}^{J} \gamma_m^j S_{m-1}^j \right). \tag{13.30}$$

As one easily sees, including fixed costs can be dealt with by rescaling ρ, thereby bringing nothing new to the analysis.

In more advanced models, *transaction costs* depend on the amount of transactions (bought and sold stocks) at each moment of time, i.e., they are given by some function

$$g(\gamma_m - \gamma_{m-1}, S_{m-1}),$$

and are paid at times when the investor changes γ_{m-1} to γ_m. In particular, the basic example presents so-called *proportional transaction costs*, where

$$g(\gamma_m - \gamma_{m-1}, S_{m-1}) = \beta \sum_{j=1}^{J} |\gamma_m^j - \gamma_{m-1}^j| S_{m-1}^j$$

(again with a fixed $\beta > 0$). We will assume only that g has the following Lipschitz property:

$$|g(\gamma_1,z) - g(\gamma_2,z)| \leq \beta|z||\gamma_1 - \gamma_2| \tag{13.31}$$

with a fixed $\beta > 0$.

To deal with transaction costs, it is convenient to extend the state space of our game, considering the states that are characterized, at time $m-1$, by $2J+1$ numbers

$$X_{m-1}, S^j_{m-1}, v_{m-1} = \gamma^j_{m-1}, \quad j = 1,\dots,J.$$

When, at time $m-1$, the investor chooses her new control parameters γ_m, the new state at time m becomes

$$X_m, \quad S^j_m = \xi^j_m S^j_{m-1}, \quad v_m = \gamma^j_m, \quad j = 1,\dots,J,$$

where the value of the portfolio is

$$X_m = \sum_{j=1}^J \gamma^j_m \xi^j_m S^j_{m-1} + \rho\left(X_{m-1} - \sum_{j=1}^J \gamma^j_m S^j_{m-1}\right) - g(\gamma_m - v_{m-1}, S_{m-1}). \tag{13.32}$$

The corresponding *reduced Bellman operator* from Sect. 13.1 takes the form

$$(\mathscr{B}f)(z,v) = \min_\gamma \max_\xi [f(\xi \circ z, \gamma) - (\gamma, \xi \circ z - \rho z) + g(\gamma - v, z)], \tag{13.33}$$

where $z, v \in \mathbb{R}^J$, or, changing the variables $\xi = (\xi^1,\dots,\xi^J)$ to $\eta = \xi \circ z$ and shifting,

$$(\mathscr{B}f)(z,v) = \min_\gamma \max_{\{\eta^j \in [z^j(d_j-\rho), z^j(u_j-\rho)]\}} [f(\eta + \rho z, \gamma) - (\gamma, \eta) + g(\gamma - v, z)]. \tag{13.34}$$

Let us assume that

$$|f(z,v_1) - f(z,v_2)| \leq \alpha|z||v_1 - v_2|$$

and α is small enough so that the requirements of Theorem 12.22 are satisfied for the right-hand side of (13.34). By Theorem 12.22,

$$(\mathscr{B}f)(z,v) = \max_\Omega \mathbf{E}_\Omega[f(\xi \circ z, \gamma_\Omega) + g(\gamma_\Omega - v, z)]. \tag{13.35}$$

Notice that since the terms with v enter additively, they cancel from the equations for γ_Ω, so that the values of γ_Ω do not depend on v. Consequently,

$$|(\mathscr{B}f)(z,v_1) - (\mathscr{B}f)(z,v_2)| \leq \max_\Omega \mathbf{E}_\Omega[g(\gamma_\Omega - v_1, z) - g(\gamma_\Omega - v_2), z)] \leq \beta|z||v_1 - v_2|. \tag{13.36}$$

Hence, if at all steps the application of Theorem 12.22 is allowed, then $(\mathscr{B}^k f)(z,v)$ remains Lipschitz in v with the Lipschitz constant $\beta|z|$ (the last-step function does not depend on v and hence trivially satisfies this condition).

Let $\varkappa_1$ and $\varkappa_2$ be the characteristics, defined before Theorem 12.22, of the set of vertices $\xi_I - \rho\mathbf{1}$ of the parallelepiped $\times_{j=1}^J [d_j, u_j] - \rho\mathbf{1}$. By Proposition 12.23, the corresponding characteristics $\varkappa_1(z)$ and $\varkappa_2(z)$ of the set of vertices of the scaled parallelepiped

$$\times_{j=1}^J [z^j d_j, z^j u_j] - \rho z$$

have the lower bounds

$$\varkappa_i(z) \geq |z| \varkappa_i \frac{1}{d\delta(z)}, \quad i = 1, 2.$$

As in each step of our process the coordinates of z are multiplied by d_j or u_j, the corresponding maximum $\delta_n(z)$ of the δ of all z that can occur in the n-step process equals

$$\delta_n(z) = \delta(z) \left(\frac{\max_j u_j}{\min_j d_j} \right)^n. \tag{13.37}$$

Thus we arrive at the following result.

Theorem 13.7. *Suppose β from* (13.31) *satisfies the estimate*

$$\beta < \min(\varkappa_1, \varkappa_2) \frac{1}{d\delta_n(z)},$$

where $\delta_n(z)$ is given by (13.37)*. Then the hedge price of a derivative security specified by a final payoff f, and with transaction costs specified previously, is given by* (13.12)*, where $\mathcal{B}$ is given by* (13.33)*. Moreover, at each step, $\mathcal{B}$ can be evaluated by Theorem 12.22, i.e., by* (13.35)*, reducing the calculations to finding a maximum over a finite set.*

Of course, for larger β, further adjustments of Theorem 12.22 are required.

Chapter 14
Rainbow Options in Discrete Time, II

14.1 Rainbow American Options and Real Options

In the world of *American options*, when an option can be exercised at any time, the operator $\mathbf{B}G(X,S^1,\ldots,S^J)$ from (13.9) changes to

$$\mathbf{B}G(X,S^1,\ldots,S^J) = \frac{1}{\rho}\max_{\gamma}\min\Big\{G(X,S^1,\ldots,S^J),\\ \frac{1}{\rho}\min_{\xi} G(\rho X + \sum_{i=1}^{J}\gamma^i\xi^i S^i - \rho\sum_{i=1}^{J}\gamma^i S^i, \xi^1 S^1,\ldots,\xi^J S^J)\Big\}, \tag{14.1}$$

so that the corresponding *reduced Bellman operator* takes the form

$$\begin{aligned}&(\mathscr{B}f)(z^1,\ldots,z^J)\\ &= \frac{1}{\rho}\min_{\gamma}\max\left[\rho f(\rho z), \max_{\xi}[f(\xi^1 z^1,\xi^2 z^2,\ldots,\xi^J z^J) - \sum_{i=1}^{J}\gamma^i z^i(\xi^i-\rho)]\right]\\ &= \max\left[f(\rho z), \frac{1}{\rho}\min_{\gamma}\max_{\xi}[f(\xi^1 z^1,\xi^2 z^2,\ldots,\xi^J z^J) - \sum_{i=1}^{J}\gamma^i z^i(\xi^i-\rho)]\right].\end{aligned} \tag{14.2}$$

Consequently, in this case the main formula (13.16) of Theorem 13.2 becomes

$$(\mathscr{B}f)(z^1,\ldots,z^J) = \max\left[f(\rho z), \frac{1}{\rho}\max_{\{\Omega\}}\mathbf{E}_\Omega f(\xi\circ z)\right], \tag{14.3}$$

which is of course not an essential increase in complexity. The hedge price for the n-step model is again given by (13.12).

Similar problems arise in the study of *real options*. We refer the reader to Dixit and Pindyck [68] for a general background and to Bensoussan et al. [31] for more

P. Bernhard et al., *The Interval Market Model in Mathematical Finance*, Static & Dynamic Game Theory: Foundations & Applications, DOI 10.1007/978-0-8176-8388-7_14,

recent mathematical results. A typical *real-option* problem can be formulated as follows. Given J instruments (e.g., commodities, assets), the value of the investment in some project at time m is supposed to be given by certain functions $f_m(S_m^1,\ldots,S_m^J)$ depending on the prices of these instruments at time m. The problem is to evaluate the price (at the initial time 0) of the option to invest in this project that can be exercised at any time during a given time interval $[0,T]$. Such a price is important since to keep the option open a firm needs to pay certain costs (say, keep required facilities ready or invest in research). We have formulated the problem in a way that makes it an example of the general evaluation of an American rainbow option, discussed above, at least when underlying instruments are tradable on a market. For practical implementation, one only has to keep in mind that the risk-free rates appropriate for the evaluation of real options are usually not those available on bank accounts used in the analysis of financial options, but rather the growth rates of the corresponding branch of industry. These rates are usually estimated via the capital asset pricing model (CAPM); see again [68].

14.2 Path Dependence and Other Modifications

The theory in Sect. 13.1 is rough, in the sense that it can be easily modified to accommodate various additional pricing mechanisms. We have already considered transaction costs and American options. Here we will discuss other modifications: path-dependent payoffs, time-dependent price jumps (including variable volatility), and nonlinear jump formations. For simplicity, we will discuss these extensions separately, but any of their combinations (including transaction costs and American versions) can be easily dealt with.

Let us start with path-dependent payoffs. That is, we generalize the context of Sect. 13.1 by making the payoff f at time m depend on the whole history of the price evolution, i.e., as defined by a function $f(S_0,S_1,\ldots,S_m)$, $S_i=(S_i^1,\ldots,S_i^J)$, on $\mathbb{R}^{J(m+1)}$. The state of the game at time m must now be specified by $(m+1)J+1$ numbers

$$X_m,\ S_i=(S_i^1,\ldots,S_i^J),\quad i=0,\ldots,m.$$

The final payoff in the n-step game is now $G=X-f(S_0,\ldots,S_n)$, and at the penultimate period $n-1$ (when $S_0,\ldots,S_{n-1}$ are known) payoff equals

$$\begin{aligned}
&\mathbf{B}G(X,S_0,\ldots,S_{n-1})\\
&\quad = X-\frac{1}{\rho}\min_{\gamma}\max_{\xi}[f(S_0,\ldots,S_{n-1},\xi\circ S_{n-1})-(\gamma,S_{n-1}\circ(\xi-\rho\mathbf{1}))]\\
&\quad = X-(\mathscr{B}_{n-1}f)(S_0,\ldots,S_{n-1}),
\end{aligned}$$

where the modified *reduced Bellman operators* are now defined as

$$(\mathscr{B}_{m-1}f)(z_0,\ldots,z_{m-1}) = \frac{1}{\rho}\min_{\gamma}\max_{\{\xi^j\in[d_j,u_j]\}}[f(z_0,\ldots,z_{m-1},\xi\circ z_{m-1}) - (\gamma,\xi\circ z-\rho z)]. \tag{14.4}$$

Consequently, by dynamic programming, the guaranteed payoff at the initial moment of time equals $X - \mathscr{B}_0(\mathscr{B}_1\cdots(\mathscr{B}_{n-1}f)\cdots)$, and the hedging price becomes

$$H^n = \mathscr{B}_0(\mathscr{B}_1\cdots(\mathscr{B}_{n-1}f)\cdots). \tag{14.5}$$

No essential changes are required if possible jump sizes are time dependent. Only the operators $\mathscr{B}_{m-1}$ from (14.4) have to be generalized to

$$(\mathscr{B}_{m-1}f)(z_0,\ldots,z_{m-1})$$
$$= \min_{\gamma}\max_{\{\xi^j\in[d_j^m,u_j^m]\}}[f(z_0,\ldots,z_{m-1},\xi\circ z_{m-1}) - (\gamma,\xi\circ z-\rho z)],$$

where the pairs (d_j^m, u_j^m), $j=1,\ldots,J$, $m=1,\ldots,n$ specify the model.

Let us turn to *nonlinear jump patterns*. Generalizing the setting of Sect. 13.1 let us assume that, instead of the stock price evolution model $S_{m+1} = \xi\circ S_m$, we are given k transformations $g_i : \mathbb{R}^J \to \mathbb{R}^J$, $i=1,\ldots,k$, which give rise naturally to two models of price dynamics: either

(1) At time $m+1$ the price S_{m+1} belongs to the closure of the convex hull of the set $\{g_i(S_m)\}$, $i=1,\ldots,k$ (*nonlinear interval model*) or
(2) S_{m+1} is one of the points $\{g_i(S_m)\}$, $i=1,\ldots,k$.

Since the first model can be approximated by the second one (by possibly increasing the number of transformations g_i), we will work with the second model.

Remark 14.1. Notice that maximizing a function over a convex polyhedron is equivalent to its maximization over the edges of this polyhedron. Hence, for convex payoffs the two preceding models are fully equivalent. However, on the one hand, not all reasonable payoffs are convex, and on the other hand, when it comes to minimization (which one needs, say, for lower prices; see Sect. 14.3), the situation becomes rather different.

Assuming for simplicity that possible jump sizes are time independent and the payoff depends only on the end value of a path, the *reduced Bellman operator* (13.10) becomes

$$(\mathscr{B}f)(z) = \frac{1}{\rho}\min_{\gamma}\max_{i\in\{1,\ldots,k\}}[f(g_i(z)) - (\gamma, g_i(z)-\rho z)], \quad z=(z^1,\ldots,z^J), \tag{14.6}$$

or, equivalently,

$$(\mathscr{B}f)(z) = \frac{1}{\rho} \min_{\gamma} \max_{\eta_i = g_i(z) - \rho z, i=1,\ldots,k} [f(\eta_i + \rho z) - (\gamma, \eta_i)]. \tag{14.7}$$

The hedge price is still given by (13.12), and operator (14.6) is calculated by Theorem 12.14, yielding

$$(\mathscr{B}f)(z) = \frac{1}{\rho} \max_{\Omega} \mathbf{E}_{\Omega} f(\eta_i + \rho z), \tag{14.8}$$

where $\mathbf{E}_{\Omega}$ denote expectations with respect to all extreme points of risk-neutral probability laws on vectors $\eta_i = g_i(z) - \rho z$.

It is worth noting that if $k = d + 1$ and $\{g_i(z)\}$ form a collection of vectors in a general position, the corresponding risk-neutral probability is unique. Consequently, our hedge price becomes fair in the strongest sense of "no arbitrage": no positive surplus is possible for all paths of the stock price evolution (if the hedge strategy is followed). In particular, the evaluation of hedge strategies can be carried out in the framework of the standard approach to option pricing, that is, by choosing as an initial (real-world) probability on jumps an arbitrary measure with full support, one concludes that there exists a unique risk-neutral equivalent martingale measure, explicitly defined via formula (12.18), and the hedge price calculated by the iterations of operator (14.7) coincides with the standard risk-neutral evaluation of derivative prices in complete markets.

14.3 Upper and Lower Values for Intrinsic Risk

The celebrated no-arbitrage property of the hedge price of an option in the Cox–Ross–Rubinstein (CRR) or Black–Scholes models means that, almost surely, with respect to the initial probability distribution on paths, the investor cannot get an additional surplus when adhering to the hedge strategy that guarantees that there could be no loss. In our setting, even though our formula in the case $J = 1$ coincides with the CRR formula, we do not assume any initial law on paths, so that the notion of "no arbitrage" is not specified either.

It is a characteristic feature of our models that picking up an a priori probability law with support on all paths leads to an incomplete market, that is, to the existence of infinitely many equivalent martingale measures, which fail to identify a fair price in a unique, consistent way. Notice that our extreme points are absolutely continuous, but usually not equivalent to a measure with full support.

Remark 14.2. The only cases of a complete market among the models discussed previously are those mentioned at the end of Sect. 14.2, that is, models with precisely $J + 1$ eligible jumps in a stock price vector S in each period.

For the analysis of incomplete markets it is of course natural to look for some subjective criteria to specify a price. Lots of work by different authors has been devoted to this endeavor; see, for example, Bielecki et al. [44]. Our approach is to search for objective bounds (price intervals), which are given by our hedging strategies, in the spirit of El Karoui and Quenez [71].

Remark 14.3. Apart from supplying the lower price (as below), one can also argue about the reasonability of our main hedging price, noting that a chance for a possible surplus can be (and actually is indeed) compensated by the inevitable inaccuracy of a model and by transaction costs (if they are not properly taken into account). Moreover, this price satisfies the so-called no strictly acceptable opportunities condition suggested in Carr et al. [53].

For completeness, let us recall the general definitions of lower and upper prices, in the game-theoretic approach to probability, given in Shafer and Vovk [136]. Assume a process (a sequence of real numbers of a fixed length, say n, specifying the evolution of an investor's capital) is specified by alternating moves of two players, an investor and Nature, with complete information (all eligible moves of each player and their results are known to each player at any time, and the moves become publicly known at the moment when a decision is made). Let us denote by $X_\gamma^\alpha(\xi)$ the last number of the resulting sequence, starting with an initial value α and obtained by applying the strategy γ (of the investor) and ξ (of Nature). By a random variable we mean just a function on the set of all possible paths.

The *upper value* (or *upper expectation*) $\overline{\mathbf{E}}f$ of a random variable f is defined as the minimal capital of the investor such that she has a strategy that guarantees that at the final moment of time, her capital will be enough to buy f, i.e.,

$$\overline{\mathbf{E}}f = \inf\{\alpha : \exists\gamma : \forall\xi, X_\gamma^\alpha(\xi) - f(\xi) \geq 0\}.$$

At the same time, the *lower value* (or *lower expectation*) $\underline{\mathbf{E}}f$ of a random variable f is defined as the maximum capital of the investor such that she has a strategy that guarantees that at the final moment of time, her capital will be enough to sell f, i.e.,

$$\underline{\mathbf{E}}f = \sup\{\alpha : \exists\gamma : \forall\xi, X_\gamma^\alpha(\xi) + f(\xi) \geq 0\}.$$

One says that the prices are *consistent* if $\overline{\mathbf{E}}f \geq \underline{\mathbf{E}}f$. If these prices coincide, we are in a kind of abstract analog of a complete market. In the general case, upper and lower prices are also referred to as *seller's and buyer's* prices, respectively.

It is seen now that in this terminology our hedging price for a derivative security is the upper (or seller's) price. The lower price can be defined similarly, that is, in the context of Sect. 13.1, the lower price is given by

$$(\mathscr{B}^n_{\mathrm{low}} f)(S_0^1, \ldots, S_0^J),$$

where

$$(\mathscr{B}_{\mathrm{low}}f)(z) = \frac{1}{\rho}\max_{\gamma}\min_{\{\xi^j\in[d_j,u_j]\}}[f(\xi\circ z)-(\gamma,\xi\circ z-\rho z)]. \tag{14.9}$$

In this simple interval model and for convex f this expression is trivial; it equals $f(\rho z)$. On the other hand, if our f is concave or, more generally, if we allow only finitely many jumps, which leads, instead of (14.9), to the operator

$$(\mathscr{B}_{\mathrm{low}}f)(z) = \frac{1}{\rho}\max_{\gamma}\min_{\{\xi^j\in\{d_j,u_j\}\}}[f(\xi\circ z)-(\gamma,\xi\circ z-\rho z)], \tag{14.10}$$

then Theorem 12.19 applies giving for the lower price the dual expression to the upper price (13.16), where maximum is turned to minimum (over the same set of extreme risk-neutral measures):

$$(\mathscr{B}_{\mathrm{low}}f)(z) = \frac{1}{\rho}\min_{\{\Omega\}}\mathbf{E}_{\Omega}f(\xi\circ z), \quad z=(z^1,\ldots,z^J). \tag{14.11}$$

The difference between lower and upper prices can be considered as a measure of the intrinsic risk of an incomplete market.

14.4 Cash-Back Methodology for Dealing with Intrinsic Risk

Apart from working with upper and lower prices, one can suggest another method for improving hedge prices in such a way that allows one to avoid arbitrage completely. This method was worked out in Lyons [108] for models with unknown volatility. It consists in specifying a possible cash back that a holder of an option should receive when the price moves (unpredictable initially) turn out to be favorable (and not the worst possible, as assumed in the minimax evaluation).

It is rather easy to specify this cash-back flow in our setting. Consider, say, the general model with nonlinear jumps described by the reduced Bellman operator (14.6), which by Theorem 12.14 can be expressed via (14.8), that is

$$(\mathscr{B}f)(z) = \frac{1}{\rho}\max_{\Omega}\mathbf{E}_{\Omega}f(\eta+\rho z),$$

where $\mathbf{E}_{\Omega}$ denote expectations with respect to all extreme points of risk-neutral probability laws on vectors η from the set $\{g_i(z)-\rho z\}$.

Suppose at some time t the jump of the price turns out to be given by a vector $g_i(z)$, which does not belong to the support S_t of the optimal risk-neutral law Ω_t (where the max is attained in the preceding expression for $\mathscr{B}$). Then the investor's unpredictable surplus (trading with his hedging strategy) equals

$$f(g_j(z))-f(g_i(z))-(\gamma_0,g_j(z)-g_i(z)), \quad i\in S_t$$

(the same value for all $i \in S_t$) or, equivalently,

$$\mathbf{E}_{\Omega_t} f(\eta + \rho z) - f(g_i(z)) + (\gamma_0, g_i(z)),$$

where $f = B^{T-t} f_T$. Thus, to avoid arbitrage, the agreement between a buyer and a seller of an option can specify payments of this amount in each case of Nature's benevolent behavior, where the price jump does not belong to the support of the optimal risk-neutral law Ω_t.

14.5 Degenerate or Random Geometry of Nonsimultaneous Jumps

The situation becomes essentially different if the jumps in various underlying stock prices cannot occur simultaneously. The absence of simultaneous jumps is a performance of certain independence of their behavior. Under this assumption, instead of 2^J possible jumps, there are only $2J$ possibilities at each moment: one of the J stock prices jumps up or down and the other just increases at the risk-free rate. Thus the *reduced Bellman operator* becomes

$$(\mathscr{B}f)(z) = \frac{1}{\rho} \min_{\gamma \in \mathbb{R}^J} \max_{i=1,\ldots,J} \max[f(\rho \check{z}^i, u_i z^i) - \gamma^i z^i (u_i - \rho), f(\rho \check{z}^i, d_i z^i) - \gamma^i z^i (d_i - \rho)], \tag{14.12}$$

where $(\rho \check{z}^i, p z^i)$, for $p \in \mathbb{R}$, denotes the vector obtained from z by multiplying its ith coordinate by p and all other coordinates by ρ, or, equivalently,

$$(\mathscr{B}f)(z) = \frac{1}{\rho} \min_{\gamma \in \mathbb{R}^J} \max_{i=1,\ldots,J} \max_{\eta \in \{\eta_{i,u}, \eta_{i,d}\}} [f(\eta + \rho z) - (\gamma, \eta)], \tag{14.13}$$

where $\eta_{i,u}$ (resp. $\eta_{i,d}$) is a vector with only one nonvanishing ith coordinate that equals $z^i(u_i - \rho)$ (resp. $z^i(d_i - \rho)$).

According to Proposition 12.9, there are precisely J extreme points of risk-neutral laws on the set of $2J$ vectors $\{\eta_{i,u}, \eta_{i,d}\}$, $i = 1, \ldots, J$, and they have support on pairs $\{\eta_{i,u}, \eta_{i,d}\}$, $i = 1, \ldots, J$. Hence these risk-neutral laws are effectively one-dimensional and are given by probabilities

$$\frac{\rho - d_i}{u_i - d_i}, \quad \frac{u_i - \rho}{u_i - d_i}.$$

Risk neutrality means that

$$\frac{\rho - d_i}{u_i - d_i}(u_i - \rho) + \frac{u_i - \rho}{u_i - d_i}(d_i - \rho) = 0.$$

Applying Theorem 12.14, we obtain the following proposition.

Proposition 14.4. *The minimax expression* (14.12) *or* (14.13) *equals*

$$(\mathscr{B}f)(z) = \frac{1}{\rho} \max_{i=1,\dots,J} \left[\frac{\rho - d_i}{u_i - d_i} f(\rho \check{z}^i, u_i z^i) + \frac{\rho - d_i}{u_i - d_i} f(\rho \check{z}^i, d_i z^i) \right]. \tag{14.14}$$

Assume now that a probability law $\{p_1, \dots, p_J\}$ on stock symbol is given specifying a name that is going to jump at each particular moment in time, that is, we find ourselves in the context of random geometry, as in Sect. 12.6. The corresponding Bellman operator takes the form of type (12.56):

$$(\mathscr{B}f)(z) = \frac{1}{\rho} \sum_{i=1}^{J} p_i \max_{\eta=\eta_{i,u},\eta_{i,d}} [f(\eta + \rho z) - (\gamma, \eta)], \tag{14.15}$$

which, similar to (14.14), is evaluated as

$$(\mathscr{B}f)(z) = \frac{1}{\rho} \sum_{i=1}^{J} p_i [\frac{\rho - d_i}{u_i - d_i} f(\rho \check{z}^i, u_i z^i) + \frac{\rho - d_i}{u_i - d_i} f(\rho \check{z}^i, d_i z^i)]. \tag{14.16}$$

This context corresponds to independent jumps (that never occur simultaneously). Allowing for correlated jumps, say with positive correlations, we will add to the family of vectors $\{\eta_{i,u}, \eta_{i,d}\}$, $i = 1, \dots, J$, the vectors $\{\eta_{jk,u}, \eta_{jk,d}\}$, $j < k$, where $\eta_{jk,u}$ (resp. $\eta_{jk,d}$) is a vector with only two nonvanishing coordinates, j and k, that equal $z^j(u_j - \rho)$ and $z^k(u_k - \rho)$ [resp. $z^j(d_j - \rho)$ and $z^k(d_k - \rho)$]. Assuming probabilities p_i on the pairs $\{\eta_{jk,u}, \eta_{jk,d}\}$ and p_{jk} on pairs $\{\eta_{jk,u}, \eta_{jk,d}\}$ leads to the Bellman operator

$$\begin{aligned} (\mathscr{B}f)(z) &= \frac{1}{\rho} \sum_{i=1}^{J} p_i \max_{\eta=\eta_{i,u},\eta_{i,d}} [f(\eta + \rho z) - (\gamma, \eta)] \\ &+ \frac{1}{\rho} \sum_{j<k} p_{jk} \max_{\eta=\eta_{jk,u},\eta_{jk,d}} [f(\eta + \rho z) - (\gamma, \eta)]. \end{aligned} \tag{14.17}$$

Applying formula (12.60) yields the following proposition.

Proposition 14.5. *Assume additionally that $u_i - \rho = \rho - d_i$ for all $i = 1, \dots, J$. Then the minimax expression* (14.17) *equals*

$$\begin{aligned} (\mathscr{B}f)(z) =& \frac{1}{2\rho} \sum_{i=1}^{J} p_i [f(\rho z + \eta_{i,u}) + f(\rho z + \eta_{i,d})] \\ &+ \frac{1}{2\rho} \sum_{j<k} p_{jk} [f(\rho z + \eta_{jk,u}) + f(\rho z + \eta_{jk,d})]. \end{aligned}$$

14.6 Stochastic Interest Rates and Stochastic Volatility

As was mentioned previously, if the distribution of certain parameters is given, it can be easily accommodated in a game-theoretic model. Here we demonstrate it on the examples of stochastic interest rates and stochastic volatilities.

Let us start with interest rates. Assume that ρ in the basic context of Sect. 13.1 is not a constant but a random variable with a given distribution. Let us denote here the corresponding expectation by $\widetilde{\mathbf{E}}$. Then, modeling the corresponding new game in the spirit of Sect. 12.6, the reduced Bellman operator (13.11) will become

$$(\mathcal{B}f)(z) = \widetilde{\mathbf{E}}\frac{1}{\rho}\min_{\gamma}\max_{\{\xi^j\in[d_j,u_j]\}}[f(\xi\circ z) - (\gamma, \xi\circ z - \rho z)]. \tag{14.18}$$

A straightforward extension of the main result of Theorem 13.2 yields

$$(\mathcal{B}f)(z) = \widetilde{\mathbf{E}}\frac{1}{\rho}\max_{\{\Omega\}}\mathbf{E}_{\Omega}f(\xi\circ z) \tag{14.19}$$

(with the same notation as in Theorem 13.2).

Next we discuss stochastic volatilities correlated with the dynamics of stock prices. For simplicity, let us consider the case of only one underlying stock S whose dynamics depends on (variable) volatility v. Assume that the changes in S and v come from systematic drifts, rS and $\Phi(v)$ respectively, and oscillating (sign changing) jumps, $\psi(v,S)$ and $\phi(v)$ respectively. Thus, at each moment in time the stock price S can change either to $\rho S = (1+r)S$ (no jumps) or to $(1+r)S \pm \psi(v,S)$, and the volatility v can change either to $v+\Phi(v)$ or to $v+\Phi(v)\pm\phi(v)$. Consequently, if we trade stocks choosing to hold γ units, our capital in a one-step game would change from X to

$$(X-\gamma S)(1+r) + \gamma((1+r)S \pm \psi(v,S)) = (1+r)X \pm \gamma\psi(v,S).$$

And if we trade on volatility choosing to hold γ units, our capital in a one-step game would change from X to

$$(X-\gamma v)(1+r) + \gamma(v+\Phi(v)\pm\phi(v)) = (1+r)X \pm \gamma(\Phi(v) - r \pm \phi(v)).$$

Assume the probability law $\{p_s, p_v, p_{sv}\}$ is given, where p_s (resp. p_v) denotes the probability that S (resp. v) will jump, and p_{sv} denotes the probability of simultaneous jumps, which we assume to occur always in the same direction (positive correlations). Then the reduced Bellman operator corresponding to a game of type (12.56) takes the form

$$\begin{aligned}(\mathcal{B}f)(z,v) = {} & \frac{p_v}{1+r}\min_{\gamma}\max_{\pm}[f(z+rz, v+\Phi(v)\pm\phi(v)) - \gamma(\Phi(v)-r\pm\phi(v))] \\ & + \frac{p_s}{1+r}\min_{\gamma}\max_{\pm}[f(z+rz\pm\psi(v,z), v+\Phi(v)) \mp \gamma\psi(v,z)]\end{aligned}$$

$$+\frac{p_{sv}}{1+r}\min_{\gamma^1,\gamma^2}\max_{\pm}[f(z+rz\pm\psi(v,z),v+\Phi(v)\pm\phi(v))\mp\gamma^1\psi(v,z)$$
$$-\gamma(\Phi(v)-r\pm\phi(v))]. \tag{14.20}$$

Applying (12.60) yields

$$(\mathscr{B}f)(z,v) = \frac{p_v}{2(1+r)}[f(z+rz,v+\Phi(v)+\phi(v))+f(z+rz,v+\Phi(v)-\phi(v))]$$
$$+\frac{p_s}{2(1+r)}[f(z+rz+\psi(v,z),v+\Phi(v))$$
$$+f(z+rz-\psi(v,z),v+\Phi(v))]$$
$$+\frac{p_{sv}}{2(1+r)}[f(z+rz+\psi(v,z),v+\Phi(v)+\phi(v))$$
$$+f(z+rz-\psi(v,z),v+\Phi(v)-\phi(v))]. \tag{14.21}$$

As we will see in Sect. 15.4, by passing to a continuous-time limit in this model, one can obtain various diffusion equations describing stock price models with stochastic volatility.

14.7 Identification of Pre-Markov Chains

This final section can be viewed as a collection of remarks on the problem of identification and calibration of *pre-Markov models* (models with a prescribed set of possible moves but unknown probabilities) from our game-theoretic perspective.

First let us explain the idea of identification of pre-Markov models on a trivial example of an interest rate model, where no theory is needed. Suppose possible interest rates, applied at various times spaced by some $\tau > 0$, can take values $a_1,\ldots,a_k$ and evolve according to a certain time-homogeneous Markov chain with transition probabilities $\{p_{ij}\}$. If the interest rate at some time t is a_i, then the price, at time t, of a bond paying one at time $t+\tau$ is $B_i(t,t+\tau) = 1/a_i$, and the price of a bond paying one at time $t+2\tau$ equals (again at time t)

$$B(t,t+2\tau)_i = \frac{1}{a_i}\mathbf{E}\frac{1}{a} = \frac{1}{a_i}\sum_{j=1}^{k} p_{ij}\frac{1}{a_j} = \frac{1}{a_i}(Q\mathbf{1})_i,$$

where we denote by Q a matrix with elements $q_{ij} = p_{ij}/a_j$ (and $\mathbf{1}$ denotes, as usual, a vector with all coordinates being 1), so that $Q\mathbf{a} = \mathbf{1}$ ($\mathbf{a}$ denotes a vector with coordinates a_i). Similarly, the price of a bond paying one at time $t+k\tau$ equals (again at time t)

$$B(t,t+k\delta)_i = \frac{1}{a_i}(Q^{k-1}\mathbf{1})_i.$$

Hence, if the prices of these bonds are given, we know $Q^k\mathbf{1}$, which allows us to identify Q if $\mathbf{1}$ is a cyclic vector (i.e., in the general position).

Now let us turn to multinomial models of stocks, where one assumes that at each moment in time the price is multiplied by one of n given positive numbers $a_1 < \cdots < a_n$. Risk neutrality for a probability law $\{p_1, \ldots, p_n\}$ on these multipliers is expressed by the equation $\sum_{i=1}^n p_i a_i = \rho$, where ρ denotes the interest rate [see (13.17)]. Suppose now that the prices of certain contingent claims (options) specified by payoffs f from a family F are given, yielding

$$\sum_{i=1}^n p_i f(a_i) = \omega(f), \quad f \in F,$$

with certain positive $\{\omega(f)\}$. If the family F is rich enough, one can expect to be able to identify a unique eligible risk-neutral probability law, so that max on the right-hand side of (12.63) disappears. The most natural f are of course the payoffs of standard European calls. Assume $n-2$ premiums of these calls (with different strike prices) are given. Since we have some flexibility in choosing a_i, we can choose $a_2, \ldots, a_{n-1}$ to coincide with strike prices of these call options. Then the probability law $\{p_1, \ldots, p_n\}$ will satisfy the system of equations

$$\begin{cases} p_1 + \cdots + p_n = 1, \\ a_1 p_1 + \cdots + a_n p_n = \rho, \\ (a_3 - a_2)p_3 + (a_4 - a_2)p_4 + \cdots (a_n - a_2)p_n = \omega_3, \\ \ldots \\ (a_{n-1} - a_{n-2})p_{n-1} + (a_n - a_{n-2})p_n = \omega_{n-1}, \\ (a_n - a_{n-1})p_n = \omega_n, \end{cases} \tag{14.22}$$

with certain ω_j. The determinant of this system is $\prod_{k=2}^n (a_k - a_{k-1})$. The system is of triangular type and, thus, explicitly solvable. To obtain a simple result, let us simplify it further by assuming that a_i are equally spaced, i.e., $a_k - a_{k-1} = \Delta$ for all $k = 2, \ldots, n$ and certain $\Delta > 0$. Then system (14.22) reduces to a system of type

$$\begin{cases} x_1 + \cdots + x_n = b_1, \\ x_2 + 2x_3 + \cdots + (n-1)x_n = b_2, \\ x_3 + 2x_4 + \cdots + (n-2)x_n = b_3, \\ \ldots \\ x_{n-1} + 2x_n = b_{n-1}, \\ x_n = b_n \end{cases} \tag{14.23}$$

(where $x_k = \Delta p_k, b_1 = \Delta, b_2 = \rho - 1$) with the explicit solution

$$\begin{cases} x_n = b_n, \\ x_{n-1} = b_{n-1} - 2b_n, \\ x_k = b_k - 2b_{k+1} + b_{k+2}, \quad k = 2, \dots, n-2, \\ x_1 = b_1 - b_2 + b_3. \end{cases} \tag{14.24}$$

Chapter 15
Continuous-Time Limits

15.1 Nonlinear Black–Scholes Equation

Our models and results are most naturally adapted to a discrete-time context, which is not a disadvantage from a practical point of view, as all concrete calculations are anyway carried out on discrete data. However, for qualitative analysis, it is desirable to be able to see what is going on in the continuous-time limit. This limit can also be simpler and, hence, be used as an approximation to a less tractable discrete model. With this in mind, let us analyze possible limits as the time between jumps and their sizes tend to zero.

Let us work with the general model of nonlinear jumps from Sect. 14.2, with the reduced Bellman operator of form (14.6). Suppose the maturity time is T. Let us decompose the planning time $[0,T]$ into n small intervals of length $\tau = T/n$ and assume

$$g_i(z) = z + \tau^{\alpha}\phi_i(z), \quad i = 1,\dots,k, \tag{15.1}$$

with some functions ϕ_i and a constant $\alpha \in [1/2, 1]$. Thus the jumps during time τ are on an order of magnitude τ^{α}. As usual, we assume that the risk-free interest rate per time τ equals

$$\rho = 1 + r\tau,$$

with $r > 0$.

From (14.6) we deduce for the one-period Bellman operator the expression

$$\mathscr{B}_{\tau} f(z) = \frac{1}{1+r\tau}\max_{I}\sum_{i\in I} p_i^I(z,\tau) f(z+\tau^{\alpha}\phi_i(z)), \tag{15.2}$$

where I are subsets of $\{1,\dots,n\}$ of size $|I| = J+1$ such that the family of vectors $z+\tau^{\alpha}\phi_i(z)$, $i\in I$, is in a general position and $\{p_i^I(z,\tau)\}$ is the risk-neutral probability law on such a family, with respect to ρz, i.e.,

$$\sum_{i\in I} p_i^I(z,\tau)(z+\tau^{\alpha}\phi_i(z)) = (1+r\tau)z. \tag{15.3}$$

P. Bernhard et al., *The Interval Market Model in Mathematical Finance*, Static & Dynamic Game Theory: Foundations & Applications, DOI 10.1007/978-0-8176-8388-7_15,

Let us deduce the Hamilton-Jacobi-Bellman (HJB) equation for the limit, as $\tau \to 0$, of the approximate cost function $\mathcal{B}_\tau^{T-t}$, $t \in [0,T]$, with a given final cost f_T, using the standard (heuristic) dynamic programming approach. That is, from (15.2) and assuming an appropriate smoothness of f we obtain the approximate equation

$$f_{t-\tau}(z) = \frac{1}{1+r\tau} \max_I \sum_{i \in I} p_i^I(z,\tau) \left[f_t(z) + \tau^\alpha \frac{\partial f_t}{\partial z} \phi_i(z) \right.$$
$$\left. + \frac{1}{2}\tau^{2\alpha} \left(\frac{\partial^2 f_t}{\partial z^2} \phi_i(z), \phi_i(z) \right) + O(\tau^{3\alpha}) \right].$$

Since $\{p_i^I\}$ are probabilities and using (15.3), this is rewritten as

$$f_t - \tau \frac{\partial f_t}{\partial t} + O(\tau^2) = \frac{1}{1+r\tau} \left[f_t(z) + r\tau \left(z, \frac{\partial f_t}{\partial z} \right) \right.$$
$$\left. + \frac{1}{2}\tau^{2\alpha} \max_I \sum_{i \in I} p_i^I(z) \left(\frac{\partial^2 f_t}{\partial z^2} \phi_i(z), \phi_i(z) \right) \right] + O(\tau^{3\alpha}),$$

where

$$p_i^I(z) = \lim_{\tau \to 0} p_i^I(z,\tau)$$

(clearly well-defined nonnegative numbers). This leads to the equation

$$rf = \frac{\partial f}{\partial t} + r(z, \frac{\partial f}{\partial z}) + \frac{1}{2} \max_I \sum_{i \in I} p_i^I(z) \left(\frac{\partial^2 f}{\partial z^2} \phi_i(z), \phi_i(z) \right) \tag{15.4}$$

in the case $\alpha = 1/2$ and to the trivial first-order equation

$$rf = \frac{\partial f}{\partial t} + r\left(z, \frac{\partial f}{\partial z} \right), \tag{15.5}$$

with the obvious solution

$$f(t,z) = \mathrm{e}^{-r(T-t)} f_T(\mathrm{e}^{-r(T-t)} z), \tag{15.6}$$

in the case $\alpha > 1/2$.

Equation (15.4) is a *nonlinear extension of the classic Black–Scholes equation*. The well-posedness of the Cauchy problem for such a nonlinear parabolic equation in the class of viscosity solutions is well known in the theory of controlled diffusions, as is the fact that the solutions solve the corresponding optimal control problem; see, for example, Fleming and Soner [72].

Remark 15.1. With this well-posedness, it should not be difficult to prove the convergence of the foregoing approximations rigorously, but we have no precise reference. Moreover, one can also be interested in pathwise approximations. For this purpose a multidimensional extension of the approach by Bick and Willinger

[43] (establishing pathwise convergence of CRR binomial approximations to the trajectories underlying the standard Black–Scholes equation in a nonprobabilistic way) would be quite relevant.

In the case $J = 1$ and the classic CCR (binomial) context with

$$\sqrt{\tau}\phi_1 = (u-1)z = \sigma\sqrt{\tau}z, \quad \sqrt{\tau}\phi_2 = (d-1)z = -\sigma\sqrt{\tau}z,$$

(15.4) becomes the usual Black–Scholes equation.

More generally, if $k = J+1$, then the corresponding market in discrete time becomes complete (as noted at the end of Sect. 14.2). In this case (15.4) reduces to

$$rf = \frac{\partial f}{\partial t} + r\left(z, \frac{\partial f}{\partial z}\right) + \frac{1}{2}\sum_{i=1}^{J+1} p_i(z)\left(\frac{\partial^2 f}{\partial z^2}\phi_i(z), \phi_i(z)\right). \tag{15.7}$$

This is a *generalized Black–Scholes equation* describing a complete market (with randomness coming from J correlated Brownian motions) whenever the diffusion matrix

$$(\sigma^2)_{jk} = \sum_{i=1}^{J+1} p_i(z)\phi_i^j(z), \phi_i^k(z)$$

is nondegenerate.

15.2 An Example with Two Colors

As a more nontrivial example, let us consider the case of $J = 2$ and a submodular final payoff f_T, so that Theorem 13.4 applies to the approximations $\mathscr{B}_\tau$. Assume the simplest (and usual) symmetric form for upper and lower jumps (other terms in Taylor expansions are irrelevant for the limiting equation):

$$u_i = 1 + \sigma_i\sqrt{\tau}, \quad d_i = 1 - \sigma_i\sqrt{\tau}, \quad i = 1,2. \tag{15.8}$$

Hence

$$\frac{u_i - \rho}{u_i - d_i} = \frac{1}{2} - \frac{r}{2\sigma_i}\sqrt{\tau}, \quad i = 1,2,$$

and

$$\kappa = -\frac{1}{2}r\sqrt{\tau}\left(\frac{1}{\sigma_1} + \frac{1}{\sigma_2}\right).$$

Because $\kappa < 0$, we find ourselves in the second case of Theorem 13.4. Hence the only eligible collection of three vectors ϕ is $(d_1, u_2), (u_1, d_2), (u_1, u_2)$, and the probability law p_i^I is $(1/2, 1/2, 0)$. Therefore, (15.4) takes the form

$$rf = \frac{\partial f}{\partial t} + r\left(z, \frac{\partial f}{\partial z}\right) + \frac{1}{2}\left[\sigma_1^2 z_1^2 \frac{\partial^2 f}{\partial z_1^2} - 2\sigma_1\sigma_2 z_1 z_2 \frac{\partial^2 f}{\partial z_1 \partial z_2} + \sigma_2^2 z_2^2 \frac{\partial^2 f}{\partial z_2^2}\right]. \tag{15.9}$$

The limiting Black–Scholes type equation is again linear in this example, but with a *degenerate second-order part.* In the analogous stochastic setting, this degeneracy would mean that only one Brownian motion governs the behavior of both underlying stocks. This is not surprising in our approach, where Nature was assumed to be a single player. One could expect uncoupled second derivatives (nondegenerate diffusion) in the limit if one chose two independent players for Nature, each playing for each stock.

Thus we are still in the context of an incomplete market. The hedge price calculated from (15.9) is actually the upper price, in the terminology of Sect. 14.3. To obtain a lower price, we will use approximations of type (14.11), leading, instead of (15.4), to the equation

$$rf = \frac{\partial f}{\partial t} + r(z, \frac{\partial f}{\partial z}) + \frac{1}{2} \min_I \sum_{i \in I} p_i^I(z) \left(\frac{\partial^2 f}{\partial z^2} \phi_i(z), \phi_i(z) \right). \tag{15.10}$$

If $J = 2$ and the payoff is submodular, then the maximum can be taken over the triple (d_1, d_2), (d_1, u_2), (u_1, d_2) or $(d_1, u_2), (u_1, d_2), (u_1, u_2)$ [under (15.8) only the second triple works]. Similarly, the minimum can be taken only over the triple $(d_1, d_2), (d_1, u_2), (u_1, u_2)$ or (d_1, d_2), (u_1, d_2), (u_1, u_2). Under (15.8) both these cases give the same limit as $\tau \to 0$, yielding for the lower price the equation

$$rf = \frac{\partial f}{\partial t} + r\left(z, \frac{\partial f}{\partial z}\right) + \frac{1}{2}\left[\sigma_1^2 z_1^2 \frac{\partial^2 f}{\partial z_1^2} + 2\sigma_1\sigma_2 z_1 z_2 \frac{\partial^2 f}{\partial z_1 \partial z_2} + \sigma_2^2 z_2^2 \frac{\partial^2 f}{\partial z_2^2}\right], \tag{15.11}$$

which differs only in the sign at the mixed derivative from the equation for the upper price.

As f was assumed to be submodular, so that its mixed second derivative is negative, we have

$$\sigma_1\sigma_2 \frac{\partial^2 f}{\partial z_1 \partial z_2} \le 0 \le -\sigma_1\sigma_2 \frac{\partial^2 f}{\partial z_1 \partial z_2}.$$

Hence, for the solution f_u of the upper value Eq. (15.9), the solution f_l of the lower value Eq. (15.11), and the solution f_c of the classic Black–Scholes equation of a complete market based on two independent Brownian motions, i.e., (15.9) or (15.11) without a term with a mixed derivative (with the same submodular initial condition f_T) we have the inequality

$$f_\mathrm{l} \le f_\mathrm{c} \le f_\mathrm{u},$$

as expected.

Equations (15.9) and (15.11) can be solved explicitly via the Fourier transform, as with the standard Black–Scholes equation, that is, changing the unknown function f to g by

$$f(z_1, z_2) = \mathrm{e}^{-r(T-t)} g\left(\frac{1}{\sigma_1} \log z_1, \frac{1}{\sigma_2} \log z_2\right),$$

so that

$$\frac{\partial f}{\partial z_i} = \mathrm{e}^{-r(T-t)} \frac{1}{\sigma_i z_i} \frac{\partial g}{\partial y_i} \left(\frac{1}{\sigma_1} \log z_1, \frac{1}{\sigma_2} \log z_2 \right)$$

transforms these equations into the equations

$$\frac{\partial g}{\partial t} + \frac{1}{2}(2r - \sigma_1)\frac{\partial g}{\partial y_1} + \frac{1}{2}(2r - \sigma_2)\frac{\partial g}{\partial y_2} + \frac{1}{2}\left[\frac{\partial^2 g}{\partial y_1^2} \mp 2\frac{\partial^2 g}{\partial y_1 \partial y_2} + \frac{\partial^2 g}{\partial y_2^2}\right] = 0 \quad (15.12)$$

(with $\mp$ respectively). Equation (15.12) has constant coefficients, and the equation for the Fourier transform $\tilde{g}(p)$ of g is obviously

$$\frac{\partial \tilde{g}}{\partial t} = \frac{1}{2}[(p_1 \mp p_2)^2 - i(2r - \sigma_1)p_1 - i(2r - \sigma_2)p_2]\tilde{g}. \quad (15.13)$$

Hence the inverse Cauchy problem for (15.12) with a given final function g_T equals the convolution of g_T with the inverse Fourier transform of the functions

$$\exp\left\{-\frac{1}{2}(T-t)[(p_1 \mp p_2)^2 - i(2r - \sigma_1)p_1 - i(2r - \sigma_2)p_2]\right\},$$

which equal (after changing the integration variables p_1 and p_2 to $q_1 = p_1 - p_2$, $q_2 = p_1 + p_2$)

$$\frac{1}{2(2\pi)^2} \int_{\mathbb{R}^2} \mathrm{d}q_1 \mathrm{d}q_2 \exp\left\{-\frac{1}{2}(T-t)q_{1,2}^2 + \frac{iq_1}{2}\left(y_1 - y_2 - \frac{(\sigma_1 - \sigma_2)(T-t)}{2}\right)\right.$$
$$\left. + \frac{iq_2}{2}\left(y_1 + y_2 + (2r - \frac{\sigma_1 + \sigma_2}{2})(T-t)\right)\right\}$$

(with $q_{1,2}$ corresponding to $\mp$), or, explicitly,

$$\frac{1}{2}\frac{1}{\sqrt{2\pi(T-t)}}\delta\left(\frac{y_1 + y_2}{2} + (r - \frac{\sigma_1 + \sigma_2}{4})(T-t)\right)$$
$$\times \exp\left\{-\frac{1}{8(T-t)}\left(y_1 - y_2 - \frac{(\sigma_1 - \sigma_2)(T-t)}{2}\right)^2\right\}$$

and

$$\frac{1}{2}\frac{1}{\sqrt{2\pi(T-t)}}\delta\left(\frac{y_1 - y_2}{2} - \frac{(\sigma_1 - \sigma_2)(T-t)}{4}\right)$$
$$\times \exp\left\{-\frac{1}{8(T-t)}\left(y_1 + y_2 + (2r - \frac{\sigma_1 + \sigma_2}{2})(T-t)\right)^2\right\}$$

respectively, where δ denotes the Dirac δ-function. Returning to (15.9) and (15.11) we conclude that the solutions f_{u} and f_{l} respectively of the inverse-time Cauchy problem for these equations are given by the formula

$$f_{\mathrm{u,l}}(t,z_1,z_2) = \int_0^\infty \int_0^\infty G^{\mp}_{T-t}(z_1,z_2;w_1,w_2) f_T(w_1,w_2)\, dw_1 dw_2, \tag{15.14}$$

with the Green functions or transition probabilities being

$$\begin{aligned} &G^{-}_{T-t}(z_1,z_2;w_1,w_2)\\ &= \frac{e^{-r(T-t)}}{2\sqrt{2\pi(T-t)}\sigma_1\sigma_2 w_1 w_2}\delta\left(\frac{1}{2\sigma_1}\log\frac{z_1}{w_1} + \frac{1}{2\sigma_2}\log\frac{z_2}{w_2} + \left(r - \frac{\sigma_1+\sigma_2}{4}\right)(T-t)\right)\\ &\exp\left\{-\frac{1}{8(T-t)}\left(\frac{1}{\sigma_1}\log\frac{z_1}{w_1} - \frac{1}{\sigma_2}\log\frac{z_2}{w_2} - \frac{(\sigma_1-\sigma_2)(T-t)}{2}\right)^2\right\} \end{aligned} \tag{15.15}$$

and

$$\begin{aligned} &G^{+}_{T-t}(z_1,z_2;w_1,w_2)\\ &= \frac{e^{-r(T-t)}}{2\sqrt{2\pi(T-t)}\sigma_1\sigma_2 w_1 w_2}\delta\left(\frac{1}{2\sigma_1}\log\frac{z_1}{w_1} - \frac{1}{2\sigma_2}\log\frac{z_2}{w_2} - \frac{(\sigma_1-\sigma_2)(T-t)}{4}\right)\\ &\exp\left\{-\frac{1}{8(T-t)}\left(\frac{1}{\sigma_1}\log\frac{z_1}{w_1} + \frac{1}{\sigma_2}\log\frac{z_2}{w_2} + (2r - \frac{\sigma_1+\sigma_2}{2})(T-t)\right)^2\right\}, \end{aligned}$$

respectively. Of course, formulas (15.14) can be further simplified by integrating over the δ-function. Singularity, presented by this δ-function, is due to the degeneracy of the second-order part of the corresponding equations.

15.3 Transaction Costs in Continuous Time

The difficulties with transaction costs are well known in the usual stochastic analysis approach; see, for example, Soner et al. [138] and Bernhard et al. [39].

In our approach, Theorem 13.7 imposes strong restrictions for incorporating transaction costs in a continuous limit. In particular, assuming jumps of size τ^α in a period of length τ, i.e., assuming (15.1), only $\alpha = 1$ can be used for the limit $\tau \to 0$ because $\delta_n(z)$ is of order $(1+\tau^\alpha)^n$, which tends to ∞, as $\tau = T/n \to 0$, whenever $\alpha < 1$. We know that for vanishing costs, assuming $\alpha = 1$ leads to the trivial limiting Eq. (15.5), which was observed by many authors (e.g., Bernhard [37], McEneaney [112], Olsder [123]). However, with transaction costs included, the model with jumps of order τ becomes not so obvious but leads to a meaningful and manageable continuous-time limit. To see this, assume that we are in the context of Sect. 15.1 and transaction costs are specified, as in Sect. 13.5, by a function g satisfying (13.31). To write a manageable approximation, we will apply the following trick: we will count at time τm the transaction costs incurred at time $\tau(m+1)$ (the latter shift in transaction cost collection does not change, of course, the limiting process). Instead of (15.2) we then get

$$\mathscr{B}_\tau f(z) = \frac{1}{1+r\tau} \max_I \sum_{i\in I} p_i^I(z,\tau)$$
$$[f(z+\tau^\alpha \phi_i(z)) + g(\gamma(z+\tau\phi_i(z),\tau) - \gamma(z,\tau), z+\tau\phi_i(z))], \tag{15.16}$$

where $\gamma(z,\tau)$ is the optimal γ chosen in the position z. Assuming g is differentiable, expanding and keeping the main terms yields the following extension of (15.5):

$$rf = \frac{\partial f}{\partial t} + r\left(z, \frac{\partial f}{\partial z}\right) + \psi(z), \tag{15.17}$$

where

$$\psi(z) = \max_I \sum_{i\in I} p_i^I(z) \sum_{m,j=1}^{J} \frac{\partial g}{\partial \gamma^m}(\gamma(z)) \frac{\partial \gamma^m}{\partial z^j} \phi_i^j(z),$$

with $\gamma(z) = \lim_{\tau\to 0} \gamma(z,\tau)$.

This is a nonhomogeneous equation, with the corresponding homogeneous equation being (15.5). Since the (time-inverse) Cauchy problem for this homogeneous equation has the explicit solution (15.6), we can write the explicit solution for the Cauchy problem of (15.17) using the standard Duhamel principle (e.g., Kolokoltsov [101]), yielding

$$f(t,z) = \mathrm{e}^{-r(T-t)} f_T(\mathrm{e}^{-r(T-t)}z) + \int_t^T \mathrm{e}^{-r(s-t)} \psi(\mathrm{e}^{-r(s-t)}z)\, \mathrm{d}s. \tag{15.18}$$

The convergence of the approximations $\mathscr{B}_\tau^{[t/\tau]} f_T$ to this solution of (15.17) follows from the known general properties of the solutions to the HJB equations (e.g., Kolokoltsov and Maslov [105]).

Of course, one can also write down the modified Eq. (15.4) obtained by introducing the transaction costs in the same way as was done previously. This is the equation

$$rf = \frac{\partial f}{\partial t} + r\left(z, \frac{\partial f}{\partial z}\right)$$
$$+\frac{1}{2} \max_I \sum_{i\in I} p_i^I(z) \left[\left(\frac{\partial^2 f}{\partial z^2} \phi_i(z), \phi_i(z) \right) + \sum_{m,j=1}^{J} \frac{\partial g}{\partial \gamma^m}(\gamma(z)) \frac{\partial \gamma^m}{\partial z^j} \phi_i^j(z) \right]. \tag{15.19}$$

However, as already mentioned, due to the restrictions of Theorem 13.7, only the solutions to a finite-difference approximation of (15.19) (with time steps τ bounded below) represent justified hedging prices. Therefore, our model suggests natural bounds for time periods between relocations of capital, when transaction costs remain amenable and do not override, so to speak, hedging strategies. Passing to the limit $\tau \to 0$ in this model (i.e., considering continuous trading) leads not to (15.19) but to the trivial strategy of keeping all the capital in risk-free bonds. This compelled triviality is, of course, well known in the usual stochastic context (e.g., Soner et al. [138]).

15.4 Models with Stochastic Volatility

As has been emphasized, our approach is mostly meant to deal with unknown volatilities. However, we would like to show here that the standard models with stochastic volatility can be easily deduced in our setting once appropriate probabilities are specified.

Let us return, then, to the model of Sect. 14.6, with the reduced Bellman operator being given by (14.21). Of course, to pass to the continuous-time limit, we must make assumptions about the jumps that are analogous to those of Sect. 15.1. That is, we scale the systematic drift rz and Φ by τ and the jumps ψ and ϕ by $\sqrt{\tau}$, changing the foregoing operator to

$$\begin{aligned}(\mathscr{B}_\tau f)(z,v) = {} & \frac{p_v}{2(1+r\tau)}\left[f(z+\tau rz, v+\tau\Phi(v)+\sqrt{\tau}\phi(v))\right.\\ & \left.+f(z+\tau rz, v+\tau\Phi(v)-\sqrt{\tau}\phi(v))\right]\\ & +\frac{p_s}{2(1+r\tau)}\left[f(z+\tau rz+\sqrt{\tau}\psi(v,z), v+\tau\Phi(v))\right.\\ & \left.+f(z+\tau rz-\sqrt{\tau}\psi(v,z), v+\tau\Phi(v))\right]\\ & +\frac{p_{sv}}{2(1+r\tau)}\left[f(z+\tau rz+\sqrt{\tau}\psi(v,z), v+\tau\Phi(v)+\sqrt{\tau}\phi(v))\right.\\ & \left.+f(z+\tau rz-\sqrt{\tau}\psi(v,z), v+\tau\Phi(v)-\sqrt{\tau}\phi(v))\right].\end{aligned}$$

Following the same procedure as at the beginning of Sect. 15.1 we obtain

$$\begin{aligned}f_t-\tau\frac{\partial f_t}{\partial t}+O(\tau^2) = {} & \frac{1}{1+r\tau}\left[f_t(z,v)++\tau rz\frac{\partial f}{\partial z}+\tau\Phi\frac{\partial f}{\partial v}\right.\\ & \left.+\frac{1}{2}p_s\tau r^2z^2\frac{\partial^2 f}{\partial z^2}+\frac{1}{2}p_v\tau\phi^2\frac{\partial^2 f}{\partial v^2}+\frac{1}{2}\tau p_{vs}\left(r^2z^2\frac{\partial^2 f}{\partial z^2}+\phi^2\frac{\partial^2 f}{\partial v^2}+2rz\phi\frac{\partial^2 f}{\partial z\partial v}\right)\right],\end{aligned}$$

yielding the partial differential equation of the *Black–Scholes type* for our *stochastic volatility model*

$$rf = \frac{\partial f}{\partial t} + rz\frac{\partial f}{\partial z} + \Phi(v)\frac{\partial f}{\partial v}$$
$$+\frac{1}{2}[p_s + p_{vs}]r^2z^2(z,v)\frac{\partial^2 f}{\partial z^2} + \frac{1}{2}[p_v + p_{vs}]\phi^2(v)\frac{\partial^2 f}{\partial v^2} + p_{vs}rz\phi(v)\frac{\partial^2 f}{\partial z\partial v}. \tag{15.20}$$

For instance, if

$$\Phi(v) = a - bv, \quad \psi(z,v) = z\sqrt{v}, \quad \phi(z,v) = \sigma\sqrt{v},$$

with certain positive constants a, b, σ, then (15.20) corresponds to the Heston stochastic volatility model.

15.5 Fractional Dynamics

So far we have analyzed models where jumps (from a given set) occur with regular frequency. However, it is natural to allow the periods between jumps to be more flexible. One can also have in mind an alternative picture of the model: instead of instantaneous jumps at fixed periods, one could think about waiting times for the distance from a previous price to reach certain levels. It is clear that these periods need not be constant. In the absence of a detailed model, it is natural to take these waiting times as random variables. In the simplest model, they can be independent and identically distributed (i.i.d.). Their intensity represents a kind of stochastic volatility. Slowing down the waiting periods is, in some sense, equivalent to decreasing the average jump size per period.

For simplicity let us deal with two-color options and submodular payoffs, so that Theorem 13.4 applies, yielding a unique eligible risk-neutral measure. Hence the changes in prices (for an investor choosing the optimal γ) follow the Markov chain $X_n^\tau(z)$ described by the recursive equation

$$X_{n+1}^\tau(z) = X_n^\tau(z) + \sqrt{\tau}\phi(X_n^\tau(z)), \quad X_0^\tau(z) = z,$$

where $\phi(z)$ is one of three points $(z^1d_1, z^2u_2), (z^1u_1, z^2d_2), (z^1u_1, z^2u_2)$ that are chosen with the corresponding risk-neutral probabilities. As was shown previously, this Markov chain converges, as $\tau \to 0$ and $n = [t/\tau]$ (where $[s]$ denotes the integer part of a real number s), to the diffusion process X_t solving the Black–Scholes type (degenerate) Eq. (15.9), i.e., a sub-Markov process with the generator

$$Lf(x) = -rf + r(z, \frac{\partial f}{\partial z}) + \frac{1}{2}\left[\sigma_1^2 z_1^2\frac{\partial^2 f}{\partial z_1^2} - 2\sigma_1\sigma_2 z_1 z_2\frac{\partial^2 f}{\partial z_1 \partial z_2} + \sigma_2^2 z_2^2\frac{\partial^2 f}{\partial z_2^2}\right]. \tag{15.21}$$

Assume that the times between jumps $T_1, T_2, \ldots$ are i.i.d. random variables with a power-law decay, that is,

$$\mathbf{P}(T_i \geq t) \sim \frac{1}{\beta t^{\beta}},$$

with $\beta \in (0,1)$ (where $\mathbf{P}$ denotes probability and $\sim$ means, as usual, that the ratio between the left-hand side and the right-hand side tends to one as $t \to \infty$). It is well known that such T_i belong to the domain of attraction of the β-stable law (e.g., Uchaikin and Zolotarev [144]), meaning that the normalized sums

$$\Theta_t^{\tau} = \tau^{1/\beta}(T_1 + \cdots + T_{[t/\tau]})$$

(where $[s]$ denotes the integer part of a real number s) converge, as $\tau \to 0$, to a β-stable Lévy motion Θ_t, which is a Lévy process on $\mathbb{R}_+$ with the fractional derivative of order β as the generator:

$$Af(t) = -\frac{\mathrm{d}^{\beta}}{\mathrm{d}(-t)^{\beta}} f(t) = -\frac{1}{\Gamma(-\beta)} \int_0^{\infty} (f(t+r) - f(t)) \frac{\mathrm{d}r}{r^{1+\beta}}.$$

We are now interested in the process $Y^{\tau}(z)$ obtained from $X_n^{\tau}(z)$ by changing the constant times between jumps to scaled random times T_i, so that

$$N_t^{\tau} = \max\{u : \Theta_u^{\tau} \leq t\}, \quad Y_t^{\tau}(z) = X_{N_t^{\tau}}^{\tau}(z).$$

The limiting process

$$N_t = \max\{u : \Theta_u \leq t\}$$

is therefore the inverse (or hitting time) process of the β-stable Lévy motion Θ_t.

By Theorem 4.2 and 5.1 of Kolokoltsov [99] (also Chap. 8 in [101]), we obtain the following result.

Theorem 15.2. *The process Y_t^{τ} converges (in the sense of distribution on paths) to the process $Y_t = X_{N_t}$, whose averages $f(T-t,x) = \mathbf{E}f(Y_{T-t}(x))$, for continuous bounded f, have the explicit integral representation*

$$f(T-t,x) = \int_0^{\infty} \int_0^{\infty} \int_0^{\infty} G_u^{-}(z_1, z_2; w_1, w_2) Q(T-t, u) \, \mathrm{d}u \mathrm{d}w_1 \mathrm{d}w_2,$$

where G^{-}, the transition probabilities of X_t, are defined by (15.15)*, and where $Q(t,u)$ denotes the probability density of the process N_t.*

Moreover, for $f \in C_{\infty}^2(\mathbb{R}^d)$, $f(t,x)$ satisfy the (generalized) fractional evolution equation (of the Black–Scholes type)

$$\frac{\mathrm{d}^{\beta}}{\mathrm{d}t^{\beta}} f(t,x) = Lf(t,x) + \frac{t^{-\beta}}{\Gamma(1-\beta)} f(x).$$

Remark 15.3. Results similar to Theorem 4.2 of Kolokoltsov [99] used previously, but with position-independent random walks, i.e., when L is the generator of a Lévy process, were obtained in Meerschaert and Scheffler [115]; see also related results in Kolokoltsov et al. [103], Henry et al. [82], and references therein. Rather general fractional Cauchy problems are discussed in Kochubei [94]. For related questions of fractional random fields see Kelbert and Leonenko [93].

A similar procedure with a general nonlinear Black–Scholes type equation (15.4) will lead, of course, to its similar fractional extension. However, a rigorous analysis of the corresponding limiting procedure is beyond the scope of the present contribution.

Chapter 16
Credit Derivatives

16.1 Basic Model with No Simultaneous Jumps

Consider a market of N securities that can default in discrete time $\delta, 2\delta, \ldots$. Let N_t denote the number of securities defaulted up to time t, and set $d_t = N - N_t$ for the number of not-yet-defaulted ones.

We will work mostly with the *instantaneous digital credit default swaps (CDSs)*. Entering an instantaneous digital CDS agreement on jth (not yet defaulted) security at time t means agreeing to pay a (protection) premium $\alpha_j(t)\delta$ and to receive back the compensation of one unit of money if j defaults during the period $(t, t+\delta]$ (and nothing otherwise). The premium $\alpha_j(t)\delta$ is chosen in such a way that there is no charge at inception at time t to enter this contract [alternatively, of course, $\alpha_j(t)\delta$ can be considered as a charge for receiving one in the case of the jth default]. Working with instantaneous digital CDSs rather than actually traded CDSs means choosing a convenient basis and is a well-accepted approach in the finance literature (e.g., Cousin et al. [55]). The infinitesimal premium $\alpha_j(t)$ is usually assumed to depend on the whole history of defaults of our basic securities until and including time t.

We start in this section with the standard (in the literature) simplifying assumption that only one default can occur in any given short period of time $(t, t+\delta]$. Thus we assume that, for any time t, the possible $(d_t + 1)$ outcomes at time $t+\delta$ are either no default or a default of only one of d_t live (not yet defaulted) securities. These outcomes can be described symbolically by $d_t + 1$ vectors in $\mathbb{R}^{d_t}$: zero vector e_0 and d_t basis vectors e_i (with the ith coordinate 1 and other coordinates vanishing), so that any short-term contingent claim starting at t for the period $(t, t+\delta]$ can be described by a function f on $\{e_0, e_1, \ldots, e_{d_t}\}$. Suppose that, to replicate any such claim, an investor, with an amount of capital X at time t, is allowed to enter an arbitrary amount γ^i of an instantaneous digital CDS agreement on the ith security, $i = 1, \ldots, d_t$ (recall that entering such an agreement is costless). Then his capital at time $t+\delta$ in case of event e_i becomes

P. Bernhard et al., *The Interval Market Model in Mathematical Finance*, Static & Dynamic Game Theory: Foundations & Applications, DOI 10.1007/978-0-8176-8388-7_16,

$$-\delta \sum_{j=1}^{d_t} \gamma^j \alpha_j(t) + f_i + \gamma^i + X,$$

where $f_i = f(e_i)$ and it assumed that $\gamma^0 = 0$ (to make this formula valid for e_0). In other words, his capital equals

$$f_i - (\gamma, \eta_i) + X,$$

where $\eta_0 = \delta(\alpha_1(t), \ldots, \alpha_{d_t}(t))$ and $\eta_i = \eta_0 - e_i$ for $i = 1, \ldots, d_t$. By *hedging price* we mean, as usual, the minimal value of X needed to be able to fulfill the obligation in any case, which is then

$$C_h = \min_{\gamma \in \mathbf{R}^{d_t}} \max_{i=0,1,\ldots,d_t} [f_i - (\gamma, \eta_i)]. \tag{16.1}$$

The vectors $\{\eta_i\}$ form a strictly positively complete family, so that Proposition 12.6 holds. Hence the corresponding market is complete and

$$C_h = \mathbf{E}\{f_\cdot\} = \sum_{j=0}^{d_t} p_j(t) f_j,$$

where $\{p_j\}$ are the corresponding unique risk-neutral probabilities. Moreover, in this case these probabilities are easily seen to have simple explicit form: $p_j(t) = \alpha_j(t)\delta$ for $j = 1, \ldots, d_t$, with

$$p_0(t) = 1 - \delta \sum_{j=1}^{d_t} \alpha_j(t).$$

One can look at the law $\{p_\cdot(t)\}$ as the transition probabilities from time t to time $t + \delta$ of a certain random process of defaults. It is, however, not Markovian, as $\alpha(t)$ can depend on the past. To make it Markovian, one must assume additionally that $\alpha_j(t)$ depend only on a current state of defaults. Let us consider three natural Markovian settings.

1. The simplest homogeneous case. The premiums $\alpha_j(t)$ are the same for all j and depend on the past only via d_t, that is, they are specified by some deterministic functions $\alpha(t,d)$. In this case a state of our process is just a number from $\{0, 1, \ldots, N\}$ (the number of not defaulted securities), and the corresponding risk-neutral transition probabilities become

$$p_{t,t+\delta}(m, m-1) = m\alpha(t,m)\delta, \quad p_{t,t+\delta}(m,m) = 1 - m\alpha(t,m)\delta, \tag{16.2}$$

with all other probabilities vanishing.

2. The general case. The state space is the set of all subsets I of $\{1,\ldots,k\}$, and premiums at time t depend on the past only via the present state $I(t)$, that is, they are given by functions $\alpha_j(t,I)$.
   The corresponding risk-neutral transition probabilities become

$$p_{t,t+\delta}(I,I\setminus j)=\alpha_j(t,I)\delta,\quad p_{t,t+\delta}(I,I)=1-\sum_{j\in I}\alpha_j(t,I)\delta. \tag{16.3}$$

3. The intermediate case. The whole set of securities is decomposed into a finite number k of classes (say, by a rating agency), the infinitesimal premiums are the same for securities from the same class and depend on the number of live securities in each class. The states are now vectors $m=(m^1,\ldots,m^k)$ (with nonnegative integers as coordinates), with the jth coordinate denoting the number of live securities in class j.

The corresponding risk-neutral transition probabilities become

$$p_{t,t+\delta}(m,m-e_i)=\alpha_i(t,m)\delta$$

for any m with $m^i>0$, and

$$p_{t,t+\delta}(m,m)=1-\delta\sum_{i:m^i\neq 0}\alpha_i(t,m),$$

with all other transitions being forbidden.

As $\delta\to 0$, these models have natural limiting Markov processes in continuous time:

1′. A Markov chain on $\{0,1,\ldots,N\}$ with transition rates

$$q_t(m,m-1)=m\alpha(t,m),$$

so that

$$\left.\frac{\mathrm{d}}{\mathrm{d}s}\right|_{s=0}\mathbf{E}(f(k_{t+s})|k_t)=(f(k_t-1)-f(k_t))k_t\alpha(t,k_t);$$

2′. A Markov chain on subsets of $\{0,1,\ldots,N\}$ with transition rates

$$q_t(I,I\setminus j)=\alpha_j(t,I);$$

3′. A Markov chain on the set of vectors with k nonnegative integer coordinates with transition rates

$$q_t(m,m-e_i)=\alpha_i(t,m),$$

so that

$$\left.\frac{\mathrm{d}}{\mathrm{d}s}\right|_{s=0}\mathbf{E}(f(m_{t+s})|m_t)=\sum_{i:m_t^i\neq 0}(f(m_t-e_i)-f(m_t))m_t^i\alpha_i(t,m_t).$$

Our exposition here was similar to that of Frey and Backhaus [75], who described the same Markov models. We have only stressed their game-theoretic origins.

16.2 Simultaneous Jumps: Completion by Tranching

Allowing for simultaneous defaults makes the preceding models incomplete. They can then be dealt with by the general techniques of Chaps. 12 and 13. On the other hand, these models can be completed again by allowing trading in *instantaneous digital CDSs*, paying compensation in the event of *simultaneous defaults*.

For definiteness, let us assume that only two simultaneous defaults can occur during each period $(t,t+\delta]$. Assume also that one can enter an instantaneous digital CDS on these events, that is, for any pair $i \neq j$, one can agree to pay a (protection) premium $\alpha_{ij}(t)\delta$ at time t and to receive back the compensation of one unit of money in case both i and j default during the period $(t,t+\delta]$ (and nothing otherwise). For convenience let us set $\alpha_{ii} = \alpha_i$, where α_i are the premium rates for the usual one-name-based CDS discussed in the previous section.

We will denote by M^d the space of symmetric $d \times d$ matrices equipped with the scalar product

$$\gamma\eta = \sum_{i=1}^{d} \gamma^{ii}\eta^{ii} + \sum_{i>j}^{d} \gamma^{ij}\eta^{ij}.$$

By e_{ij} we denote a matrix with one at the intersection of the ith column and jth row and at the intersection of the jth column and ith row and with other elements vanishing.

Arguing now as in the previous section we can conclude that for a contingent claim paying f_{ij}, $i \neq j$, in the case of simultaneous default of i and j, paying f_i i defaults, and paying f_0 in the case of no default during the period $(t,t+\delta]$ the hedge price at time t becomes

$$C_h = \min_{\gamma \in M^{d_t}} \max_{\eta \in \{\eta_0, \eta_{ij}\}} [f(\eta) - (\gamma, \eta)], \tag{16.4}$$

where $\eta_0 \in M^{d_t}$ has elements $\delta\alpha_{ij}$, $\eta_{ii} = \eta_0 - e_{ii}$, and $\eta_{ij} = \eta_0 - e_{ij} - e_{ii} - e_{ij}$ for $i \neq j$. We have $1 + d_t(d_t+1)/2$ vectors η in $d_t(d_t+1)/2$-dimensional space. These vectors are strongly positively complete, so that Proposition 12.12 applies, specifying unique risk-neutral probabilities (complete market setting!) that can be looked at as transition probabilities for a Markov chain, just as in the previous section. Similar limiting continuous-time Markov chains can be identified.

16.3 Mean-Field Limit, Fluctuations, and Stochastic Law of Large Numbers

In this section we touch briefly on the *dynamic law of large numbers (LLN) limit* for Markov chains arising from the evaluation of CDSs. These chains are, after all, just certain death processes. The simplest LLN limit, the mean-field limit, is well studied for them. Thus, applying the results of Sect. 5.11 from Kolokoltsov [101] (see alternatively Benaïm and Le Boudec [20] or Darling and Norris [64]) we can conclude that this limit for the Markov model (3′) of Sect. 16.1 is a deterministic process on $\mathbb{R}_+^N$ with the evolution described by the ordinary differential equation

$$\dot{\omega}^j = -\omega_j \alpha_j(t, \omega_1, \ldots, \omega_k). \tag{16.5}$$

Similarly, the results of Kolokoltsov [100, 101] allows one to specify the limiting Gaussian process for the fluctuations of this Markov model (3′) around its mean-field limit (16.5).

The deterministic mean-field limit is definitely not the only possible dynamic LLN for processes of type 3′. Stochastic LLNs are natural sometimes. The most popular limits of this kind are super processes that come from pure branching (no interaction) processes. Similar limits that include the interaction of a rather general type (including processes of type 3′ above) are described in Kolokoltsov [98, 101]. They are given by Markov processes on $\mathbb{R}_+^N$ with pseudodifferential generators of the Lévy–Khintchine type.

Part V
Viability Approach to Complex Option Pricing and Portfolio Insurance

Authors:
Patrick Saint-Pierre
Université Paris-Dauphine, France

Jean-Pierre Aubin
Société VIMADES (Viabilité, Marchés, Automatique et Décision), France

In this part, we use the tools of viability theory, and more specifically the Guaranteed Capture Basin Algorithm, to solve several problems in mathematical finance: option pricing for a large class of complex options and, in the second chapter, portfolio insurance.

The first chapter, authored by Patrick Saint-Pierre, summarizes results obtained with the collaboration of Jean-Pierre Aubin and Dominique Pujal from LEDA-SDFi, Université Paris-Dauphine. See [17]. The theory of hedging under the NGARCH model of Sect. 17.4.4 was developed with the collaboration of Michèle Breton from HEC Montreal and GERAD, Quebec.

The second chapter, authored by Jean-Pierre Aubin, results from a joint work with Luxi Chen and Olivier Dordan. It was partially supported by the Commission of the European Communities under the 7th Framework Programme Marie Curie Initial Training Network (FP7-PEOPLE-2010-ITN), Project SADCO, Contract 264735.

The authors thank Pierre Bernhard, Pierre Cardaliaguet, Marc Quincampoix, and Georges Haddad for their contributions to differential game theory; Giuseppe Da Prato, Halim Doss, Hélène Frankowska, and Jerzy Zabczyk for their work with stochastic and tychastic viability; Daniel Gabay, Nadia Lericolais, Dominique Pujal, and Francine Roure for their initiation to finance; Marie-Hélène Durand regarding uncertainty; and Sophie Martin concerning resilience and other tychastic indicators.

Notation

- T: Exercise time
- $t \in [0,T]$: Current time
- ρ: Time step of discrete-time theory
- $t_n = n\rho$: Current time at step n
- $N = T/\rho$: Total number of steps
- t^n: Time of nth impulse
- $\psi(x)$: Reachable states by an impulse from x
- K: Exercise price
- U: Terminal payment
- δ: Transaction cost ratio
- S_0: Riskless bond price
- r_0: Riskless return rate
- p_0: Number of riskless bonds in portfolio
- S: Risky asset price
- S^n: Risky asset price at t_n
- r: Risky asset return
- $r^\flat, r^\sharp$: Minimum and maximum risky asset return rate
- p: Number of risky shares in portfolio
- $E = pS$: Portfolio exposure
- $E^\heartsuit$: Optimal hedging strategy
- W: Portfolio worth
- W^n: Portfolio worth at t_n.
- $W^\heartsuit$: Optimal portfolio worth

Chapter 17
Computational Methods Based on the Guaranteed Capture Basin Algorithm

17.1 Guaranteed Capture Basin Method for Evaluating Portfolios

17.1.1 Classical Option Evaluation

The fundamental problem arising in the framework of dynamic replicating portfolios is to determine a hedging strategy $p(\cdot)$ such that, whatever the uncertain evolution of the underlying asset price $S(\cdot)$ is, a payoff is realized at the exercise time or at any time before, depending on the type of option. This can be formalized in terms of viability and target capturability in the presence of uncertainty, guaranteeing the capture of a target. This is the main issue of viability theory. Therefore, one can deduce from the mathematical and geometrical properties of capture basins the optimal rules for managing complex financial instruments. Furthermore, once discretized in this natural formulation, the Guaranteed Capture Basin Algorithm provides the valuation of an optimal portfolio and its management under different representations of uncertainty. We refer the reader to Chap. 18 by Jean-Pierre Aubin, Luxi Chen, and Olivier Dordan.

A put or call is an agreement conferring the right to sell (put) or buy (call) a quantity of an asset at a given date (European put or call) or at any date before a fixed date T (American put or call). We aim at determining the value of the agreement at the start. This value is the price the seller should ask for to protect herself against risk. It measures the cost of risk covering. Facing the risks inherent in her position, the seller builds up a theoretical portfolio by investing in the underlying asset through self-financing. This permanently adjusted portfolio yields the same losses and profits as the put or call. It is said to replicate the put or call.

For European "vanilla" options we know that the Guaranteed Capture Basin Algorithm provides both approximated values of the option depending on the refinement level of the discretization, on the maturity and price of the underlying asset, and on the option hedging strategy. In cases where uncertainty is characterized

P. Bernhard et al., *The Interval Market Model in Mathematical Finance*, Static & Dynamic Game Theory: Foundations & Applications, DOI 10.1007/978-0-8176-8388-7_17,

by a geometric Brownian motion leading to the Black–Scholes option pricing formula and considering approximation of this Brownian motion by random walk following the Cox, Ross, and Rubinstein (CRR) discrete approach, the Guaranteed Capture Basin Algorithm returns classic valuation results but also provides valuation and hedging strategies when considering other types of uncertainty such as tychastic uncertainty.

17.1.2 Limits of Classic Evaluation Methods

Taking into account transaction costs in the framework of the Black–Scholes approach becomes intricate [23,126]. The Guaranteed Capture Basin Method can be applied to handle any constraints on asset prices or share quantities or any financial restrictions and in the presence of transaction costs. In addition, when studying barrier options or Bermuda options, the extension of the capture basin method to impulse dynamics authorizes the evaluation of such options where discontinuity appears.

After all, since numerical processes necessarily require discretization of both space and time variables, uncertainty representation must be matched to the discrete models being considered [147].

However, since, due to the complexity of models, a more analytic formula cannot be developed, unlike the Black–Scholes formula in the classic case, the "cost to pay" will come from computation limitations.

17.2 The Dynamic System Underlying Financial Instruments

The viability/capturability algorithm constitutes a helpful tool to evaluate and manage a portfolio. The usual framework of stochastic control is superseded by the dynamic game approach, where the notion of guaranteed viable-capture basin plays a prior role in selecting the appropriate portfolio in due time. In this section we introduce the mathematical concepts that will be needed to embed the option valuation problem in the framework of viability theory, allowing an implementation of the Guaranteed Capture Basin Algorithm that extends the Viability Kernel Algorithm. We define the state and control variables, the constraint set and the target to be reached, the uncertainty parameterization, and the dynamic system governing the evolution of all variables. Even if we just need to consider discrete (in time) dynamic systems for numerical issues, we present both continuous and discrete formulations of the dynamics.

17.2.1 State and Control Variables

Let S_0 and p_0 denote the price and the quantity of a riskless asset and S and p the price and the quantity of a risky asset, which constitute the portfolio. Let $W := p_0S_0 + pS$ be the value of the portfolio. Time t evolves in the interval $[0,T]$, where T is the maturity time. We denote by $\tau := T - t$ the time left before maturity. In the time discretization process we will introduce the number N of time steps $\rho = \frac{T}{N}$.

17.2.2 Viability Constraints and Target

Constraints and targets are formalized by a couple of functions, $(\mathbf{b},\mathbf{c})$, $\mathbf{b} \le \mathbf{c}$, where

- $\forall (t,S) \in \mathbb{R}_+ \times \mathbb{R}^2_+$, $\mathbf{b}(t,S)$ describes the financial constraints;
- $\forall (t,S) \in \mathbb{R}_+ \times \mathbb{R}^2_+$, $\mathbf{c}(t,S)$ describes the objective to reach at maturity time, which serves as the target.

For instance, the intrinsic value of an option is associated with the payoff function $U(S) = (S-K)^+$ for a call or $U(S) = (K-S)^+$ for a put. In that case we express the two functions $\mathbf{b}$ and $\mathbf{c}$ in terms of U:

$$\mathbf{b} : (t,S) \mapsto \mathbf{b}(t,S) = \begin{cases} U(S) & \text{for American options,} \\ 0 & \text{for European options,} \end{cases} \tag{17.1}$$

and

$$\mathbf{c} : (t,S) \to \mathbf{c}(t,S) = \begin{cases} U(S) & \text{if } t = 0, \\ +\infty & \text{otherwise.} \end{cases} \tag{17.2}$$

Managing a portfolio requires that the constraint $\mathbf{b}(t,S(t)) \le W(t)$, or $\mathbf{b}(n\rho,S^n) \le W^n$ in the discrete case, be satisfied.

The condition characterizing the target reads $\mathbf{c}(t,S) \le W$, or $\mathbf{c}(n\rho,S^n) \le W^n$ in the discrete case.

One already notices that maps $\mathbf{b}$ and $\mathbf{c}$ are only lower semicontinuous since their epigraphs are closed. Sometimes their values may coincide, but their roles are different: the constraint must not be violated, while reaching the target stops the process.

We define the constraint set

$$\begin{aligned} \mathscr{K} := \{(\tau,S,W) \in \mathbb{R}^3 \text{ such that } & \mathbf{b}(T-\tau,S) \le W,\ \tau \ge 0, \\ & S^\flat(T-\tau) \le S(T-\tau) \le S^\sharp(T-\tau),\ \mathscr{P}(S) \cap [0,1] \ne \emptyset\}, \end{aligned} \tag{17.3}$$

where $\mathscr{P}(S)$ is the set of asset quantities available on the market if the price is S. The property $\mathscr{P}(S) \cap [0,1] \ne \emptyset$ means that there exists $p \in \mathscr{P}(S)$ with $p \in [0,1]$.

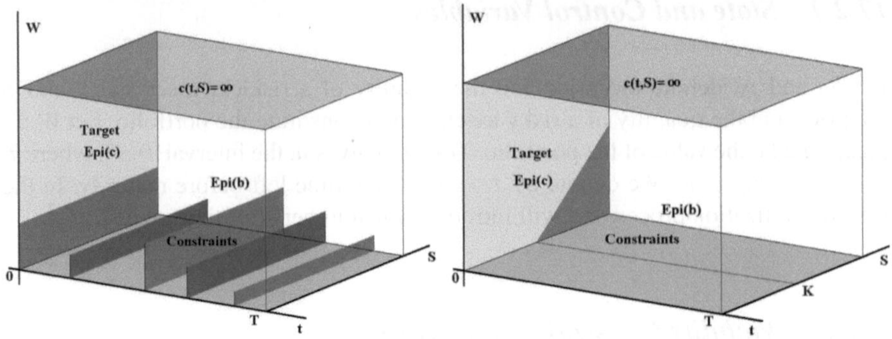

Fig. 17.1 Examples of constraints and targets for guaranteeing cash flows

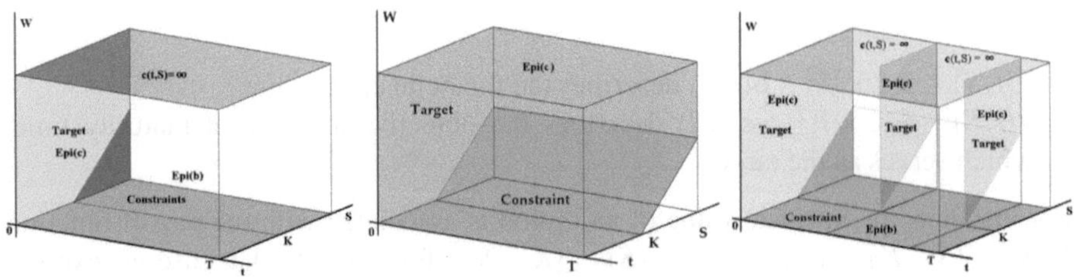

Fig. 17.2 Three examples of constraints and targets: European, American, and capped-style constraint and target specifications. In the case of capped-style options, the target is a union of targets so that the function **c** is defined as a function whose epigraph is the union of the epigraphs of functions $\mathbf{c}_i$

Indeed, the seller of an option looks to secure the best price. One can always insure against risk by buying the asset, choosing $p_0 = 0$ and $p = 1$, but better covering will be p less than 1. Also, we can restrict the set of available controls to $\mathscr{P}(S) \cap [0,1]$. For option evaluation, usually $\mathscr{P}(S) \cap [0,1]$ is set to $[0,1]$, assuming that market liquidity holds.

We define the target set

$$\mathscr{T} := \{(\tau, S, W) \in \mathbb{R}^3 \text{ such that } \mathbf{c}(T-\tau, S) \leq W\}. \qquad (17.4)$$

Figures 17.1 and 17.2 illustrate the situation where constraints are imposed, depending on the financial context. In the first example, some predetermined cash flows *could* be paid, but not necessarily, at a fixed date: the value of the financial instrument must always remain above the series of upright frames. This problem arises when aiming at guaranteeing capital. In the second example, the value of the portfolio must remain at any time greater than the payoff value, but for American options this occurs at any time $t \in [0,T]$. This amounts to choosing $\mathbf{b}(t,S) = U(S) := \max(0, S-K)$, $\forall t \in [0,T]$ instead of $\mathbf{b}(t,S) = 0$, $\forall t \in [0,T[$ and $\mathbf{b}(T,S) = \max(0, S-K)$ for European options. For Bermuda options, we must choose $\mathbf{b}(t_i,S) = \max(0, S-K_i)$, $\forall i \in I$ and $\mathbf{b}(t_i,S) = 0$, $\forall t \neq t_i$, where $(t_i)_{i \in I}$ is a finite set of dates when the option can be exercised. For capped-style options, we must choose $\mathbf{c}(t_i, S) = \max(0, S - K_i)$, $\forall i \in I$ and $\mathbf{c}(t_i, S) = +\infty$, $\forall t \neq t_i$, where $(t_i)_{i\in I}$ is a finite set of dates when the option, if the price of the underlying asset closes at or above K_i, is automatically exercised since, in that case, the option expires.

17.2.3 Uncertainty of the Environment

The velocity of the price evolution $t \to S(t)$ is, in the framework of continuous models, governed by its return $r(t) := \frac{S'(t)}{S(t)}$. Uncertainty arises from the lack of full knowledge of the return evolution, for instance, when uncertainty is described using an observable *tychastic* measure, $r(t) \in \frac{D\Sigma(t,S(t))}{S(t)}$ (see Chap. 18 by Jean-Pierre Aubin, Luxi Chen, and Olivier Dordan.) This is the core of financial modeling for continuous models covering the two main concepts of uncertainty: the stochastic approach based on statistical measurement and the tychastic approach closely related to differential game theory, games against nature, and robust control. In this chapter dealing with numerical issues, we will consider discrete-time models with returns of the form

$$r(t_n, t_{n+1}, S)) \in Q_n(S) := \frac{1}{S}\frac{\mathrm{Succ}(S, t_n, t_{n+1}) - S}{t_{n+1} - t_n},$$

where uncertainty is wrapped up in the definition of the set-valued map $(S, t^-, t^+) \rightsquigarrow \mathrm{Succ}(S, t^-, t^+)$ describing all the possible evolutions of the asset price S between dates t^- and t^+. Uncertainty is parameterized by a variable υ belonging to a set Q or Q_ρ generating the set $\mathrm{Succ}(S, t^-, t^+)$. For instance, in the CRR discrete binomial model that approximates the stochastic uncertainty appearing in Wiener processes, $\mathrm{Succ}(S, t^-, t^+) = \{(1 + r_{\mathrm{d}}(t^+ - t^-))S, (1 + r_u(t^+ - t^-))S\}$, where r_{d} and r_u are the "up" and "down" return rates. In this case, $\mathrm{Succ}(S, t^-, t^+) = \{(1 + \upsilon(t^+ - t^-))S,\ \upsilon \in \mathcal{Q} := \{r_{\mathrm{d}}, r_u\}\}$, where $\{r_{\mathrm{d}}, r_u\} = \{\mathrm{e}^{-\sigma\sqrt{\rho}} - 1, \mathrm{e}^{+\sigma\sqrt{\rho}} - 1\}$, $\rho = t^+ - t^-$, denotes the time step and σ the volatility.

Without loss of generality, we will assume that the returns r_0, r_{d}, and r_u are independent of S_0.

17.2.4 Differential and Discrete Games Describing Portfolio Evolution

Riskless asset and risky asset are governed by differential or discrete equations

$$\begin{cases} S_0'(t) &= r_0 S_0(t), \\ S'(t) &= r(t)S(t), \end{cases} \quad \text{or} \quad \begin{cases} S_0^{n+1} = S_0^n(1 + \rho r_0), \\ S^{n+1} = S^n(1 + r_\rho^n). \end{cases}$$

The self-financing principle of the portfolio reads

$$\begin{aligned}&\forall t \geq 0,\ p_0'(t)S_0(t) + p'(t)S(t) = 0\\ &\text{or}\\ &\forall n \in \{0,\ldots,N\},\ (p_0^{n+1} - p_0^n)S_0^n + (p^{n+1} - p^n)S^n = 0,\end{aligned}$$

so that the value of the portfolio satisfies

$$\begin{aligned}&W'(t) = p_0(t)S_0(t)r_0 + p(t)S(t)r(t)\\ &\text{or}\\ &W^{n+1} = W^n + \rho\, p_0^n S_0^n r_0 + p^n S^n r_\rho^n.\end{aligned}$$

We can definitely choose $S_0(0) = 1$ and $r_0(S_0) = r_0$ so that the evolution of S_0^n is completely determined and the value of the portfolio becomes

$$\begin{aligned}&W'(t) = W(t)r_0(S(t)) - p(t)S(t)(r_0 - r(t))\\ &\text{or}\\ &W^{n+1} = (1+\rho r_0)W^n - p^n S^n(\rho r_0 - r_\rho^n).\end{aligned}$$

To summarize, the time $\tau(t)$ left before maturity, the asset price $S(t)$, and the capital $W(t)$ are solutions to the dynamic or discrete system

$$\begin{cases}\tau'(t) &= -1,\\ S'(t) &= S(t)r(t),\\ W'(t) &= W(t)r_0 - p(t)S(t)(r_0 - r(t)),\end{cases}$$

or

$$\begin{cases}\tau^{n+1} &= \tau^n - \rho,\\ S^{n+1} &= S^n(1 + r_\rho^n),\\ W^{n+1} &= (1+\rho r_0)W^n - p^n S^n(\rho r_0 - r_\rho^n),\end{cases} \tag{17.5}$$

where $p(t) \in \mathscr{P}(S(t))$ or $p^n \in \mathscr{P}(S^n)$ and $r(t) \in \mathscr{Q}(t,S(t))$ or $r_\rho^n \in \mathscr{Q}_\rho^n$.

Our aim is to determine the set of initial conditions $(S(0),W(0)) \in \mathbb{R}_+^2$ or $(S^0,W^0) \in \mathbb{R}_+^2$ for which there exists a map $(\tau,S,W) \to \widehat{p}(\tau,S,W)$ or a sequence of maps $((S,W) \to \widehat{p}^n(S,W))_{n=0,\ldots,N-1}$ such that, whatever the realization $\upsilon(\cdot)$ or sequence $(\upsilon^n)_{n=0,\ldots,N-1}$ of uncertainty, the evolution $t \to (\tau(t),S(t),W(t))$ or $(\tau^n,S^n,W^n)_{n=0,\ldots,N}$ solution to

$$\begin{cases}\tau'(t) &= -1,\\ S'(t) &= S(t)r(t),\\ W'(t) &= W(t)r_0 - \widehat{p}(\tau(t),S(t),W(t))S(t)(r_0 - r(t))\end{cases}$$

or

$$\begin{cases}\tau^{n+1} &= \tau^n - \rho,\\ S^{n+1} &= S^n(1 + r_\rho^n),\\ W^{n+1} &= (1+\rho r_0)W^n - \widehat{p}^n(S^n,W^n)S^n(\rho r_0 - r_\rho^n)\end{cases} \tag{17.6}$$

remains in the constraint set before reaching the target.

We refer the reader to [7, 134] for the convergence theorems of approximated solutions to (17.5) when $\rho \to 0$ and to [13] for the stochastic case.

Let us denote $x := (\tau, S, W)$, $u = p$, $v = r$ and define the maps

$$\begin{aligned} &\Phi(x,u,v) := (-1, Sv, Wr_0 - pS(r_0 - v)) \\ &\quad \text{and} \\ &\Phi_\rho(x,u,v) := (\tau - \rho, S(1+v), (1+\rho r_0)W - p(\rho r_0 - v)), \end{aligned} \tag{17.7}$$

which are the right-hand sides of (17.5).

An evolutionary game, continuous or discrete, is defined by a retroaction (feedback) map $x \mapsto \mathscr{U}(x)$, a perturbation map $x \mapsto \mathscr{Q}(x)$, and a dynamic system

$$\begin{cases} x'(t) \in \Phi(x(t), u(t), v(t)), \\ u(t) \in \mathscr{U}(x(t)), \\ v(t) \in \mathscr{Q}(x(t)), \end{cases} \quad \text{or} \quad \begin{cases} x^{n+1} \in \Phi_\rho(x^n, u^n, v^n), \\ u^n \in \mathscr{U}(x^n), \\ v^n \in \mathscr{Q}_\rho^n. \end{cases}$$

Definition 17.1. A guaranteed capture domain of $\mathscr{T}$ in $\mathscr{K}$ for the evolutionary game $(\Phi, \mathscr{U}, \mathscr{Q})$ (resp. $(\Phi_\rho, \mathscr{U}, \mathscr{Q}_\rho)$) is a subset $\mathscr{D}$ of elements $x_0 \in \mathscr{K}$ (resp. $x_0 \in \mathscr{K}$) such that there exists a feedback $\widetilde{u} : x \to \widetilde{u}(x) \in \mathscr{U}(x)$ (resp. $(\widetilde{u}^n)_n$) such that all the solutions to $x'(t) \in \Phi(x(t), \widetilde{u}(x(t)), \mathscr{Q}(x(t)))$ (resp. $x^{n+1} \in \Phi_\rho(x^n, \widetilde{u}^n, Q_\rho(x^n))$ reach the target $\mathscr{T}$ before leaving $\mathscr{K}$. The guaranteed capture basin of $\mathscr{T}$ in $\mathscr{K}$ is the largest guaranteed capture domain of $\mathscr{T}$ in $\mathscr{K}$ contained in $\mathscr{K}$. It is denoted by $\mathrm{Capt}(\mathscr{K}, \mathscr{T}, (\Phi, \mathscr{U}, \mathscr{Q}))$ (resp. $\mathrm{Capt}(\mathscr{K}, \mathscr{T}, (\Phi_\rho, \mathscr{U}, \mathscr{Q}_\rho)))$.

Obviously we are interested in initial positions in the guaranteed capture basin with $t = 0$ ($n = 0$). We refer to [52] for more details about Differential Games and discriminating kernels, and to [17, 128] about evolutionary games.

17.2.5 Guaranteed Capture Basin Algorithm

To compute approximations of the guaranteed capture basin we introduce an extension of the Viability Kernel Algorithm to the case of dynamic games. It consists in the construction of a decreasing sequence of subsets $\mathscr{K}_\rho^i$ defined by

$$\begin{aligned} &\mathscr{K}_\rho^0 := \mathscr{K} \cup \mathscr{T}, \\ &\mathscr{K}_\rho^{i+1} := \{x \in \mathscr{K}_\rho^i \text{ such that } \exists u \in \mathscr{U}(x),\ \Phi_\rho(x, u, \mathscr{Q}_\rho) \subset \mathscr{K}_\rho^i\}. \end{aligned} \tag{17.8}$$

This algorithm is strongly related to so-called set-valued numerical analysis since it consists in the construction of sets converging in the sense of Painlevé-Kuratowski to a set of elements sharing a given property. It is beyond the scope of this chapter to study the full discretization in space, with respect to x, of this algorithm, but we simply state that, using the Painlevé–Kuratowski definition of set limit, one can prove that $\mathrm{Lim}_{\rho\to 0}\mathrm{Lim}_{i\to\infty}\mathscr{K}_\rho^i$ exists and coincides with the guaranteed capture basin of $\mathscr{T}$ in $\mathscr{K}$ of $(\Phi, \mathscr{U}, \mathscr{Q})$.

17.2.6 Approximation of Valuation Function

By construction, since the time component τ of x belongs to the interval $[0, T = N\rho]$, we are interested in the initial positions $(\tau^0, S^0, W^0) \in \mathscr{K}_\rho^N$ with $\tau^0 = T$. The question becomes ones of determining the subset of initial capital W such that the capture condition respecting the financial constraints is satisfied for a given initial price S and then the right decision rule that will ensure the capture, i.e., the payoff requirement. Let us define the function $V_\rho : \mathbb{R}_+ \times \mathbb{R}_+ \mapsto \mathbb{R}_+$ by

$$V_\rho(\tau, S) := \inf\{W \text{ such that } (\tau, S, W) \in \mathrm{Capt}(\mathscr{K}, \mathscr{T}, (\Phi_\rho, \mathscr{U}, \mathscr{Q}_\rho))\},$$

which associates with each couple (τ, S) the cheapest value of capital such that, if τ is the time to maturity and if S is the current asset price for any initial portfolio value $W^0 \geq V_\rho(T, S^0)$, then there exists a feedback control that guarantees the objectives to be reached while preserving the constraints, for any realization of the uncertainties. It is immediate to prove that $\mathrm{Epi}(V_\rho) := \mathrm{Capt}(\mathscr{K}, \mathscr{T}, (\Phi_\rho, \mathscr{U}, \mathscr{Q}_\rho))$. Let us associate to the sequence of sets $\mathscr{K}_\rho^i$ the sequence of functions V_ρ^i *epigraphically* defined by

$$\mathrm{Epi}(V_\rho^i) := \mathscr{K}_\rho^i.$$

The following result makes clear the relation between viability, capturability, and financial instrument evaluation.

Theorem 17.2. *For any $\tau = n\rho \in \{0, \ldots, N\rho\}$ and $S \in [S^\flat(T-\tau), S^\sharp(T-\tau)]$, let us define recursively the functions V_ρ^i by*

$$\begin{aligned} V_\rho^0(\tau, S) &= \min(\mathbf{b}(T-\tau, S), \mathbf{c}(T-\tau, S)), \\ V_\rho^{i+1}(\tau, S) &= \max\Bigg([V_\rho^i(\tau, S), \\ &\qquad \inf_{u\in\mathscr{U}(S)} \sup_{v\in\mathscr{Q}_\rho} \frac{V_\rho^i(\tau-\rho, S(1+v)) + uS(\rho r_0 - v)}{1+\rho r_0}\Bigg). \end{aligned} \tag{17.9}$$

Then $V_\rho^n(n\rho, S)$ coincides with the value of an option for the discrete model if the time to maturity is $\tau = n\rho$ and the price at time $t = T - n\rho$ of the underlying asset is S.

Proof. From Algorithm 17.8, since $\mathrm{Epi}(V_\rho^0) = \mathscr{K} \cup \mathscr{T}$ and since

$$\mathrm{Epi}(V_\rho^{i+1}) = \{x \in \mathrm{Epi}(V_\rho^i) \mid \exists u \in \mathscr{U}(S),\ \forall v \in \mathscr{Q}_\rho,\\ \Phi_\rho(x,u,v) \in \mathrm{Epi}(V_\rho^i)\},$$

from the definition of K_ρ^{i+1} given by (17.8), $x \in K_\rho^{i+1}$ if and only if

$$\begin{cases} x = (\tau,S,W) \in K_\rho^i \text{ and } \exists u \in U(x) \text{ such that } \forall v \in \mathscr{Q}_\rho, \\ \Phi_\rho(x,u,v) := x + \rho\varphi(x,u,v) \\ \qquad = (\tau-\rho, S(1+v), W(1+\rho r_0) - uS(\rho r_0 - v)) \in K_\rho^i. \end{cases}$$

Using the epigraphic characterization of the sets

$$\begin{cases} W \geq V_\rho^i(\tau,S) \text{ and } \exists u \in \mathscr{U}(x) \text{ such that } \forall v \in \mathscr{Q}_\rho, \\ W(1+\rho r_0) - uS(\rho r_0 - v)) \geq V_\rho^i(\tau-\rho, S(1+v)), \end{cases}$$

or, equivalently,

$$\begin{cases} W \geq V_\rho^i(\tau,S) \text{ and } \exists u \in U(x) \text{ such that } \forall v \in \mathscr{Q}_\rho, \\ W \geq \dfrac{V_\rho^i(\tau-\rho, S(1+v)) + uS(\rho r_0 - v))}{1+\rho r_0}, \end{cases}$$

so that necessarily

$$W \geq V_\rho^{i-1}(\tau,S) \\ := \max\left[V_\rho^i(\tau,S),\ \inf_{u\in U(x)} \sup_{v \in Q_\rho} \frac{V_\rho^i(\tau-\rho, S(1+v)) + uS(\rho r_0 - v))}{1+\rho r_0}\right].$$

□

17.2.7 *Implementing the Guaranteed Capture Basin Method to Evaluate a European Call*

The following results illustrate that, when uncertainty is modeled with $\mathscr{Q}_\rho = [e^{-\sigma\sqrt{\rho}} - 1, e^{+\sigma\sqrt{\rho}} - 1]$, $K = 100$, and $T = 1$ (year), as in the CRR binomial model, the Guaranteed Capture Basin Algorithm provides the same classic results. Computations were made with $\rho = 1/320$.

Figures 17.3 and 17.4 show a graph of the evaluation function and the optimal strategy p, which represents the amount of risky asset necessary to compose the replicating portfolio. The capture basin is the upper domain lower bounded by this function, the *epigraph* of the evaluation function.

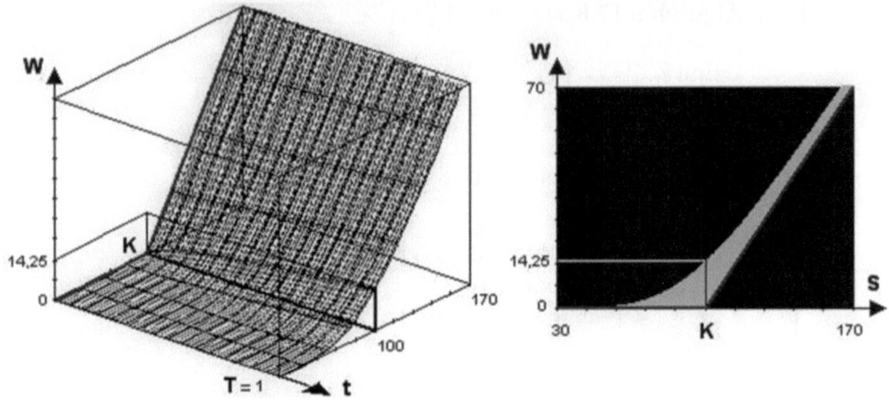

Fig. 17.3 Guaranteed evaluation function of a European call

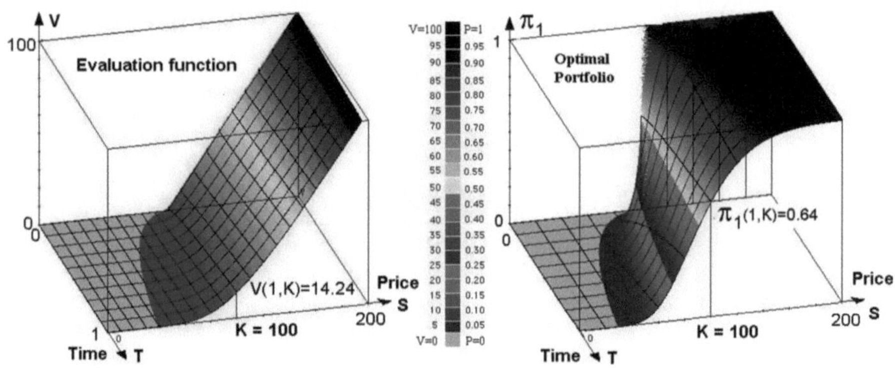

Fig. 17.4 Optimal strategy p corresponding to guaranteed evaluation function of a European call

Figure 17.5 presents different values of a call for different maturity times and different values for volatility obtained by the capture basin method. They practically coincide with the values obtained by the CRR binomial approximation method.

17.3 Extension of Capture Basin Methods to Evaluate Complex Instruments

We now design models and algorithms that allow us to evaluate and to provide the optimal strategy in the following situations:

1. When transaction costs hold for standard options;
2. When some conditions must be fulfilled at known dates, like for Bermuda or capped-style options.

Maturity	Volatility	Cox, Ross & Rubinstein	Capture Basin Algorithm W	π_0	π_1
1	10%	**6.72**	**6.82**	-64.06	0.7078
0.5	10%	**4.15**	**4.21**	-60.95	0.6507
0.1	10%	**1.52**	**1.53**	-55.47	0.5699
1	20%	**10.45**	**10.46**	-53.23	0.6368
0.5	20%	**6.89**	**6.89**	-52.91	0.5979
0.1	20%	**2.77**	**2.79**	-51.68	0.5446
1	25%	**12.27**	**12.34**	-50.41	0.6274
0.5	25%	**8.22**	**8.26**	-50.84	0.5910
0.1	25%	**3.39**	**3.42**	-50.72	0.5413
1	30%	**14.23**	**14.23**	-48.20	0.6242
0.5	30%	**9.65**	**9.64**	-49.24	0.5887
0.1	30%	**4.03**	**4.04**	-49.97	0.5401
1	50%	**21.73**	**21.76**	-41.69	0.6344
0.5	50%	**15.09**	**15.12**	-44.66	0.5977
0.1	50%	**6.53**	**6.54**	-47.87	0.5441

Fig. 17.5 Comparison of numerical results obtained by the capture basin method and the CRR binomial method (*Source*: [129])

We illustrate this approach through numerical applications. For all these applications we present the continuous and discrete dynamic systems underlying the corresponding model to which we apply the Guaranteed Capture Basin Algorithm or its extension to impulse systems.

17.3.1 Taking into Account Transaction Costs and Constraints

We assume that $S(t) \in [S^\flat(t), S^\sharp(t)]$ and that the quantity $P(t)$ of the risky asset is also constrained to remain in a given interval $[0, P^\sharp(t)]$. The quantity P (uppercase) is no longer a control but a state variable. Its variation, which is the quantity of shares bought or sold at each transaction, denoted by u, becomes the control of the system. In the presence of transaction costs, the self-financing assumption is superseded by

$$\forall t \in [0,T],\ P_0'(t)S_0(t) + P'(t)S(t) + \delta|P'(t)|S(t) + \delta_0(P'(t)) = 0$$

or

$$\forall n \in \{0,\ldots,N-1\},$$
$$(P_0^{n+1} - P_0^n)S_0^n + (P^{n+1} - P^n)S^n + \delta|P^{n+1} - P^n|S^n + \delta_0(P^{n+1} - P^n) = 0,$$

where δ is the transaction cost rate, which can possibly depend on $|P'|S$ and $\delta_0(u) = 0$ if $u = 0$ and $\delta_0(u) = \delta_0$ if $u \neq 0$, where δ_0 represents the fixed cost. The function $\delta_0(\cdot)$ is discontinuous.

So that, with the same foregoing simplifications, the evolution of the value of the portfolio becomes

$$W'(t) = W(t)r_0 - P(t)S(t)(r_0 - r(t)) - \delta|u(t)|S(t) - \delta_0(u(t))$$

or

$$W^{n+1} = (1+\rho r_0)W^n - P^n S^n(\rho r_0 - r^n_\rho) - \delta|u^n|S^n - \delta_0(u^n).$$

To summarize, the time $\tau(t)$ to maturity, the asset price $S(t)$, the capital $W(t)$, and the number of shares $P(t)$ are solutions to the nonlinear dynamic or discrete system

$$\begin{cases} \tau'(t) &= -1,\\ S'(t) &= S(t)r(t),\\ P'(t) &= u(t),\\ W'(t) &= W(t)r_0 - P(t)S(t)(r_0 - r(t)) - \delta|u(t)|S(t) - \delta_0(u(t)), \end{cases}$$

or

$$\begin{cases} \tau^{n+1} &= \tau^n - \rho,\\ S^{n+1} &= S^n(1+\rho r^n_\rho),\\ P^{n+1} &= P^n + \rho u^n,\\ W^{n+1} &= (1+\rho r_0)W^n - P^n S^n(\rho r_0 - r^n_\rho) - \delta|u^n|S^n - \delta_0(u^n), \end{cases}$$

where $u(t) \in \mathbb{R}$ or $u^n \in \mathscr{U}_\rho(S^n, P^n)$ (defined subsequently) and $r(t) \in \mathscr{Q}(t, S(t))$ or $r^n_\rho \in \mathscr{Q}^n_\rho$.

The state variable (τ, S, P, W) must evolve, before reaching the target $\mathscr{T}$, in a constrained set $\mathscr{K}$, including *trading constraints* that bound the variation of the share of the risky asset so as to maintain $P(t) \in \mathscr{P}(S(t))$ or $P^n \in \mathscr{P}(S^n)$. For instance, if $\mathscr{P}(S) = [0,1]$, then $\mathscr{U}_\rho(S,P) = \left[-\frac{P}{\rho}, 1-\frac{P}{\rho}\right]$. The sets $\mathscr{K}$ and $\mathscr{T}$ are defined by

$$\begin{aligned} \mathscr{K} := \{(\tau,S,P,W) \in \mathbb{R}^4 \text{ such that } \mathbf{b}(T-\tau,S,P) \le W,\ \tau \ge 0,\ P \in \mathscr{P}(S),\\ S_\flat(T-\tau) \le S(T-\tau) \le S_\sharp(T-\tau),\ P \in [0,P^\sharp]\},\\ \mathscr{T} := \{(\tau,S,P,W) \in \mathbb{R}^3 \text{ such that } \mathbf{c}(T-\tau,S,P) \le W\}, \end{aligned}$$

where $\mathbf{c}(T-\tau,S,P) = \begin{cases} (S-K)^+ & \text{if } \tau = T,\\ +\infty & \text{if not.} \end{cases}$

Our aim is to determine the set of initial conditions $(S(0),P(0),W(0)) \in \mathbb{R}^3_+$ or $(S^0,P^0,W^0) \in \mathbb{R}^3_+$ for which there exists a map $(\tau,S,P,W) \to \widehat{u}(\tau,S,P,W)$ or a sequence of maps $((S,P,W) \to \widehat{u}^n(S,P,W))_{n=0,\ldots,N-1}$ such that, whatever the realization $t \mapsto v(t)$ or the sequence $(v^n)_{n=0,\ldots,N-1}$ of uncertainty happens to be, the evolution $t \to (\tau(t),S(t),P(t),W(t))$ or $(\tau^n,S^n,P(t),W^n)_{n=0,\ldots,N}$ solution to

$$\begin{cases} \tau'(t) &= -1, \\ S'(t) &= S(t)r(t), \\ P'(t) &= \widehat{u}(\tau(t),S(t),P(t),W(t)), \\ W'(t) &= W(t)r_0 - P(t)S(t)(r_0 - r(t)) \\ & \quad -\delta|\widehat{u}(\tau(t),S(t),P(t),W(t))|S(t) - \delta_0(\widehat{u}\tau(t),S(t),P(t),W(t)) \end{cases}$$

or

$$\begin{cases} \tau^{n+1} &= \tau^n - \rho, \\ S^{n+1} &= S^n(1+\rho r_\rho^n), \\ P^{n+1} &= P^n + \rho\widehat{u}^n(S^n,P^n,W^n), \\ W^{n+1} &= (1+\rho r_0)W^n - P^n S^n(\rho r_0 - r_\rho^n) \\ & \quad -\delta|\widehat{u}^n(S^n,P^n,W^n)|S^n - \delta_0(\widehat{u}^n S^n,P^n,W^n) \end{cases}$$

remains in the constraint set $\mathscr{K}$ before reaching the target $\mathscr{T}$. To that end, we apply the Guaranteed Capture Basin Algorithm setting $x = (\tau,S,P,W)$, $u = P'$, $v = r$, and

$$\begin{aligned} \Phi(x,u,v) &:= (-1, Sv(S,v), u, Wr_0 - PS(r_0 - v) - \delta|u|S), \\ &\text{and} \\ \Phi_\rho(x,u,v) &:= (\tau-\rho, S(1+v), P+\rho u, (1+\rho r_0)W - P(\rho r_0 - v) \\ &\quad -\rho\delta|u|S - \delta_0(u)). \end{aligned} \tag{17.10}$$

17.3.2 Approximation of Valuation Function in the Presence of Transaction Costs

Applying the Guaranteed Capture Basin Algorithm we determine the subset of initial positions (τ^0,S^0,P^0,W^0) for which there exists a "buy and sell" strategy such that the financial constraints are satisfied until the payoff is realized, whatever the considered uncertainty. Let us define the function $V : \mathbb{R}_+ \times \mathbb{R}_+ \times \mathbb{R}_+ \mapsto \mathbb{R}_+$ by

$$V_\rho(\tau,S,P) := \inf\{W \text{ such that } (\tau,S,P,W) \in \mathrm{Capt}(\mathscr{K},\mathscr{T},(\Phi_\rho,\mathscr{U},\mathscr{Q}_\rho))\},$$

which associates to each triple (τ,S,P) the cheapest value of capital needed to guarantee the objectives to be reached while preserving the constraints. We still have $\mathrm{Epi}(V_\rho) := \mathrm{Capt}(\mathscr{K},\mathscr{T},(\Phi_\rho,\mathscr{U},\mathscr{Q}_\rho))$, so that we can compute recursively the functions V_ρ^i:

$$\begin{aligned} V_\rho^0(\tau,S,P) &= \min(\mathbf{b}(T-\tau,S),\mathbf{c}(T-\tau,S)),\ \forall\tau = n\rho \in \{0,\ldots,N\rho\}, \\ V_\rho^{i+1}(\tau,S,P) &= \max\Bigg[V_\rho^i(\tau,S,P), \\ \inf_{u\in\mathscr{U}(S,P)}\sup_{v\in\mathscr{Q}_\rho} & \frac{V_\rho^i(\tau-\rho,S(1+v),P+\rho u)+PS(\rho r_0 - v)+\rho\delta|u|S+\delta_0(u)}{1+\rho r_0}\Bigg]. \end{aligned} \tag{17.11}$$

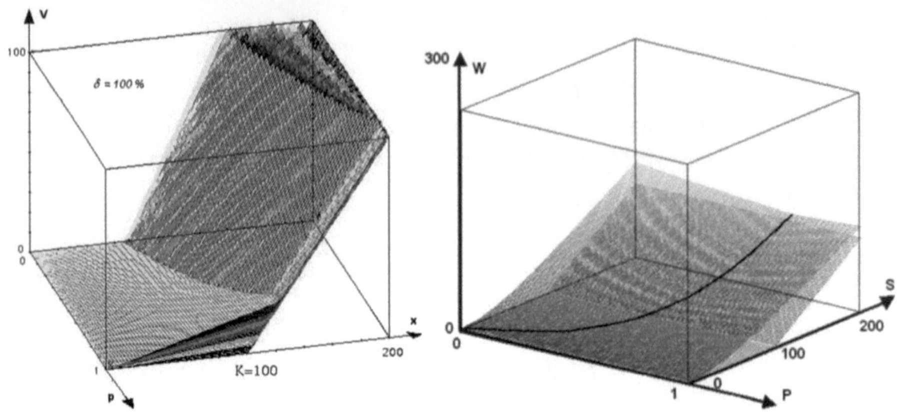

Fig. 17.6 Evaluation of a European call with transaction costs

Then the approximate value of the option in the presence of transaction costs is given by

$$V_\rho(S,P) = V_\rho^N(T,S,P),$$

and the initial portfolio, determined by a quantity of shares $\widehat{P}$ depending on the value S at the initial time, is given by

$$\widehat{P} = \Pi(S) := \text{Argmin}_P \left(V_\rho(\tau,S,P) + \rho\delta PS\right).$$

We illustrate this method in the same context as above for evaluating European calls in the presence of transaction costs. Figure 17.6a shows the superimposed sequence of valuation functions for different values of time. In Fig. 17.6b, only the first and last valuation functions are superimposed. Shown in the figure is the curve corresponding to the best choice $\widehat{P}$ of quantity of shares that one must buy to initialize the portfolio if the price of the risky asset is S.

One interesting question is the following one: taking into account transaction costs while running the replicating process is obviously different from running the replicating process without transaction costs and calculating a posteriori the cost of all the effective transactions done during the life of the option. The difference becomes very sensitive when the time step – like, for instance, in the CRR procedure – becomes smaller and smaller since the a posteriori cost becomes infinite, whereas in our approach, it remains bounded. Figure 17.7 shows the difference clearly. On the left, the upper figure shows the evolution of the value of the replicating portfolio

- When there are no transaction costs,
- When the transaction costs are endogenized in the model where the control is the variation of the quantity of the asset bought or sold, and

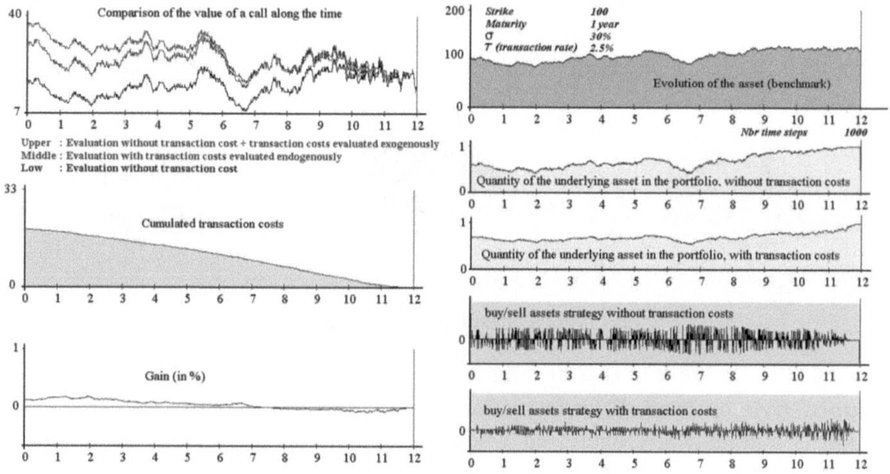

Fig. 17.7 Comparison of vanilla option valuation with and without transaction costs *(with the permission of VIMADES Cie)*

- When the evaluation of transaction costs is determined ex post on the basis of the evolution of the quantity of asset of the replicating portfolio as if there were no transaction costs.

17.3.3 Bermuda and Capped-Style Options: Two Examples of Constrained and Multitarget Problems

Bermuda options belong to the nonstandard American option family with the following extended features:

- Early exercise is restricted to certain dates $(T_i)_{i\in\mathscr{I}}$, $T_i \in [0,T]$.
- Early exercise is allowed only for certain parts of the life of the option.
- The strike price may change during the life of the option.

In addition, Bermuda options are characterized by the data of pairs $(T_i, K_i)_{i\in I}$ when exercise is possible at certain fixed dates T_i within the maturity period $[0,T]$ with a strike price K_i. In our formalization, this can simply be written through the definition of the constraint function. Let

$$\mathbf{b}_i(t,S,P) = \begin{cases} [S-K_i]^+ & \text{if } t = T_i, \\ 0 & \text{if } t \neq T_i, \end{cases} \quad \text{and set } \mathbf{b}(t,S,P) = \max_i \mathbf{b}_i(t,S,P), \text{with}$$

$$\mathbf{c}(t,S,P) = \begin{cases} [S-K]^+ & \text{if } t = T, \\ 0 & \text{if } t \neq T. \end{cases}$$

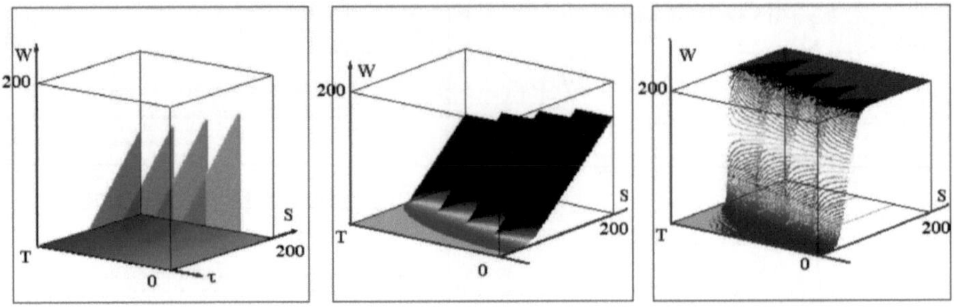

Fig. 17.8 Evaluation of a Bermuda option

Capped-style options differ from Bermuda options in the sense that the capped option is automatically exercised when the underlying asset closes at or above the option strike. This can simply be written through the definition of the target function. Let

$$\mathbf{c}_i(t,S,P) = \begin{cases} S - K_i \text{ if } \ t = T_i \text{ and } S \geq K_i, \\ +\infty \quad \text{ if not,} \end{cases}$$

and set

$$\mathbf{c}(t,S,P) = \min_i \mathbf{c}_i(t,S,P) \text{with}$$
$$\mathbf{b}(t,S,P) = 0.$$

This amounts to considering the target $\mathscr{T}$ as the union of targets $\mathscr{T}_i$, as shown in Fig. 17.2.

Then, applying the Guaranteed Capture Basin Algorithm defined previously in the presence of transaction costs, we obtain an approximation of the evaluation of these options satisfying $\forall i \in \mathscr{I}$, $\forall S \in S^\flat(T_i) \leq S \leq S^\sharp(T_i)$, $W(T_i) \geq [S - K_i]^+$, which means that at each time T_i when the option can or must be exercised, the value of the replicating portfolio is sufficient to ensure the payoff.

Figure 17.8 illustrates the computation of a Bermuda option with three specific dates between issue date and expiry date. On Fig. 17.8a, a (discontinuous) graph of the constraint function $\mathbf{b}$ is depicted. The target is the epigraph of $\mathbf{c}$. On Fig. 17.8b is represented a graph of the evaluation function computed with $\delta = 0$ (without transaction costs). On Fig. 17.8c is represented the optimal policy, that is to say, the quantity of shares $\Pi(t,S)$ of the risky asset in the replicating portfolio.

Figure 17.9 shows the impact of transaction costs when evaluating a Bermuda call taking or not taking into account a priori the transaction costs.

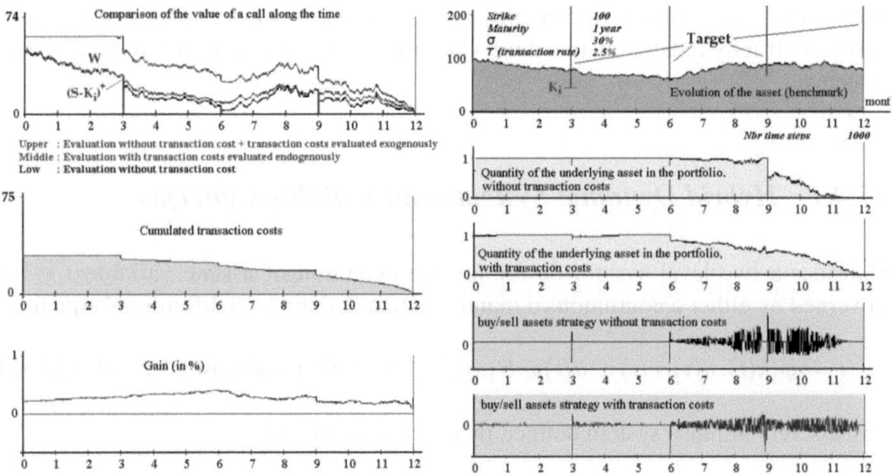

Fig. 17.9 Comparison of Bermuda option valuation with and without transaction costs (with permission of VIMADES Cie)

17.4 Evaluation of Complex Financial Instruments Using Impulse Systems

Impulse dynamic systems refer to processes whereby the state can "jump" to a new position following some given rules. Hybrid systems refer to systems that can change from one mode to another following some change rules depending on several parameters. Both hybrid and impulse systems can be represented interbreeding continuous and discrete dynamic systems.

Mathematical and numerical tools developed in the framework of hybrid dynamic systems make it possible to solve various control problems in the presence of uncertainty. Here we integrate uncertainty and impulse thanks to the Guaranteed Capture Basin Algorithm applied to evolutionary systems. This was developed in [60] in the context of nonanticipative strategies. We refer mainly to [16, 58, 59] for theorical studies and numerical issues related to the viability approach to impulse systems and to [47, 48] for applications to finance (barrier options) and economics (managing the life cycle). In this context we will be able to evaluate options in two additional complex situations:

1. When certain conditions must be fulfilled or dismissed at unknown dates. This is the case for barrier options, which are options that can have four modes – "in-up" or "in-down" and "out-up" or "out-down" modes – depending on the asset price evolution;
2. When the "volatility" is periodically adjusted depending on the past and present asset prices.

To illustrate how viability techniques can be extended to these nonlinear complex cases, we briefly recall how to extend viability concepts in the framework of hybrid systems.

17.4.1 Hybrid Dynamic Systems and Viability Concepts

Consider a two-level system describing the evolution of a state variable $x \in \mathbb{R}^n$ governed by either a continuous dynamic system defined by a differential equation

$$x'(t) = \varphi(x(t), u(t), v(t)),\ \ u(t) \in U(x(t)),\ \ v(t) \in \mathscr{Q}, \text{ for almost all } t \geq 0, \quad (17.12)$$

or a discrete impulse system defined by the reset equation

$$x^+ \in \psi(x^-), \quad (17.13)$$

where $\psi(x^-)$ is the set of all available states x^+ attainable from the position x^-. Resetting is possible when x belongs to a prescribed subset $\mathscr{R}$ of $\mathbb{R}^N$, called the *reset set*, so that $\psi(x) = \emptyset$ whenever $x \notin \mathscr{R}$. We assume that the map ψ has no fixed point and the map φ is bounded:

$$\begin{aligned} &(1) \quad \exists m > 0,\ \inf_{x \in K} d(\psi(x), x) \geq m, \\ &(2) \quad \exists M > 0,\ \sup_{x \in K} \sup_{u \in \mathscr{U}(x)} \sup_{v \in Q} ||\varphi(x,u,v)|| \leq M. \end{aligned} \quad (17.14)$$

Definition 17.3. We call a *run* of impulse system (φ, ψ), with initial condition x_0 and given pair $(u(\cdot), v(\cdot))$ of measurable functions, a collection of finite or infinite sequences

$$\{\delta_i, x_i, \xi_i(\cdot)\}_{i \in \mathbb{N}}$$

in $\mathbb{R}^+ \times \mathbb{R}^n \times \mathscr{S}_F(\mathbb{R}^n, u(\cdot), v(\cdot))$, where $\{\delta_i\}_{i \in \mathbb{N}}$ is a sequence of durations such that for all $i \in \mathbb{N}$,

$$\xi_i'(t) = \varphi(\xi_i(t), u(t+t_i), v(t+t_i)),\ \ \xi_i(0) = x_i, \quad \xi_i(\delta_i) \in \mathscr{R},\ \ x_{i+1} \in \psi(\xi_i(\delta_i)).$$

Let us set $t_i := \sum_{j=0}^{i-1} \delta_j$.

We call a *trajectory* associated with a run $\{\delta_i, x_i, \xi_i(\cdot)\}_{i \in \mathbb{N}}$ the function $xi(\cdot)$ given by

$$\xi(t) = \begin{cases} x_0 & \text{if } t < 0, \\ \xi_i(t - t_i) & \text{if } t \in [t_i, t_i + \delta_i). \end{cases} \quad (17.15)$$

We are interested in trajectories that remain in a constraint set $\mathscr{K}$, eventually resetting if necessary within the reset rule, until they satisfy at a prescribed time

T, or eventually before, some specific condition summarized in the definition of the target $\mathscr{T}$, and this regardless of the evolution of the uncertainty parameter $v(\cdot)$. We search the largest domain of initial positions from which starts at least one run switching between continuous motives and impulses but remaining in $\mathscr{K}$ until it reaches the target $\mathscr{T}$, whatever $v(\cdot)$ is. We can assume without loss of generality that $\mathscr{R} \subset \mathscr{K} \cap \psi^{-1}(\mathscr{K})$, and we denote by $S_{(\varphi,\psi,\mathscr{K},\mathscr{T})}(x_0)$ the set of runs starting from x_0 and viable in $\mathscr{K}$.

We need to know whether the successor of ξ^n is a consequence of a jump or if it derives from the approximation of the continuous process. For that purpose we add two variables and their evolution rules: the time t and the total number of resets or jumps j. If ξ^{n+1} is a consequence of a jump, then $d(\xi^{n+1}, \xi^n) \geq m$, whereas if it derives from a continuous process, then $d(\xi^{n+1}, \xi^n) \leq \rho M$. Choosing the time step sufficiently small, $\rho \leq \frac{m}{2M}$, then, if ξ^{n+1} is a consequence of a jump, we necessarily have $d(\xi^{n+1}, \xi^n) \geq 2\rho M$. Let us define the discrete map (jump counter)

$$j_\rho(\xi^n) = \begin{cases} j^n & \text{if } d(\xi^{n+1}, \xi^n) \leq \rho M, \\ j^n + 1 & \text{if } d(\xi^{n+1}, \xi^n) \geq 2\rho M, \end{cases}$$

and we consider the following extended semi-implicit discrete dynamic system:

$$\begin{aligned} \xi^{n+1} &\in \{\xi^n + \rho\varphi(\xi^n, u^n, v^n)\} \cup \{\psi(\xi^n)\}, \\ j^{n+1} &= \begin{cases} j^n & \text{if } d(\xi^{n+1}, \xi^n) \leq \rho M, \\ j^n + 1 & \text{if } d(\xi^{n+1}, \xi^n) \geq 2\rho M, \end{cases} \\ t^{n+1} &= t^n + (j_p^{n+1} - j_p^n)\rho. \end{aligned} \tag{17.16}$$

17.4.2 Guaranteed Hybrid Capture Basin Algorithm

To compute approximations of the guaranteed hybrid capture basin, we construct a decreasing sequence of subsets $\mathscr{K}_\rho^i$ defined by

$$\begin{aligned} \mathscr{K}_\rho^0 &:= \mathscr{K} \cup \mathscr{T}, \\ \mathscr{K}_\rho^{i+1} &:= \{(\xi, j, t) \in \mathscr{K}_\rho^i,\ \exists u \in \mathscr{P}(\xi) \text{ such that} \\ &\qquad \forall v \in \mathscr{Q}_\rho,\ (\xi + \rho\varphi(\xi, u, v), j^n, t + \rho) \text{ or } (\psi(\xi), j+1, t) \in \mathscr{K}_\rho^i\}. \end{aligned} \tag{17.17}$$

We simply point out that $\mathrm{Lim}_{\rho \to 0} \mathrm{Lim}_{i \to \infty} \mathscr{K}_\rho^i$ exists and coincides with the guaranteed hybrid capture basin of $\mathscr{T}$ in $\mathscr{K}$ for $(\varphi, \psi, \mathscr{U}, \mathscr{Q})$.

An element $(\xi, j, t) \in \mathscr{K}_\rho^{i+1}$ if and only if $(\xi, j, t) \in \mathscr{K}_\rho^i$ and either there exists a control $u \in \mathscr{U}(x)$ such that for any $v \in \mathscr{Q}_\rho$, $S_\rho(x + \rho f(x, u, v)) \in \mathscr{K}_\rho^p$ or $(\psi(\xi), j + 1, t) \in \mathscr{K}_\rho^i$.

17.4.3 Evaluation of Barrier Options

Barrier options also belong to nonstandard American or European options depending on unpredictable events that change in an impulse manner the nature of the agreement. The barrier mechanism complicates the evaluation of the replicating portfolio. A barrier is a particular value S^* for a price $S(t)$ beyond which the agreement changes. There are four types of options with barriers:
"Up and in": the agreement becomes effective at the first time t^* when $S(t) < S^*, \forall t < t^*, S(t^*) = S^*$;
"Down and in": the agreement becomes effective at the first time t^* when $S(t) > S^*, \forall t < t^*, S(t^*) = S^*$;
"Up and out": the agreement ceases at the first time t^* when $S(t) < S^*, \forall t < t^*, S(t^*) = S^*$;
"Down and out": the agreement ceases at the first time t^* when $S(t) > S^*, \forall t < t^*, S(t^*) = S^*$.

The challenge is to evaluate today an agreement that can vanish or become effective at some unknown date in the future. For the sake of simplicity we only consider here the "up and in" case of a call absent transaction costs, focusing our attention on the impulse complexion.

To this end, we introduce a discrete variable $L \in \{0,1\}$ that labels the state of the agreement – effective for $L = 1$ or noneffective for $L = 0$ – and consider the hybrid dynamic system:

$$\text{Continuous level}\quad \begin{cases} \tau'(t) &= -1, \\ S'(t) &= S(t)r(t), \\ L'(t) &= 0, \\ W'(t) &= \begin{cases} W(t)r_0 - P(t)S(t)(r_0 - r(t)) & \text{if } L(t) = 1, \\ 0 & \text{if } L(t) = 0; \end{cases} \end{cases}$$

$$\text{Impulse level}\quad \begin{cases} \tau^+ &= \tau^-, \\ S_1^+ &= S_1^-, \\ L^+ &= \begin{cases} 1 & \text{if } S_1^- \geq S^*, \\ L^- & \text{if } S_1^- < S^*, \end{cases} \\ W^+ &= W^-. \end{cases}$$

Note that, as formulated, the evolution of the label's variable is increasing. This corresponds to the case of "in" options.

As for previous option evaluations, we wish to determine the guaranteed capture basins of the epigraph of $\mathbf{c}$ while remaining in the epigraph of $\mathbf{b}$ and $\tau > 0$.

Let us remark that, for this problem, the right-hand side is governed by conditional operations. Let $\rho = \frac{T}{N}$ and $n \leq N$. The discrete system then becomes

$$\begin{cases} \tau^{n+1} &= \tau^n - \rho, \\ S^{n+1} &= S^n(1+r_\rho^n), \\ L^{n+1} &= \begin{cases} 1 & \text{if } S^{n+1} \geq S^*, \\ L^n & \text{if not,} \end{cases} \\ W^{n+1} &= \begin{cases} (1+\rho r_0)W^n - p^n S^n(\rho r_0 - r_\rho^n) & \text{if } L^{n+1} = 1, \\ W^n & \text{if } L^{n+1} = 0, \end{cases} \end{cases} \tag{17.18}$$

under the constraints

$$\begin{aligned} \mathscr{K} := \{(\tau, S, W) \in \mathbb{R}^3 \text{ such that } \mathbf{b}(T-\tau, S) \leq W,\ \tau \geq 0, \\ S^\flat(T-\tau) \leq S(T-\tau) \leq S^\sharp(T-\tau),\ \mathscr{P}(S) \cap [0,1] \neq \emptyset\}, \end{aligned}$$

with the target $\mathscr{T} := \{(\tau, S, W) \in \mathbb{R}^3 \text{ such that } \mathbf{c}(T-\tau, S) \leq W\}$.

Let r_ρ^M and r_ρ^m denote the upper and lower bounds between which the return r varies over each interval of length ρ. Control and uncertainty are henceforth assumed to vary within $p \in [0,1]$ and $r_\rho \in \mathscr{Q}_\rho := [r_\rho^m, r_\rho^M]$.

We define the switching function

$$\Lambda(L, S, S^*) = \begin{cases} 1 & \text{if } S \geq S^*, \\ L & \text{if not,} \end{cases}$$

and we denote

$$\begin{aligned} &\Phi_\rho(\tau, S, L, W, p, r_\rho) \\ &\quad := (\tau - \rho, S(1+r_\rho), \Lambda(L, S(1+r_\rho), S^*), \\ &\qquad W + \Lambda(L, S(1+r_\rho), S^*)\left(W r_{\rho 0}(t)) + pS(r_\rho - \rho r_0)\right)). \end{aligned}$$

The Guaranteed Hybrid Capture Basin Algorithm leads to the construction of a decreasing sequence of subsets K_ρ^i as in (17.8), which can be rewritten as epigraphs of functions $(\tau, S, L) \to V_\rho^i(\tau, S, L)$ defined recursively by

$$\begin{aligned} &V_\rho^0(\tau, S, L) = \begin{cases} \min(\mathbf{b}(\tau, S), \mathbf{c}(\tau, S)) & \text{if } L = 1, \\ 0 & \text{if } L = 0, \end{cases} \\ &V_\rho^{i+1}(\tau, S, L) = \max\Bigg(V_\rho^i(\tau, S, L), \\ &\inf_{p \in \mathscr{P}(S) \cap [0,1]} \sup_{r_\rho \in [r_\rho^m, r_\rho^M]} \frac{1}{1 + r_{\rho 0}(t)} \Big[V_\rho^i(\tau - \rho, S(1+r_\rho), \Lambda(L, S(1+r_\rho), S^*)) \\ &\qquad\qquad + p\Lambda(L, S(1+r_\rho), S^*) S(\rho r_0 - r_\rho) \Big] \Bigg), \\ &\forall \tau = n\rho \in \{0, \ldots, N\rho\}. \end{aligned} \tag{17.19}$$

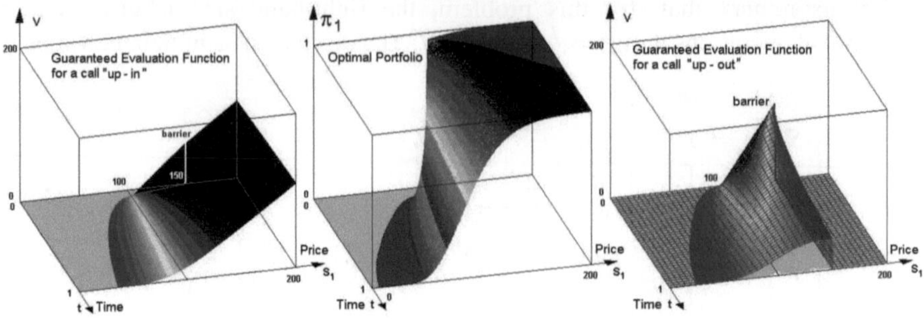

Fig. 17.10 Evaluation of barrier options

Note that for $L = 1$ the approximated guaranteed evaluation function coincides with the approximated guaranteed evaluation function corresponding to the case of a plain call without barriers. Also the *true* function is not trivial for $L = 0$ and for $S < S^*$.

Figure 17.10 shows the graph of the evaluation function $V_\rho(t,S,0) = V_\rho^N(t,S,0)$ of a European call with an "up and in" barrier with $s^* = 150$ and the optimal strategy p. On the right is a graph of the evaluation function of an "up and out" call with $s^* = 150$. In both case, the uncertainty is of CRR-type $[r_\rho^m, r_\rho^M] = [\mathrm{e}^{-\sigma\sqrt{\rho}} - 1, \mathrm{e}^{+\sigma\sqrt{\rho}} - 1]$.

17.4.4 Evaluation of Options Using NGARCH Uncertainty Correction and in the Presence of Transaction Costs

We have seen that the formulation of barrier options in the framework of hybrid dynamic systems using the Hybrid Capture Basin Algorithm leads to the definition of algorithms allowing for an evaluation of options that are subject to impulse events translating unpredictable events that change the structure of the agreement in a discontinuous way. In the barrier option mechanism, the status of the agreement changes when some risky asset price reaches or leaves a given range set; the date is unknown. Garch or Ngarch models also belong to the class of impulse systems where parameters characterizing uncertainty are modified according to the unpredictable result of an observation that occurs at some prescribed dates (see [26]). The characterization of epigraphs of evaluation functions arising in NGARCH modelization using the Guaranteed Capture Basin Algorithm was developed with the collaboration of Michèle Breton (HEC Montreal and GERAD, Quebec).

One can take into account transaction costs taht fixed or dependent on the amount of monetary exchange between the risky and the nonrisky parts of a portfolio. Fixed costs again generate a discontinuous evolution of the value of the portfolio. Other transaction costs are usually nondifferentiable. Thus taking into account transaction costs prevents the implementation of algorithms based on numerical methods for approximating continuous valuation functions.

In what follows we will consider the case of a single risky asset. However, there are no mathematical difficulties when considering a set of different risky assets except when attempting to implement the algorithm on computer since the required memory size grows exponentially with the number of these assets. We now consider the evaluation of a vanilla put in the presence of transaction costs that is associated with the dynamic – continuous or discrete – system

$$\begin{aligned}
\tau'(t) &= -1,\\
S'(t) &= S(t)r(t),\\
P'(t) &= u(t),\\
W'(t) &= W(t)r_0 - P(t)S(t)(r_0 - r(t)) - \delta|u(t)|S(t) - \delta_0(u(t)).
\end{aligned} \tag{17.20}$$

17.4.4.1 Garch Process

In our model, uncertainty is formalized by the condition $v \in \mathcal{Q}_\rho$, where $\mathcal{Q}_\rho$ denotes the uncertainty range, which depends on the time step ρ. In the Garch mechanism we introduce a sequence of dates of observations $(t_0 = 0, t_1, t_2, \dots, t_j, \dots, t_p = T)$ at which the set $\mathcal{Q}_\rho$ is reevaluated such that for all $t \in [t_j, t_{j+1}[$, $r \in \mathcal{Q}_j(S^{j-1}, S^j)$, where S^j is the observed price or the risky asset at date t_j. The set $\mathcal{Q}_j(S^{j-1}, S^j)$ is constant during the time interval $[t_j, t_{j+1}[$. This amounts to correcting the "volatility" according to the "past price" and the "actual price" of the risky asset, "past price" being understood in the sense of a single value of S observed at the beginning of the previous period $[t_{j-1}, t_j[$ and "actual price" being understood as the price observed at the beginning of the actual period $[t_j, t_{j+1}[$. We will detail the Garch process when presenting the discrete process.

Let us introduce two new variables denoted by S_a representing the value of the risky asset observed at the beginning of the previous period and σ representing the uncertainty range. In our application, σ^2 will correspond to the classic "volatility," but it can parameterize any uncertainty representation. Let us denote $\alpha := S/S_a$.

The hybrid dynamic system associated with this problem is transformed as follows:

$$\left\{\begin{array}{ll}
\multicolumn{2}{c}{\text{Continuous dynamics}}\\
\tau'(t) &= -1,\\
S'(t) &= S(t)r(t),\\
S_a'(t) &= 0,\\
P'(t) &= u(t),\\
\sigma'(t) &= 0,\\
W'(t) &= W(t)r_0 - P(t)S(t)(r_0 - r(t))\\
&\quad -\delta_1|u(t)S(t)| - \delta_0(u(t)),
\end{array}\right.
\quad \text{and} \quad
\left\{\begin{array}{ll}
\multicolumn{2}{c}{\text{Impulse dynamics}}\\
\tau^+ &= \tau^-,\\
S^+ &= S^-,\\
S_a^+ &= S_a^-,\\
P^+ &= P^-,\\
\sigma^+ &= g(S^-, S_a^-, \sigma^-),\\
W^+ &= W^-,
\end{array}\right. \tag{17.21}$$

with $u(t) \in \mathscr{U}(S(t),P(t))$ and $r(t) \in \mathscr{Q}(\sigma(t))$ and where $\sigma \to g(S^-,S_a^-,\sigma)$ describes the Garch volatility correction.

Let us introduce the variable $\alpha := \frac{S}{S_a}$. Since all along each period $[t_j,t_{j+1}]$, $S_a(t)$ is constant, we have $\alpha'(t) = \frac{S'(t)}{S_a} = v(t)\alpha(t)$. To simplify the notations, we suppose that the volatility reevaluation map g depends only on α. Thus we can rewrite system (17.21) describing the dynamics of the state $(\tau,\alpha,S_a,P,\sigma,W)$ as follows:

$$\begin{cases} \text{Continuous dynamics} \\ \tau'(t) = -1, \\ \alpha'(t) = \alpha(t)r(t), \\ S_a'(t) = 0, \\ P'(t) = u(t), \\ \sigma'(t) = 0, \\ W'(t) = W(t)r_0 - P(t)S(t)(r_0 - r(t)) \\ \qquad -\delta_1|u(t)S(t)| - \delta_0(u(t)), \end{cases} \text{and} \begin{cases} \text{Impulse dynamics} \\ \tau^+ = \tau^-, \\ \alpha^+ = \alpha^-, \\ S_a^+ = S_a^-, \\ P^+ = P^-, \\ \sigma^+ = g(\tau,\alpha^-,\sigma^-), \\ W^+ = W^-, \end{cases} \tag{17.22}$$

with $u(t) \in \mathscr{U}(S(t),P(t))$ and $r(t) \in Q(\sigma(t))$.

Let us denote the state variable $x := (\tau,\alpha,S_a,P,\sigma,W)$, $v := r$ and the maps

$$\Phi(x,u,v) := (-1,\alpha v,0,u,0,Wr_0 - PS(r_0 - v) - \delta_1|uS| - \delta_0(u)),$$
$$\Psi(x) := (\tau,\alpha,S_a,P,g(\tau,\alpha,\sigma),W).$$

The reset set associated with the hybrid system is that of $(\tau,\alpha,S_a,p,\sigma,W)$ on dates when observations of the risky asset price occurs generating a reevaluation of the uncertainty estimation. It is given by

$$\mathscr{R} := \{(\tau,\alpha,S_a,p,W) \text{ such that } \tau = \tau_j := T - t_j,\ j = 0,\ldots,p\}.$$

The mathematical assumptions for applying viability tools in the context of hybrid systems need to prove an upper semicontinuity property of the set-valued map $(x,v) \to \{\Phi(x,u,v),\ u \in \mathscr{U}(x)\}$, the convexity and compactness of the images, and that the graph of the reset map Ψ is closed, which are true here.

To implement the Hybrid Guaranteed Capture Basin Algorithm, we need to coordinate the time discretization and the observation dates. If $\vartheta := \frac{T}{p}$ denotes the duration of each time interval between two consecutive observations, then we assume that observations are done periodically and that $t_j = j\vartheta$. Let us set $\rho := \frac{T}{N}$ and let us define $J := \frac{\rho}{\theta}$ the number of time steps between two observations. Then we can now write the associated discrete dynamic system associated with (17.22) as follows:

$$\left\{\begin{array}{ll} \text{Continuous dynamics} & \\ \tau^{n+1} & = \tau^n - 1, \\ \alpha^{n+1} & = \alpha^n(1+r^n_\rho), \\ S_a^{n+1} & = S_a^n, \\ P^{n+1} & = P^n + \rho u^n, \\ \sigma^{n+1} & = \sigma^n, \\ W^{n+1} & = W^n(1+\rho r_0) - P^n\alpha^n S_a^n(\rho r_0 - r^n_\rho) \\ & \quad -\delta_1|u^n\alpha^n S_a^n| - \delta_0(u^n), \end{array}\right. \quad \text{and} \quad \left\{\begin{array}{ll} \text{Impulse dynamics} & \\ \tau^+ & = \tau^-, \\ \alpha^+ & = \alpha^-, \\ S_a^+ & = S_a^-, \\ P^+ & = P^-, \\ \sigma^+ & = g(\tau,\alpha^-,\sigma^-), \\ W^+ & = W^-, \end{array}\right. \tag{17.23}$$

where $u^n \in \mathscr{U}(P^n) := \{u \in \mathbb{R} \text{ such that } P^n + \rho u \in [0.1]\}$ and $r^n_\rho \in \mathscr{Q}_\rho(\sigma^n) := [-\sigma^n\sqrt{\rho}+(\sigma^n)^2, \sigma^n\sqrt{\rho}+(\sigma^n)^2]$.

The NGarch models specify the impulsive change of the uncertainty set. For instance, in the framework of stochastic processes, this change is described through the volatility adjustment given by the following statements:

1. For all $n \in \{0,\dots,N\}$, $n \neq 0 \bmod J$, $\sigma^n = \sigma^{n-1}$,
2. For all $n \in \{0,\dots,N\}$, $n = 0 \bmod J$, $\sigma^n = g(\alpha^{n-1},\sigma^{n-1})$,

$$\text{where } g(\alpha,\sigma) = \sqrt{\beta_0 + \beta_1\sigma^2 + \beta_2\sigma^2\left(\frac{\ln(\alpha)-r+\frac{\sigma^2}{2}}{\sigma} - \theta\right)^2}.$$

As previously, the Guaranteed Hybrid Capture Basin Algorithm leads to the construction of a decreasing sequence of subsets K^i_ρ as in (17.8), which can be rewritten as epigraphs of the functions $(\tau,\alpha,S_a,P,\sigma) \to V^i_\rho(\tau,\alpha,S_a,P,\sigma)$ defined recursively by

$$\left\{\begin{array}{l} V^0_\rho(\tau,\alpha,S_a,P,\sigma) = \min(\mathbf{b}(\tau,\alpha S_a), \mathbf{c}(\tau,\alpha S_a)), \\ V^{i+1}_\rho(\tau,\alpha,S_a,P,\sigma) = \max\Bigg(V^i_\rho(\tau,\alpha,S_a,P,\sigma), \\ \inf\limits_{u\in\mathscr{U}(P)} \sup\limits_{v\in Q_\rho(\sigma)} \dfrac{1}{1+\rho r_0}\Big[V^i_\rho(\tau-\rho,\alpha(1+v),S_a,P+\rho u,\sigma) \\ \qquad\qquad + P\alpha S_a(\rho r_0 - v) + \delta_1|u\alpha^n S_a^n| + \delta_0(u)\Big]\Bigg), \\ \forall \tau = n\rho \in \{0,\dots,N\rho\}. \end{array}\right.$$

This provides the minimal value of the replicating portfolio and its composition, which guarantees that, if the initial price of the risky asset is S_a and the initial volatility is σ^2, then one can find a buying/selling strategy u such that, whatever the uncertainty, in the presence of transaction costs and applying regularly the Garch correction rule g at prescribed times, the payoff function will be superseded.

Numerical implementation of the Guaranteed Capture Basin Algorithm for evaluating options in the presence of transaction costs using the NGARCH model is an ongoing work with Michèle Breton, HEC Montréal.

Chapter 18
Asset and Liability Insurance Management (ALIM) for Risk Eradication

18.1 Introduction

Asset and liability management (ALM) deals with approaches that allow a company to manage the composition of its risky assets or underlying in such a way that they are *always larger than its liabilities*. Choosing a management rule is a choice under contingent uncertainty (choosing an exposition of the portfolio) and tychastic uncertainty (valid for risky returns above a forecasted lower bound).

The objective of portfolio insurance is to enable investors to participate opportunistically in market performance while providing protection of any type of liability against all evolutions of risky prices consistent with a prediction model, at each date up to maturity. The liability is understood in the largest sense, from the standard protection at exercise time or allowing for variable annuities, as in life insurance or retirement pension plans. In other words, portfolio insurance's main objective is

1. Either to evaluate risk
2. Or to eliminate or limit the loss of portfolio value

while allowing the portfolio to benefit opportunistically, to some extent, from a rise in the "market." Here, we propose to eradicate losses.

The motivation for such a strategy is based on the simple observation that during, for example, the crashes of October 1987 and October 1989, strategies such as *buy and hold* (which fixes the risky part of the portfolio once and for all) or the *constant proportion portfolio insurance* (CPPI) method (which amounts to keeping constant the "cushion multiplier," which is the ratio of the expectation over the cushion, or surplus, defined as the difference between the value of the portfolio and the liability). CPPI is a widely used fund management technique and a capital guarantee derivative security that provide participation in the performance of the underlying asset [125] but "*...could result in very significant losses*," violating the "I" in CPPI. See, for instance, the studies of [50, 54], among several authors. Cont and Tankov point out the fact that the CPPI does not eliminate risk: *Yet the possibility of going below the floor, known as "gap risk," is widely recognized by CPPI managers: there is a*

P. Bernhard et al., *The Interval Market Model in Mathematical Finance*, Static & Dynamic Game Theory: Foundations & Applications, DOI 10.1007/978-0-8176-8388-7_18,

nonzero probability that, during a sudden downside move, the fund manager will not have time to readjust the portfolio, which then crashes through the floor. In this case, the issuer has to refund the difference, at maturity, between the actual portfolio value and the guaranteed amount. It is therefore important for the issuer of the CPPI note to quantify and manage this "gap risk."

In their paper [50], the authors describe the CPPI in the following way: *[...] An alternative approach [...] based on the following two ideas: first, the portfolio is* always maintained *above a certain minimum level called the floor, the difference or the "surplus" being called the "cushion" – the floor is assumed to grow at a fixed rate (for example, at the riskless rate of interest) such that* at the maturity of the fund, it is at least equal to the guaranteed amount*; second, the exposure to the market at any moment is determined as a (non-decreasing) function of the cushion, usually a constant multiple of the cushion. [...] The CPPI is a technique easy to understand and implement, and independent of time. [...] There is a small risk of the portfolio crashing through the floor in between two rebalancements, as happened with some assured portfolios during the 1987 crash. In such a case, it is impossible even to meet the guarantee. Therefore, one objective of management might be to* minimize *this possibility.*

Indeed, CPPI-type models suffer from (at least) two shortcomings: they do not *compute* but simply *evaluate* by statistical methods, optimizing several criteria including the initial investment (cushion) and the fixed weights or multipliers (used as controls), and they do not provide the value of the hedging portfolio, but only its expectation.

Take advantage of highs while protecting against lows is our aim in designing a management rule, the *Viabilist Portfolio Performance and Insurance* (VPPI) rule, which *eradicates the gap risk.*

The VPPI robot trader of VIMADES (http://$v$imades.com) is indeed a *"risk-eradication" software for the insurance and management of a portfolio* consisting of a risky asset to hedge a liability up to a given exercise date. It calculates both

1. At the date of investment, the *minimum guaranteed investment* (MGI) needed to insure liabilities (serving as a *solvency capital requirement* (SCR)).[1]
2. The VIMADES robot trader also calculates the *VPPI management rule*, which, on each date until the exercise date, provides the number of shares of the risky asset (or the exposure of the portfolio) according to the return of the risky asset *known at that time*.

The aim of doing this is to ensure that the value of the portfolio to "always" exceed liabilities during the exercise period.

[1]The SCR should reflect a level of eligible proprietary funds that enables insurance and reinsurance undertakings to absorb significant losses and that gives reasonable assurance to policyholders and beneficiaries that payments will be made as they fall due, as required in "Pillar I" of the Solvency II framework of the European Directive 2009/138/EC. Unfortunately, it stipulates that "the Solvency Capital Requirement is calculated using Value-at-Risk techniques" and not risk-eradication techniques!

The word *"always"* carries with it some requirements.

1. *On the date of investment*, the manager uses a predictive model of his choice providing the *lower bounds of future returns on the risky asset* (or underlying), which can be calculated in various ways:

   - With probabilistic models requiring the knowledge of a "volatilimeter".
   - Using extrapolation methods based on history-dependent (path-dependent) dynamic systems as a prediction mechanism. The *VIMADES Extrapolator* we suggest in this study is an example that takes into account the velocity, acceleration, and jerk of the history of an evolution to capture its trend [8, 14, 15].

2. In addition, that at *each date*, the actual return is assumed to be greater than the forecasted lower bound of the risky asset (worst case), so that the number of shares calculated by the portfolio VPPI management rules provides a value higher than the liability.

The problem is to insure and manage a portfolio consisting of a risky asset to hedge a liability flow up to a given exercise date. This is a *dynamic mechanism ensuring that, at each date, the value of the portfolio "always" exceeds liabilities.*

This is a very simple *viability* problem that has prompted the emergence of concepts and mathematical and algorithmic results gathered under the name of "viability theory" [see [7] for a recent mathematical presentation or [6] for a essay (in French)].

Denote by 0 the *investment date*, by $T > 0$ the *exercise time* (or the term, the horizon, etc.), by $t \geq 0$ the *current (or spot) time*, and by $T - t \geq 0$ the *time to maturity*. We set:

$$\begin{cases} L(t) & \textit{floor} \text{ (of liabilities);} \\ S_0(t) & \textit{price} \text{ of riskless asset;} \\ S(t) & \textit{price} \text{ of underlying;} \\ r_0(t) & = \dfrac{dS_0(t)}{S_0(t)dt} \; \textit{return} \text{ on riskless asset;} \\ r(t) & = \dfrac{dS(t)}{S(t)dt} \; \textit{return} \text{ on the underlying;} \\ p_0(t) & \text{number of shares of riskless asset;} \\ p(t) & \text{number of shares of underlying;} \\ W(t) & = p_0(t)S_0(t) + p(t)S(t) \; \textit{portfolio value;} \\ E^0(t) & = p_0(t)S_0(t) \; \textit{liquid component} \text{ of portfolio.} \\ E(t) & = p(t)S(t) \text{ the } \textit{exposure} \text{ (risky component) of the portfolio} \end{cases}$$

The insurance requirement can be written in the form

$$\forall t \in [0, T], \;\; W(t) \geq L(t) \;\; \text{and} \; W(T) = L(T). \tag{18.1}$$

Since the VPPI algorithm provides the management rules, few words are needed to define them: in the simple case of self-financed two-asset portfolios, the management is done at each instant t by choosing and acting on the exposure $E(t)$ for investing in more or fewer underlying shares.

The main problem is the choice of a management rule for hedging the portfolio and, thus, for assessing it. This is a part of uncertainty, called *contingent uncertainty*, because we do not know what management rule to choose (in his *Essais de théodicée*, *Leibniz* defines it thus: *"Contingency is a non-necessity, a characteristic attribute of freedom"*).

The first step requires looking for such rules in a "contingent reservoir" of "management rules" $(t,W) \mapsto \widetilde{E}(t,W)$ available to investors (or mechanisms of regulation). In the last analysis, given the time and value of a portfolio, the management rule dictates the exposure. In this study, contingent uncertainty is described by *financial constraints on exposures*, which require that each period t

$$\forall t \in [0,T],\ \forall W \geq 0,\ E \in [E^\flat(t,W), E^\sharp(t,W)].$$

Example: If

$$\begin{cases} \qquad B(t) \geq 0 & \text{is the target allocation and} \\ A(t) := -\frac{-p_0(t)S_0(t)}{W(t)} \geq B(t)-1 & \text{is the maximum cash allocation,} \end{cases}$$

then we set

$$E^\flat(t,W) := B(t)W(t) \text{ and } E^\sharp(t,W) := (1+A(t))W(t).$$

If $A(t) = -|A(t)| < 0$, then the condition $-p_0(t)S_0(t) \leq A(t)W$ can be written as $p_0(t)S_0(t) \geq |A(t)|W$, which means that the portfolio should *include a minimum share of the monetary value of the portfolio*.

These bounds $A(t)$ describe more or less severe prudential constraints. The default values are $A(t) = 0$ and $B(t) = 0$ (so that the contingent reservoir is in this case the interval $[0,W]$).

What we do know is, for each asset, the lower prices $S^\flat(t)$ (*LOW*) and upper prices $S^\sharp(t)$ (*HIGH*) defining the interval $\Sigma(t) := [S^\flat(t), S^\sharp(t)]$ in which the price $S(t)$ evolves. They are provided by brokerage firms.

The "tychastic" map $S(t) \in \Sigma(t)$ is an example of a tube, i.e., a set-valued (here, an interval-valued) map. One can use the (graphical) derivatives of set-valued maps introduced in [3] to deduce the range of derivatives $S'(t)$ of viable evolutions $t \mapsto S(t) \in \Sigma(t)$; see also among other references the monographs [12, 130]. Introducing the derivative $\mathrm{D}\Sigma(t,S)$ of this set-valued map σ defined at (t,S), where $S \in \Sigma(t)$, we derive from the viability theorem that the viable evolutions are governed by the *differential inclusion*

$$\forall t \geq 0,\ S'(t) \in \mathrm{D}\Sigma(t,S(t)).$$

In this very simple case of interval-valued tubes, we can compute it under adequate assumptions.

The value of the portfolio is governed by a (very simple) *tychastic controlled system*, where the controls are the exposures $E(t,W) \in [E^\flat(t,W), E^\sharp(t,W)]$ of the portfolio and the "tyches" (see Sect. 18.3.2, p. 332) are the prices $S(t) \in [S^\flat(t), S^\sharp(t)]$ of the underlying:

$$\begin{cases} \forall t \in [0,T], \\ (i) \;\; W'(t) \; = \; r_0(t)W(t) + E(t)(r(t) - r_0(t)) \text{ (evolutionary engine)}, \\ (ii) \;\; E(t) \; \in \; [E^\flat(t,W(t)), E^\sharp(t,W(t))] \text{ (controls)}, \\ (iii) \; r(t) \; \in \; \frac{D\Sigma(t,S(t))}{S(t)} \text{ (tyches)}, \end{cases} \tag{18.2}$$

subject to the floor constraints (18.1) (p. 321).

It may happen that the returns do not satisfy the requirement $S(t) \in \Sigma(t)$ for some t due to prediction errors whenever the forecasted lower bound $S^\flat(t)$ is too large. There exists an impulse ratchet mechanism that can be integrated into the regulated tychastic system (18.2) (p. 323).

Definition 18.1 Minimum Guaranteed Investments. Suppose that the floor $t \mapsto L(t)$ and the bounds $[E^\flat(t,W), E^\sharp(t,W)]$ describing the contingent uncertainty are given.

Assume that the lower bounds $r^\flat(t)$ of the returns on the underlying describing the tychastic uncertainty are known. The problem is to find at each date t

1. The (exposure) *management rule* $E^\heartsuit(t,W) \in [E^\flat(t,W), E^\sharp(t,W)]$;
2. The *minimum guaranteed investment* (MGI) $W^\heartsuit(t)$.
3. And especially the initial minimum guaranteed investment ("viability insurance") $W^\heartsuit(0)$

such that

1. Starting at investment date 0 from $W_0 \geq W^\heartsuit(0)$, then *regardless of the evolution of the tyches* $r(t) \geq r^\flat(t)$, the value $W(t)$ of the portfolio governed by the *management module*

$$W'(t) \; = \; r_0(t)W(t) + E^\heartsuit(t,W(t))(r(t) - r_0(t)) \text{ (VPPI management module)}$$

is always above the floor, and, actually, to the MGI;

2. Starting at investment date 0 from $W_0 < W^\heartsuit(0)$, regardless of the management rule $\widehat{E}(t,W) \in [E^\flat(t,W), E^\sharp(t,W)]$ (including the CPPI management rule and its variants), there exists *at least* one evolution of returns $r(t) \geq r^\flat(t)$ for which the value of the portfolio managed by

$$W'(t) \; = \; r_0(t)W(t) + \widehat{E}(t,W(t))(r(t) - r_0(t))$$

pierces the floor.

The epigraph $\mathscr{E}p(L)$ of the floor function $L(\cdot)$ is defined by

$$\mathscr{E}p(L) := \{(t,W) \text{ such that } W \; \geq \; L(t)\}.$$

Hence inequality constraint (18.1) (p. 321) can be written as the environment

$$\begin{cases} \forall t \in [0,T], \ (t,W(t)) \in \mathscr{E}p(L) \subset \mathbb{R}^2_+ \text{ (environment)}, \\ (T,W(T)) \in \text{Graph}(L) \text{ (target)}. \end{cases}$$

This viability formulation suggests the use of viability theory, and the concept of *guaranteed capture basin* of this target viable in the epigraph of L under an adequate auxiliary system (Chap. 17 or [8] or [7]). The MGI function is the south boundary of this guaranteed capture basin, which can be computed with the *capture basin algorithm* (Fig. 18.4).

18.2 VPPI in Summer 2011 Crisis

We describe the assumptions of the problem and their consequences using the Euro OverNight Index Average (EONIA). The European cousin of the American *Fed Funds Effective (Overnight Rate)* and the British *London Inter-Bank Offered Rate* as a riskless asset and the Cotation Assistée en Continu (CAC 40) as an underlying for two example floors, the classic one and a variable-annuities one. We chose the period from 2 June to 16 September 2011 as a 75-day exercise period, which is short enough for the readability of the graphics.

The following figures were automatically provided by the demonstration version of the VPPI robot trader of VIMADES in a PDF report and are reproduced here for two types of floor.

18.2.1 *Inputs*

18.2.1.1 Floor (Liabilities)

The *floor* is defined by a function $L : t \geq 0 \mapsto L(t) \geq 0$. The *liability flow* to insure plays the role of a constraint: the floor must never be "pierced" by the value of the portfolio. The *capital to guarantee (at exercise time)* is the final value $L(T)$ of the floor on the exercise date T.

The *cushion* (analogous to the *surplus* in ALM) is the difference between the portfolio value and the floor (to be guaranteed).

18.2.1.2 Forecast Mechanism

At each date t, the forecast mechanism chosen by the investor must provide the *lower bounds* $S^\flat(\tau)$ and *upper bounds* of the prices of the risky asset up to exercise time ($\tau \in [t,T]$).

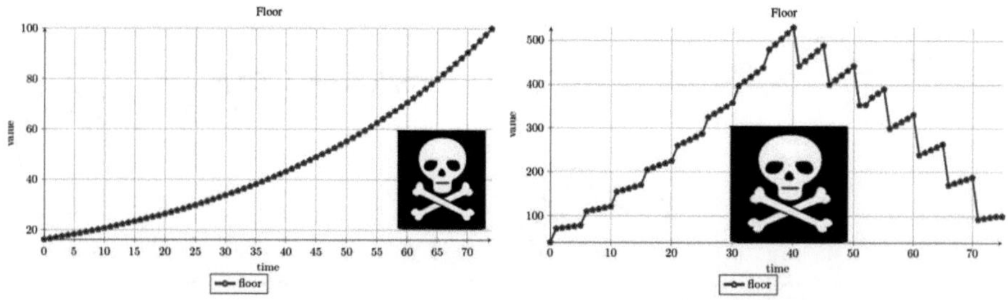

Fig. 18.1 *Floor examples. Left*: example of a classic floor growing at a fixed rate (for example, at the riskless rate of interest). *Right*: example of a "variable annuities floor": the insurer makes periodic payments during an accumulation phase and receives periodic payments for the payout phase. It is no longer continuous but punctuated by "jumps" on scheduled dates. In both cases, the value of the portfolio must *always* be above the floor, or, in mathematical terms, be viable in the epigraph of the floor function $L(\cdot)$ up to exercise time

In this case, we can define the lower bound $d^{\flat}\Sigma(t)$ of the returns of the prices by the formula

$$\mathrm{d}^{\flat}\Sigma(t) := \liminf_{h\to 0+} \frac{S^{\flat}(t+h) - S^{\sharp}(t)}{h}$$

because for any differentiable function $S(t) \in \Sigma(t)$,

$$\frac{S^{\flat}(t+h) - S^{\sharp}(t)}{h} \leq \frac{S(t+h) - S(t)}{h}$$

and, consequently, $\mathrm{d}^{\flat}\Sigma(t) \leq S'(t)$. We thus infer that returns satisfy

$$r(t) := \frac{S'(t)}{S(t)} \geq r^{\flat}(t) := \frac{\mathrm{d}^{\flat}\Sigma(t)}{S^{\sharp}(t)}.$$

Hence, knowing extrapolations or forecasts or the upper and lower bounds of prices, we can forecast the lower bounds of their returns.

The VIMADES Extrapolator is an example of history-dependent differential inclusion that depends on the history of the evolution, as well as its derivatives up to a given order, in order to capture the trends. Here, we used the velocity, the acceleration, and the jerk of the past evolution during the four preceding dates (Figs. 18.2 and 18.3).

Using the VIMADES Extrapolator and knowing the last four dates, we can extrapolate the upper and lower bounds of the prices and check that the extrapolated intervals $[S^{\flat}(t), S^{\sharp}(t)]$ are very close to the actual ones.

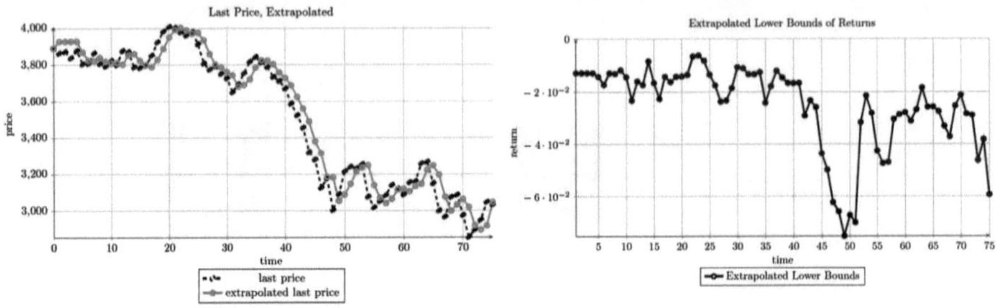

Fig. 18.2 *Extrapolation of an evolution and of the lower bounds of the risky return. Left*: Example of extrapolation of evolution of last-price series by VIMADES Extrapolator. *Right*: The VIMADES Extrapolator was used to compute forecasts of the lower bounds of the risky asset

Fig. 18.3 *VIMADES Extrapolator.* The VIMADES Extrapolator is used to extrapolate the upper and lower bounds of the price of the risky asset. *Left*: On the initial date, we assume that the values of the upper and lower bounds of the prices and the last prices are known for the four preceding dates. *Middle*: This figure displays the series of actual prices. *Right*: This figure displays the extrapolation of the upper and lower bounds of prices, which are used to compute the forecasted future lower bounds of the risky asset

18.2.2 Outputs

If the floor $L(\cdot)$ and the forecasted lower bounds $r^{\flat}(t)$ of the risky asset are given, then the VPPI robot trader provides eradication insurance.

1. On the date of investment, the insurance (Fig. 18.4):
   (a) The MGI,
   (b) The VPPI management gule.
2. On each date, the performance (Fig. 18.5):
   (a) Value of the portfolio,
   (b) Number of shares,
   (c) Or, equivalently, the exposure (Figs. 18.4 and 18.5).

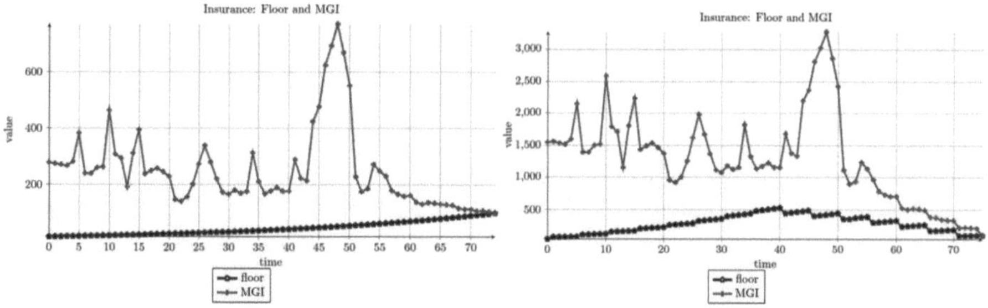

Fig. 18.4 *The Floor and the MGI.* For each floor, the bottom curve represents the floor $L(t)$ that should never be pierced by the portfolio value. The *top curve* $W^{\heartsuit}(t)$ represents the MGI, which depends only on the floor, the riskless return, and lower bounds of underlying returns obtained by the VIMADES Extrapolator

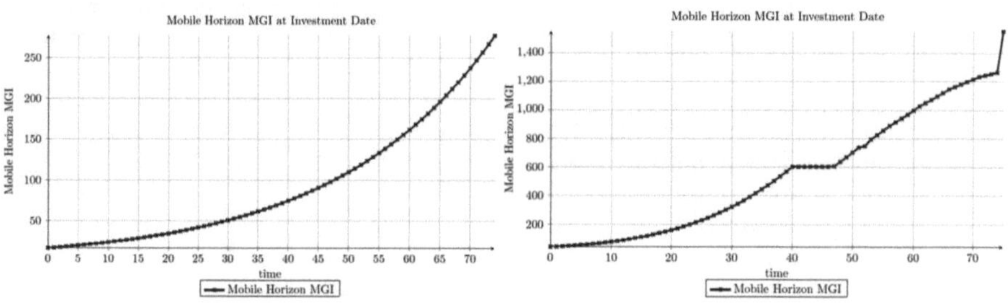

Fig. 18.5 *Mobile minimum guaranteed investments.* For each floor, the graphs display the minimum guaranteed investment up to each date prior to the exercise date

18.2.2.1 Minimum Guaranteed Investment

Although eradication of risk requires an amount of the MGI that may seem too high, it provides valuable information for investors who wish to measure the risk of choosing a smaller investment: even if the worst is not certain, it can happen during a crisis.

The *mobile minimum guaranteed investment* measures this kind of risk: it calculates the minimum guaranteed investment up to each date prior to the exercise date. By inverting this relation, it associates with any investment the final date of the period during which the hedging of the liability is guaranteed.

We observe that if the reluctance to immobilize a guaranteed investment is due to the idea of allocating this amount to invest it in other assets hoping that *diversification will be beneficial*, it is sufficient to conclude that in this case, *the allocated investments will even be lower than the required mMGI of each of these assets, worsening the risk taken.* This is a strong motivation for either classifying assets by their required minimum guaranteed investment or studying a portfolio with many assets and computing the MGI of each asset, which could be used by credit rating agencies to rate assets.

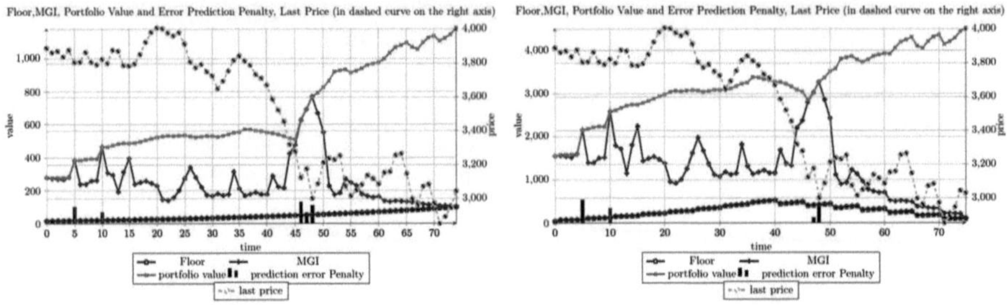

Fig. 18.6 *Value of the hedging portfolio.* For each floor, the *bottom curve* represents the floor $L(t)$ that should never be pierced by the portfolio value. The graphs of the MGI are still displayed. The *top curve* is the graph of the value of the portfolio managed by the VPPI management rule when, on each date, the price of the underlying is known; loans to correct prediction errors, represented as a histogram, compensate for the difference between the value of the MGI and the portfolio value when it is lower than that of the MGI. The *dotted line* represents the evolution of the underlying price (right scale) to visualize together the behavior of the MGI and of the value of the portfolio depending on the price

minimum guaranteed investment (MGI)	277.82	minimum guaranteed investment (MGI)	1551.91
minimum guaranteed cushion (MGC)	261.66	minimum guaranteed cushion (MGC)	1512.28
actualized exercice value	1177.3	actualized exercice value	4495.05
cumulated prediction penalties	471.03	cumulated prediction penalties	1413.3
liquidating dividend (in %)	4.12	liquidating dividend (in %)	2.91
net liquidating dividend (in %)	2.33	net liquidating dividend (in %)	1.98
riskless rate during period (in %)	0.63	riskless rate during period (in %)	0.63

Fig. 18.7 *Synoptic tables.* For each floor, the synoptic tables are represented. Despite the crisis, the spreads between the net liquid dividends and the riskless rates during the exercise period are equal respectively to 1.70% and 0.35% over a 75-day exercise period: *takes advantage of highs while protect against lows*, as it was announced, and the "I" in *(Viabilist Portfolio Performance and Insurance)* is a real insurance

18.2.2.2 Portfolio Values

The following table summarizes the principal numerical characteristic features of the portfolio (Figs. 18.6 and 18.7):

1. On the *investment date*, the *information*:
   (a) The value $W^{\heartsuit}(0)$ of the *initial MGI* (e.g., solvency capital requirement, economic capital);
   (b) The value $W^{\heartsuit}(0) - L(0)$ of the *initial minimal guaranteed cushion* (MGC).
2. On *each date*, when the asset prices are known, the *management of the portfolio* is summarized by
   (a) The *actualized value* $e^{-\int_0^T r_0(\tau)d\tau} W(T)$ of the value $W(T)$ at exercise time.
   (b) The cumulative value $\widehat{\mathbb{B}}(T)$ of *actualized prediction error penalties* obtained by the ratchet mechanism during the exercise period.

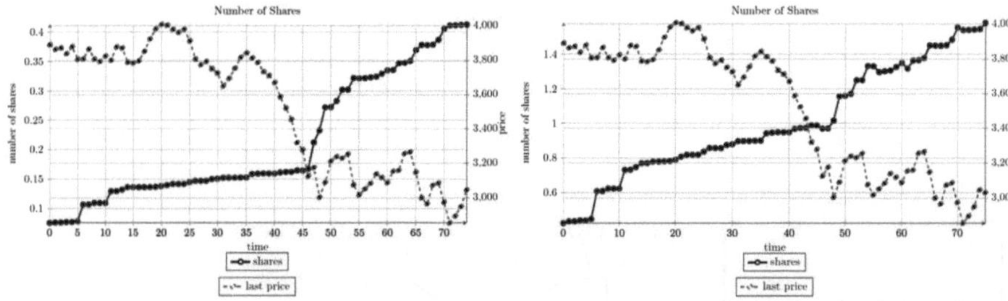

Fig. 18.8 *Number of shares.* Once the VPPI management rule is calculated and stored in a computer's memory, the management module provides the number of shares of the underlying within its portfolio based on the realization of prices. For each floor, the figure represents the evolution of the number of shares

(c) The net *return* (including error predictions corrected by the ratchet mechanism) $\dfrac{e^{-\int_0^T r_0(\tau)d\tau}W(T)+\widehat{\mathbb{B}}(T)}{W(0)}$, as a percentage.

(d) The net *liquidating dividend* (including error predictions corrected by the ratchet mechanism), $\dfrac{e^{-\int_0^T r_0(\tau)d\tau}(W(T)-L(T))+\widehat{\mathbb{B}}(T)}{W(0)-L(0)}$, as a percentage.

(e) The *riskless rate during the exercise period,* $\dfrac{e^{\int_0^T r_0(\tau)d\tau}}{W(0)-L(0)}$, as a percentage.

18.2.2.3 Number of Risky Shares in Portfolio

The knowledge of the initial MGI, which serves as a pricer, is only a part of the solution to the problem and needs to be complemented by a knowledge of the management rule to give it meaning. The VPPI robot trader provides on each date the amount of shares to manage the portfolio taking advantage of highs while protecting against lows by remaining viable (above the floor).

This VPPI version does not assume that there are scarcity constraints on the number of available shares or brokerage fees. A current version taking into account constraints on shares, brokerage fees, and minimizing transactions (for minimizing brokerage fees) in process, which will be extended to the case of a portfolio with several assets.

18.2.3 Flow Chart of VPPI Software

A flow chart of the VPPI software provides a concise way to summarize the functioning of the robot trader, divided into the computation of the MGI and the management rule on the date of investment, and then the computation of the shares of the portfolio along with its values.

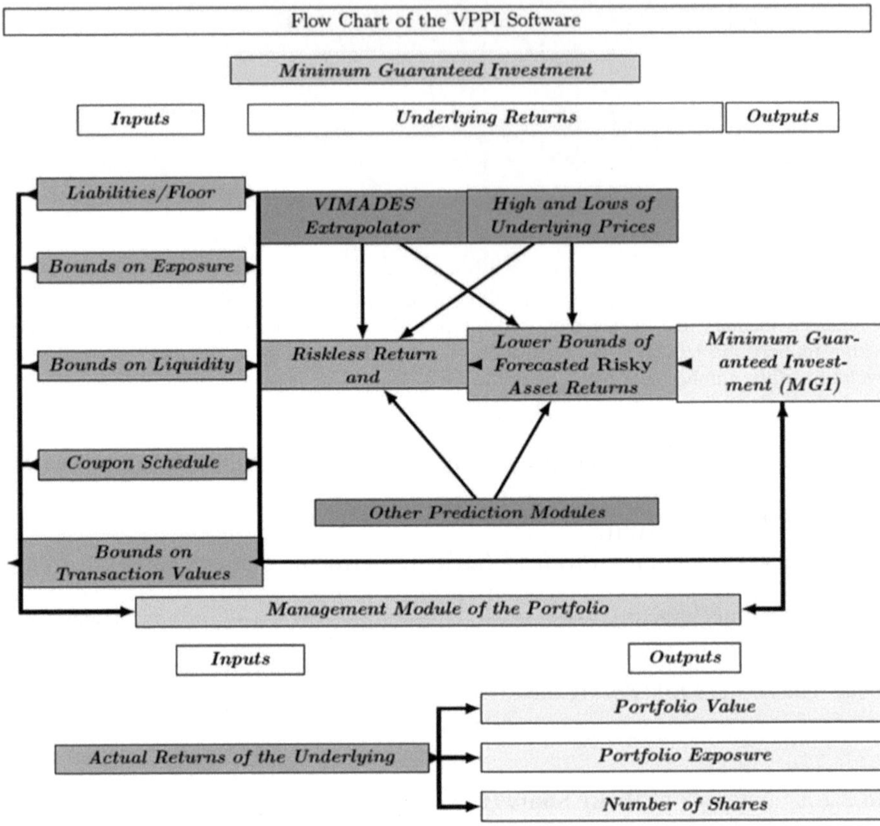

Fig. 18.9

18.3 Uncertainties

Choosing a management rule represents a choice under contingent uncertainty, but a choice that the investor or the manager can make.

Evaluation or eradication of losses hides behind the fact that there are uncertainties on which the investor cannot act. In this particular insurance problem, these uncertain variables are the returns, which are assumed to be unknown, whereas economic theory is concerned with the analysis and computation of supply and demand adjustment laws, among which is *Walras's tâtonnement*, or *groping*, in the hope of explaining the mechanisms of price formation (e.g., [5] or Chap. 5 of [6]). In the last analysis, the choice of the prices is made by the invisible hand of the "Market," the new deity in which many economists and investors believe. But it is a hidden deity that leaves the investor with the task of forecasting it.

In most financial scenarios, investors take into account their ignorance of the pricing mechanism. They assume instead that prices *evolve under uncertainty* and that they can master this uncertainty. They still share the belief that the "Market knows best" how to regulate the prices, above all without human or political regulations. The question becomes to know how to master this uncertainty. For that, many of them "trade" the Adam Smith invisible hand against a Brownian process, since it seems that this unfortunate hand is shaking the asset price like a particle on the surface of a liquid. It should then be enough to assume that average returns and volatilities are known for the purpose of managing portfolios.

Hence the design of the management rule takes into account not only the floor but also some "measure" of this kind of uncertainty:

1. Either we choose an a priori management rule and we exchange the Adam Smith invisible hand on the formation of asset prices against stochastic uncertainty to derive management rules of the portfolio evaluating losses through a battery of measures mentioned earlier
2. Or we build a management rule a posteriori that would allow us to eradicate loss against tychastic uncertainty: the value of the portfolio is "always" greater than or equal to the liabilities whatever the returns ranging over a "tychastic reservoir" (composed of risky returns higher than their forecasted lower bounds) in which returns appear unexpectedly.

The problem is no longer one of assessing the probability of risky return realizations but *determining the subset in which such returns can emerge*. Prediction models or extrapolation techniques no longer consist in determining trends and volatilities but, in the tychastic case, the *lower bounds of risky returns defining the tychastic reservoir at each future date.*

18.3.1 Stochastic Uncertainty

Stochastic uncertainty on returns is described by a space Ω, filtration $\mathscr{F}_t$, the probability $\mathbb{P}$, a Brownian process $B(t)$, a drift $\gamma(S)$, and a volatility $\sigma(S)$:

$$r(t)\mathrm{d}t := \frac{\mathrm{d}S(t)}{S(t)} = \gamma(S(t))\mathrm{d}t + \sigma(S(t))\mathrm{d}B(t). \tag{18.3}$$

1. The random events are not explicitly identified. The set Ω is not described explicitly (one can always choose the space of all evolutions or the interval $[0,1]$ in the proofs of theorems). Only the drift and volatility are assumed to be explicitly known. Random variables are hidden in their laws.
2. Stochastic uncertainty does not study the "package of evolutions" (depending on $\omega \in \Omega$), but *functionals over this package*, such as the different moments and their statistical consequences (e.g., averages, variance) used in the evaluation of risk (they deal with both the space of evolutions and spaces of functionals on these evolutions). Even though, in some cases, Monte Carlo methods provide an approximation of the set of evolutions (for constant ω), there is no mechanism used for selecting those satisfying a particular property.

3. Required properties are valid for *almost all constants* ω.
4. Stochastic differential equations provide only the expectation of the package of evolutions but do not allow for the selection of the right one whenever, for every time $t > 0$, the effective ω (which then depends on time) is known.

We cite a few references among many: [28–30, 49, 120] and, in the discrete case, [147].

18.3.2 Tychastic Uncertainty

What we do know is, for each asset, the range $\Sigma(t) := [S^\flat(t), S^\sharp(t)]$ in which the price $S(t)$ evolves, which is an observable measure of "tychastic uncertainty." The first question we should ask when using a stochastic model is as follows: (18.3) is the consistency with the price constraint; so do (almost) all evolutions $S(\cdot)$ governed by this stochastic differential equation satisfy $S(t) \in \Sigma(t)$ for all $t \in [0,T]$? This is a stochastic viability problem that can be solved and that shows that it is not the case for lack of a known "volatilitimeter."

The motivation for tychastic uncertainty is to start from the constraint $S(t) \in \Sigma(t)$, which is an example of a tube, i.e., a set-valued (here, an interval-valued) map. One can use the (graphical) derivatives of set-valued maps introduced in [3] to deduce the range of derivatives $S'(t)$ of viable evolutions $t \mapsto S(t) \in \Sigma(t)$: see also among other references the monographs [4, 7, 12]. Introducing the derivative $\mathrm{D}\Sigma(t,S)$ of this set-valued map Σ defined at (t,S) where $S \in \Sigma(t)$, we derive from the viability theorem that the viable evolutions are governed by the *differential inclusion*

$$\forall t \geq 0, \ S'(t) \in \mathrm{D}\Sigma(t,S(t))$$

In this very simple case of interval-valued tubes, we can compute it under adequate assumptions,

$$\mathrm{D}\Sigma(t,S(t)) = \left[\liminf_{h\to 0+}\frac{S^\flat(t+h)-S^\sharp(t)}{h}, \limsup_{h\to 0+}\frac{S^\sharp(t+h)-S^\flat(t)}{h}\right]$$

and is easily accessible.

Therefore, *it is natural to trade stochastic differential equations (18.3), which do not necessarily provide viable evolutions* $S(t) \in \Sigma(t)$, *for the differential inclusion* $S'(t) \in \mathrm{D}\Sigma(t,S(t))$ *governing viable evolutions in this tube.*

In this study, tyches are the prices of the underlying over which the investor has no influence ["Mrs. Market" is the common name given to the deity (Tyche) that sets prices, and, nowadays, "markets," which are opaque disguises to hide plain speculators]. The uncertainty is described by the tychastic map $t \mapsto \Sigma(t)$.

1. Tyches are identified (prices of the underlying, in our case) and can then *be used* in dynamic management systems when they are actually observed and known on each date during the exercise period.

2. For this reason, the results are computed in the worst case (*eradication of risk* instead of its *statistical evaluation*).
3. Required properties are valid for *all evolutions* of tyches $t \mapsto S(t) \in \Sigma(t)$ instead of *constant* ω.

The investor is supposed to provide a tychastic map $\Sigma(t) := [S^{\flat}(t), S^{\sharp}(t)]$ of the forecasted prices. This amounts to forecasting the lower and higher bounds of prices. There are a myriad of ways for forecasting the upper and lower bounds of prices, from chartist to the most sophisticated econometric methods, which include the VIMADES Extrapolator.

Contingent uncertainty "offsets" tychastic uncertainty: In fact, the MGI decreases when the contingent reservoir increases and increases when the "tychastic reservoir" $\nabla(t)$ increases, that is to say, in this study, when the lower bound of the underlying return decreases.

Note at this point that the stochastic viability of the tube$\Sigma(t)$ under the stochastic differential equation

$$\mathrm{d}S(t) = \rho(t)S(t)\mathrm{d}t + \sigma(t)S(t)\mathrm{d}B(t) \tag{18.4}$$

is a particular case of its tychastic viability under the Stratonovitch tychastic differential inclusion

$$S'(t) = \rho(t)S(t) - \frac{\sigma(t)S^2(t)}{2} + \sigma(t)S(t)v(t) \text{ where } v(t) \in \mathbb{R},$$

where $\rho(t)S(t) - \frac{\sigma(t)S^2(t)}{2}$ is the Stratonovitch drift, thanks to the Stroock–Varadhan support theorem (see [139, 140] and [9–11, 61–63]). Since the tyches $v(\cdot)$ of the Stratonovitch tychastic differential equation range over the whole space $\mathbb{R}$, the stochastic viability kernel of $\Sigma(t)$ is empty! In other words, the stochastic differential equation (18.4) governs the evolution of prices, not all of which are viable.

18.3.3 Historic Differential Inclusions

The articles [14, 15] propose to replace the use of stochastic differential equations for forecasting uncertain future evolutions by history-dependent (or path-dependent, memory-dependent, functional) control systems. At each instant, they associate with the history of the evolution up to each time t a set of velocities.

Histories are evolutions $\varphi \in \mathscr{H}(\mathbb{R}^n) := \mathscr{C}(-\infty, 0, \mathbb{R}^n)$ defined for negative times that play the role of a state space. They require a specific "differential" calculus. For instance, let a history-dependent functional $v : \varphi \in \mathscr{H}(\mathbb{R}^n) \mapsto v(\varphi) \in \mathbb{R}$. The addition operator $\varphi \mapsto \varphi + h\psi$ is replaced by the *concatenation operator* $\Diamond_h$ associating with each history $\varphi \in \mathscr{C}(-\infty, 0; \mathbb{R}^n)$ the function $\varphi \Diamond_h \psi \in \mathscr{C}(-\infty, 0; \mathbb{R}^n)$ defined by

$$(\varphi \Diamond_h \psi)(\tau) := \begin{cases} \varphi(\tau+h) & \text{if } \tau \in [-\infty, -h], \\ \varphi(0) + \psi(\tau+h) & \text{if } \tau \in [-h, 0]. \end{cases}$$

This allows us to define the concept of *Clio*[2] *derivatives* by taking the limits of "differential quotients"

$$\nabla_h v(\varphi)(\psi) := \frac{v((\varphi \Diamond_h \psi)) - v(\varphi)}{h}$$

to obtain

$$Dv(\varphi)(\psi) := \lim_{h \to 0+} \nabla_h v(\varphi)(\psi)$$

if this limit exists and is linear and continuous on $\mathscr{H}(\mathbb{R}^n)$. Then the gradient of v at φ is an element of the dual $\mathscr{H}(\mathbb{R}^n)^*$ of $\mathscr{H}(\mathbb{R}^n)$, i.e., a vector measure.

Actually, the first "general" viability theorem was proved by Georges Haddad in the framework of history-dependent differential inclusions in the late 1970s (see [4, 79–81]). Since these investigations, motivated by the evolutionary systems in the life sciences, including economics and finance, are much more involved than that of differential inclusions, most of the viability studies rested on the case of differential inclusions.

Actually, one can also use history-dependent differential equations or inclusions depending on the functionals of past evolutions, such as their derivatives up to a given order.

Hence, these history-dependent differential inclusions can be used for extrapolating asset prices, in particular their upper and lower bounds, providing the lower bounds of the returns.

The *VIMADES Extrapolator* (based on Laurent Schwartz distributions and on [15]) is an example of a history-dependent differential inclusion that bypasses the use of a "volatilitimeter" by extrapolating each history-dependent on (past) evolutions of upper (*HIGH*) and lower bounds (*LOW*) of the underlying prices, provided by brokerage firms, from which we can forecast the lower bounds of the future returns of the underlying [8].

18.3.4 Nature and Measures of Insurance

In the stochastic case, risk is "measured" by real numbers through statistical evaluation (value at risk, for example). These numbers are *abstractions*, which differ according to the methods and techniques used.

---

[2]Clio, muse of history, was, like all the muses, the daughter of Zeus and Mnemosyne, goddess of memory.

In the tychastic case, risk is "measured" by numbers, providing at each date t the MGI $W^{\heartsuit}(t)$, ensuring that the floor will *never* be pierced later. The guaranteed cash flow $t \mapsto W^{\heartsuit}(t)$ associates the MGI with each date.

18.4 Comparing VPPI and CPPI

The CPPI and VPPI management rules of a portfolio for hedging a floor are quite different, the first one belonging to the class of *a priori* management rules, where the potential losses are statistically evaluated, the second one belonging to the class of *a posteriori* management rules designed to eradicate losses. Hence they are difficult to compare because they answer different questions, related to either risk evaluation or risk eradication. The question that the CPPI management rule leaves open is the empirical determination of the multiplier and the initial cushion, whereas the VPPI management rule "encapsulates" the cushion multipliers and computes the guaranteed minimum cushion:

1. *CPPI*: The floor, the initial investment, and a prediction mechanism are given, as are multipliers (constant as the CPPI or not as in other rules).
2. *VPPI*: The MGI and the VPPI management rule are deduced from the floor and lower bounds of future risky returns.

In both cases, however, the floor is fixed, and the choice of forecasting mechanism is open, from chartist methods to statistical and stochastic techniques to extrapolation mechanisms based on history-dependent differential inclusions.

Whenever the CPPI management rule is chosen and "integrated" into the stochastic differential equation governing the values of the portfolio, the statistical evaluations depend naturally on this choice (see [54], for instance).

The same is true for the VPPI management rule, where the MGI (or cushion) depends on the flow of the lower bounds of forecasted returns for losses to be permanently eradicated.

Comparison of VPPI and CPPI

	VPPI	CPPI
Multipliers	Computed	Given
Regulation	Yes	Given
Insurance	Yes (MGI)	Given (loss evaluation)
Prediction	Yes	Yes
errors	(impulse management)	(jump processes)
Forecasting	Any method for predicting lower bounds	Stochastic
methods	of returns, e.g., VIMADES Extrapolator	processes

References

1. Ahn, H., Dayal, M., Grannan, E., Swindle, G.: Option replication with transaction costs: general diffusion limits. Ann. Appl. Probab. **8**, 676–707 (1998)
2. Artzner, P., Delbaen, F., Eber, J.M., Heath, D.: Coherent measures of risk. Math. Finance **9**, 203–228 (1999)
3. Aubin, J.P.: Contingent derivatives of set-valued maps and existence of solutions to nonlinear inclusions and differential inclusions. In: Nachbin, L. (ed.) Mathematical Analysis and Applications, Advances in Mathematics, vol. 7a, pp. 159–229 (1981)
4. Aubin, J.P.: Viability Theory. Birkhaüser, Boston (1991)
5. Aubin, J.P.: Dynamic Economic Theory: A Viability Approach. Springer, Berlin (1997)
6. Aubin, J.P.: La mort du devin, l'émergence du démiurge. Essai sur la contingence et la viabilité des systèmes. Éditions Beauchesne (2010)
7. Aubin, J.P., Bayen, A., Saint-Pierre, P.: Viability Theory. New Directions. Springer, Berlin (2011)
8. Aubin, J.P., Chen, L., Dordan, O., Saint-Pierre, P.: Viabilist and tychastic approaches to guaranteed ALM problem. Risk Decis. Anal. (2011)
9. Aubin, J.P., Da Prato, G.: Stochastic Nagumo's viability theorem. Stoch. Anal. Appl. **13**, 1–11 (1995)
10. Aubin, J.P., Da Prato, G.: The viability theorem for stochastic differential inclusions. Stoch. Anal. Appl. **16**, 1–15 (1998)
11. Aubin, J.P., Da Prato, H., Frankowska, H.: Stochastic invariance for differential inclusions. J. Set-Valued Anal. **8**, 181–201 (2000)
12. Aubin, J.P., Frankowska, H.: Set Valued Analysis. Birkhaüser, Boston (1990)
13. Aubin, J.P., H., D.: Characterization of stochastic viability of any non-smooth set involving its generalized contingent curvature. Stoch. Anal. Appl. **25**, 951–981 (2003)
14. Aubin, J.P., Haddad, G.: Path-dependent impulse and hybrid systems. In: Benedetto, D., Sangiovanni-Vincentelli (eds.) Hybrid Systems: Computation and Control, pp. 119–132. Springer, Berlin (2001)
15. Aubin, J.P., Haddad, G.: History (path) dependent optimal control and portfolio valuation and management. J. Positivity **6**, 331–358 (2002)
16. Aubin, J.P., Lygeros, J., Quincampoix, M., Sastry, S., Seube, N.: Impulse differential inclusions: a viability approach to hybrid systems. IEEE Trans. Automat. Control **47**, 2–18 (2002)
17. Aubin, J.P., Pujal, D., Saint-Pierre, P.: Dynamic management of portfolios with transaction costs under tychastic uncertainty. In: Ben Hammeur, H., Breton, M. (eds.) Numerical Methods in Finance, pp. 59–89. Springer, New York (2005)

P. Bernhard et al., *The Interval Market Model in Mathematical Finance*, Static & Dynamic Game Theory: Foundations & Applications, DOI 10.1007/978-0-8176-8388-7,

18. Avellaneda, M., Levy, A., Parás, A.: Pricing and hedging derivative securities in markets with uncertain volatilities. Appl. Math. Finance **2**, 73–88 (1995)
19. Bachelier, L.: Théorie de la spéculation. Annales scientifiques de l'É.N.S., 3ème série **Tome 17**, 21–86 (1900)
20. Banaïm, M., Le Boudec, J.Y.: A class of mean field interaction models for computer and communication systems. Perform. Eval. **65**, 823–838 (2008)
21. Bardi, M., Capuzzo-Dolcetta, I.: Optimal Control and Viscosity Solutions of Hamilton–Jacobi–Bellman Equations. Birkhaüser, Boston (1997)
22. Barles, G.: Solutions de viscosité des équations de Hamilton–Jacobi. Springer, Berlin (1994)
23. Barles, G., Sonner, H.M.: Option pricing with transaction costs and a nonlinear Black and Scholes equation. Finance Stoch. **2** (1998)
24. Basel Committee on the Global Finance System: Stress testing by large financial institutions: current practice and aggregation issues (2000)
25. Bellman, R.E.: Dynamic Programming. Princeton University Press, Princeton (1957)
26. Ben Ameur, H., Breton, M., Martinez, J.: A dynamic programming approach for pricing derivatives in the garch model. Manage. Sci. **55**, 252–266 (2009)
27. Ben-Tal, A., El-Ghaoui, L., Nemirovski, A.: Robust Optimization. Princeton University Press, Princeton (2009)
28. Benaïm, M., El Karoui, N.: Promenade alatoire: Chaînes de Markov et simulations; martingales et stratégies. École Polytechnique, Palaiseau, France (2005)
29. Bensoussan, A., Crouhy, M., Galai, D.: Stochastic equity volatility and the capital structure of the firm. Philos. Trans. R. Soc. Lond. **A347** (1994)
30. Bensoussan, A., Crouhy, M., Galai, D.: Stochastic volatility related to the leverage effect. II: Valuation of european equity options and warrants. Appl. Math. Finance **2**, 43–60 (1995)
31. Bensoussan, A., Diltz, J.D., Sing Ru, H.: Real options games in complete and incomplete markets with several decision makers. SIAM J. Finan. Math. **1**, 666–728 (2010)
32. Bensoussan, A., Lions, J.L.: Contrôle impulsionnel et inéquations quasi-variationnelles. Dunod, Paris (1982)
33. Bernhard, P.: Singular surfaces in differential games, an introduction. In: Haggedorn, P., Olsder, G., Knobloch, H. (eds.) Differential Games and Applications. Lecture Notes in Information and Control Sciences, vol. 3, pp. 1–33. Springer, Berlin (1977)
34. Bernhard, P.: Une approche déterministe de l'évaluation d'options. In: Menaldi, J.L., Rofman, E., Sulem, A. (eds.) Optimal Control and Partial Differential Equations, volume in honor of Professor Alain Bensoussan's 60th birthday, pp. 511–520. IOS Press (2001)
35. Bernhard, P.: A robust control approach to option pricing. In: Salmon, M. (ed.) Applications of Robust Decision Theory and Ambiguity in Finance. City University Press, London (2003)
36. Bernhard, P.: The robust control approach to option pricing and interval models: an overview. In: Breton, M., Ben-Ameur, H. (eds.) Numerical Methods in Finance, pp. 91–108. Springer, New York (2005)
37. Bernhard, P.: A robust control approach to option pricing including transaction costs. In: Nowak, A.S., Szajowski, K. (eds.) Advances in Dynamic Games, Applications to Economics, Finance, Optimization, and Stochastic Control. Annals of the ISDG, also 9th ISDG International Symposium on Dynamic Games and Applications, Adelaide, South Australia, 2000, vol. 7, pp. 391–416. Birkhaüser, Boston (2005)
38. Bernhard, P., El Farouq, N., Thiery, S.: An impulsive differential game arising in finance with interesting singularities. In: Haurie, A., Muto, S., Petrosjan, L.A., Raghavan, T. (eds.) Advances in Dynamic Games. Annals of the ISDG, also 10th ISDG International Symposium on Dynamic Games and Applications, Saint Petersburg, 2002, vol. 8, pp. 335–363. Birkhaüser, Boston (2006)
39. Bernhard, P., El Farouq, N., Thiery, S.: Robust control approach to option pricing: representation theorem and fast algorithm. SIAM J. Control Optim. **46**, 2280–2302 (2007)
40. Bernhard, P., Thiery, S., Deschamps, M.: La tarification d'options. Proposition pour une approche déterministe. In: 56ème Congrès annuel de l'Association Française des Sciences Économiques. Paris (2007)

41. Bertsimas, D., Bandi, C., Chen, A.: Robust option pricing: An ε-arbitrage approach. Eur. J. Oper. Res. (submitted, 2010)
42. Bertsimas, D., Brown, D., Caramanis, C.: Theory and applications of robust optimization. SIAM Rev. **53**, 464–501 (2011)
43. Bick, A., Willinger, W.: Dynamic spanning without probabilities. Stoch. Process. Appl. **50**, 349–374 (1994)
44. Bielecki, T.R., Jeanblanc, M., Rutkowski, M.: Hedging of defaultable claims. In: Paris-Princeton Lectures on Mathematical Finance. Springer Lecture Notes in Mathematics, vol. 1847, pp. 1–132. Springer, Berlin (2003)
45. Bingham, N.H., Kiesel, R.: Risk-Neutral Valuation. Springer, Berlin (2004)
46. Black, F., Scholes, M.: The pricing of options and corporate liabilities. J. Polit. Econ. **81**, 229–263 (1973)
47. Bonneuil, N., Saint-Pierre, P.: The hybrid guaranteed viability capture basin algorithm in economics. In: Alur, R., Pappas, G. (eds.) Hybrid Systems: Computation and Control. Springer Lecture Notes in Computer Science, vol. 2993, pp. 187–202. Springer, Berlin (2004)
48. Bonneuil, N., Saint-Pierre, P.: Beyond optimality: managing children, assets, and consumption over the life cycle. J. Math. Econ. **44**, 227–241 (2008)
49. Bouchaud: Theory of Financial Risk and Derivative Pricing: From Statistical Physics to Risk Management. Cambridge University Press, London (2009)
50. Boulier, J.F., Kanniganti, A.: Expected performance and risk of various portfolio insurance strategies. http://www.actuaries.org/AFIR/colloquia/Brussels/Boulier-Kanniganti.pdf (2005)
51. Capuzzo-Dolcetta, I.: On a discrete approximation of the Hamilton Jacobi equation of dynamic programming. Appl. Math. Optim. **10**, 367–377 (1983)
52. Cardaliaguet, P., Quincampoix, M., Saint-Pierre, P.: Set-valued numerical methods for optimal control and differential games. In: Nowak, A. (ed.) Stochastic and Differential Games. Theory and Numerical Methods. Annals of the International Society of Dynamic Games, pp. 177–247. Birkhaüser, Boston (1999)
53. Carr, P., Geman, H., Madan, D.B.: Pricing and hedging in incomplete markets. J. Finan. Econ. **32**, 131–167 (2001)
54. Cont, R., Tankov, P.: Constant proportion portfolio insurance in presence of jumps in asset prices. Math. Finance **19**, 379–401 (2009)
55. Cousin, A., Jeanblanc, M., Laurent, J.P.: Hedging CDO tranches in a markovian environment. In: Paris-Princeton Lectures on Mathematical Finance. Springer Lecture Notes in Mathematics, vol. 2003, pp. 1–63. Springer, Berlin (2010)
56. Cox, J.C., Ross, A.: The valuation of options for alternative stochastic processes. J. Financ. Econ. **3**, 145–166 (1976)
57. Cox, J.C., Ross, S.A., Rubinstein, M.: Option pricing: a simplified approach. J. Finan. Econ. **7**, 229–263 (1979)
58. Crück, E.: Target problems under state constraints for nonlinear controlled impulsive systems. J. Math. Anal. Appl. **270**, 636–656 (2002)
59. Crück, E., Saint-Pierre, P.: Nonlinear impulse target problems under state constraints: a numerical analysis based on viability theory. Set-Valued Anal. **12**, 383–416 (2004)
60. Crück, E., Quincampoix, M., Saint-Pierre, P.: Pursuit-Evasion Games with Impulsive Dynamics, in Advances in Dynamic Game Theory, Numerical Methods, Algorithms, and Applications to Ecology and Economics, Annals of the International Society of Dynamic Games, Vol. 9. Jorgensen, Steffen; Quincampoix, Marc; Vincent, Thomas L. (Eds.) XXII, 718 p. 190 illus (2007)
61. Da Prato, G., Frankowska, H.: A stochastic Filippov theorem. Stoch. Calculus **12**, 409–426 (1994)
62. Da Prato, G., Frankowska, H.: Stochastic viability for compact sets in terms of the distance function. Dyn. Syst. Appl. **10**, 177–184 (2001)
63. Da Prato, G., Frankowska, H.: Invariance of stochastic control systems with deterministic arguments. J. Differ. Equat. **200**, 18–52 (2004)

64. Darling, R.W.R., Norris, J.R.: Differential equation approximations for Markov chains. Probab. Surv. **5**, 37–79 (2008)
65. Davis, M.H.A., Norman, A.R.: Portfolio selection with transaction costs. Math. Oper. Res. **15**, 676–713 (1990)
66. De Meyer, B., Moussa Saley, H.: On the strategic origin of Brownian motion in finance. Int. J. Game Theory **31**, 285–319 (2002)
67. Derman, E., Kani, I.: Riding on a smile. RISK, pp. 32–39 (1994)
68. Dixit, A., Pindyck, R.S.: Investment under Uncertainty. Princeton University Press, Princeton (1994)
69. El Farouq, N., Barles, G., Bernhard, P.: Deterministic minimax impulse control. Appl. Math. Optim. **61**, 353–378 (2010)
70. El Farouq, N., Bernhard, P.: Option pricing with zero lower bound of impulse cost. In: 14th International Symposium on Dynamic Games and Applications, Banff, Canada (2010)
71. El Karoui, N., Quenez, M.C.: Dynamic programming and pricing of contingent claims in an incomplete market. SIAM J. Control **33**, 29–66 (1995)
72. Fleming, W.H., Soner, H.M.: Controlled Markov Processes and Viscosity Solutions. Springer, Berlin (2006)
73. Föllmer, H.: Calcul d'Itô sans probabilités. In: Azéma, J., Yor, M. (eds.) Séminaire de probabilités XV. Lecture Notes in Mathematics, vol. 850. Springer, Berlin (1981)
74. Föllmer, H., Kramkov, D.: Optional decompositions under constraints. Probab. Theory Related Fields **109**, 1–25 (1997)
75. Frey, R., Backhaus, J.: Pricing and hedging of portfolio credit derivatives with interacting default intensities. J. Theor. Appl. Finance **11**, 611–634 (2008)
76. Friedman, M.: Essays in Positive Economics. University of Chicago Press, Chicago (1953)
77. Geske, R.: The valuation of compound options. J. Finan. Econ. **7**, 63–81 (1979)
78. Greenspan, A.: Greenspan's plea for stress testing. Risk **13** (2000)
79. Haddad, G.: Monotone trajectories of differential inclusions with memory. Isr. J. Math. **39**, 83–100 (1981)
80. Haddad, G.: Monotone viable trajectories for functional differential inclusions. J. Differ. Equat. **42**, 1–24 (1981)
81. Haddad, G.: Topological properties of the set of solutions for functional differential inclusions. Nonlinear Anal. Theory Methods Appl. **5**, 1349–1366 (1981)
82. Henry, B.I., Langlands, T.A.M., Straka, P.: Fractional fokker-planck equations for subdiffusion with space- and time-dependent forces. Phys. Rev. Lett. **105**, 170602 (2010)
83. Hobson, D.: Comparison results for stochastic volatility models via coupling. Finance Stoch. **14**, 129–152 (2010)
84. Hobson, D.G.: Volatility misspecification, option pricing and super-replicatin via coupling. Ann. Appl. Probab. **8**, 193–205 (1998)
85. Hobson, D.G.: The skorohod embedding problem and model-independent bounds for option prices. In: Paris-Princeton Lectures on Mathematical Finance. Springer Lecture Notes in Mathematics, vol. 2003, pp. 267–318. Springer, Berlin (2010)
86. Hucki, Z., Kokoltsov, V.: Pricing of rainbow options: game theoretic approach. Int. Game Theory Rev. **9**, 215–242 (2007) (Preprint: Nottingham Trent University, 2003)
87. Hull, J., White, A.: The pricing of options on assets with stochastic volatilities. J. Finance **42**, 281–300 (1987)
88. Hull, J.: Options, Futures, and Other Derivatives, 8th edn. Pearson, London (2011)
89. Isaacs, R.: Differential Games, a Mathematical Theory with Applications to Optimization, Control and Warfare. Wiley, New York (1965)
90. Jayne, J.E., Rogers, C.A.: Selectors. Princeton University Press, Princeton (2002)
91. Joshua: The book of Joshua. In: The Bible (circa 500 B.C.)
92. Jumarie, G.: Derivation and solutions of some fractional Black–Scholes equations in coarse-grained space and time. Comp. Math. Appl. **59**, 1142–1164 (2010)
93. Kelbert, L.Y., Leonenko, N.N., Ruiz-Medina, M.D.: Fractional random fields associated with stochastic fractional heat equations. Adv. Appl. Probab. **37**, 108–133 (2005)

94. Kochubei, A.N.: Distributed order calculus: an operator-theoretic interpretation. Ukrainian Math. J. **60**, 551–562 (2008)
95. Kolokoltsov, V.N.: Nonexpansive maps and option pricing theory. Kybernetica **34**, 713–724 (1998)
96. Kolokoltsov, V.N.: Idempotent structures in optimisation, translated from the International Conference for the 90th anniversary of L. S. Pontryagin, Moscow 1999. J. Math. Sci. **104**, 847–880 (2001)
97. Kolokoltsov, V.N.: Idempotent structures in optimization theory. J. Math. Sci. **104**, 847–880 (2001)
98. Kolokoltsov, V.N.: Measure-valued limits of interacting particle systems with k-nary interactions ii. Stoch. Stoch. Rep. **76**, 45–58 (2004)
99. Kolokoltsov, V.N.: Generalized continuous-time random walks, Subordination by hitting times and fractional dynamics. Theor. Probab. Appl. **53**, 594–609 (2009)
100. Kolokoltsov, V.N.: Nonlinear Markov Processes and Kinetic Equations. Cambridge University Press, London (2010)
101. Kolokoltsov, V.N.: Markov processes, semigroups and generators. Studies in Mathematics, vol. 38. De Gruyter, Berlin (2011)
102. Kolokoltsov, V.N.: Game theoretic analysis of incomplete markets: emergence of probabilities, nonlinear and fractional Black-Scholes equations. Risk Decis. Anal. http://arxiv.org/abs/1105.3053 (to appear)
103. Kolokoltsov, V.N., Korolev, V., Uchaikin, V.: Fractional stable distributions. J. Math. Sci. **105**, 2570–2577 (2001)
104. Kolokoltsov, V.N., Malafeyev, O.A.: Understanding Game Theory. World Scientific, Singapore (2010)
105. Kolokoltsov, V.N., Maslov, V.P.: Idempotent Analysis and Its Application. Kluwer, Dordrecht (1997)
106. Kramkov, D.O.: Optional decomposition of supermartingales and hedging contingent claims in incomplete security markets. Probab. Theory Related Fields **105**, 459–479 (1996)
107. Laubsch, A.: Stress testing. In: R.M. Group (ed.) Risk Management – A Practitioner's Guide, pp. 21–37 (1999)
108. Lyons, T.: Uncertain volatility and the risk-free synthesis of derivatives. Appl. Math. Finance **2**, 177–133 (1995)
109. Mandelbrot, B., Gomory, R.E., Cootner, P.H., Fama, E.F., Morris, W.S., Taylor, H.M.: Fractals and Scaling in Finance. Springer, New York (1997)
110. Markowitz, H.M.: Mean-Variance Analysis in Portfolio Choice and Capital Markets. Blackwell, Boston (1987)
111. Maslov, V.P.: Nonlinear averages in economics. Math. Notes **78**, 347–363 (2005)
112. McEneaney, W.: A robust control framework for option pricing. Math. Oper. Res. **22**, 22–221 (1997)
113. Meerschaert, M.M., Nane, E., Xiao, Y.: Correlated continuous time random walks. Stat. Probab. Lett. **79**, 1194–1202 (2009)
114. Meerschaert, M.M., Scala, E.: Coupled continuous time random walks in finance. Physica A **370**, 114–118 (2006)
115. Meerschaert, M.M., Scheffler, H.P.: Limit theorems for continuous-time random walks with infinite mean waiting times. J. Appl. Probab. **41**, 623–638 (2004)
116. Melikyan, A., Bernhard, P.: Geometry of optimal trajectories around a focal singular surface in differential games. Appl. Math. Optim. **52**, 23–37 (2005)
117. Merton, R.C.: Option pricing when underlying stock returns are discontinuous. J. Finan. Econ. **3**, 125–144 (1976)
118. Merton, R.C.: Continuous-Time Finance. Blackwell, Boston (1992)
119. Morgan, J.: Value at risk. Risk Man Special Supplement, June 1996, 68–71 (1996)
120. Neftci, S.: Principles of Financial Engineering. Academic Press, New York (2009)
121. Neftci, S.: An Introduction to the Mathematics of Financial Derivatives. Academic Press, London (1996)

122. Ng, S.: Hypermodels in Mathematical Finance: Modeling via Infinitesimal Analysis. World Scientific, River Edge (2003)
123. Olsder, G.: Control theoretic thoughts on option pricing. Int. Game Theory Rev. **2**, 209–228 (2000)
124. Palczewski, J., Schenk-Hoppé, K.R.: From discrete to continuous time evolutionary finance models. J. Econ. Dyn. Control **34**, 913–931 (2010)
125. Perold, A.F., Black, F.: Theory of constant proportion portfolio insurance. J. Econ. Dyn. Control **16**, 403–426 (1992)
126. Perrakis, S., Lefoll, J.: Derivative asset pricing with transaction costs: an extension. Comput. Econ. **10** (1997)
127. Pliska, S.: Introduction to Mathematical Finance: Discrete Time Models. Blackwell, Oxford (1997)
128. Pujal, D.: Évaluation et gestion dynamiques de portefeuille. Ph.D. thesis, Université Paris-Dauphine (2000)
129. Pujal, D., Saint-Pierre, P.: L'algorithme du bassin de capture appliqué pour évaluer des options européennes, américaines ou exotiques. Revue de l'Association Française de Finance (2004)
130. Rockafellar, R.T., Wets, R.: Variational Analysis. Springer, Berlin (1997)
131. Roorda, B., Engwerda, J., Schumacher, H.: Coherent acceptability measures in multiperiod models. Math. Finance **15**, 589–612 (2005)
132. Roorda, B., Engwerda, J., Schumacher, H.: Performance of hedging strategies in interval models. Kybernetica (Preprint: 2000) **41**, 575–592 (2005)
133. Rubinstein, M.: Displaced diffusion option pricing. J. Finance **38**, 213–217 (1983)
134. Saint-Pierre, P.: Approximation of the viability kernel. Appl. Math. Optim. **29**, 187–209 (1994)
135. Samuelson, P.A.: Lifetime portfolio selection by dynamic stochastic programming. Rev. Econ. Stat. **51**, 239–246 (1969)
136. Shafer, G., Vovk, V.: Probability and Finance: It's Only a Game! Wiley, New York (2001)
137. Shaiju, A.J., Dharmatti, S.: Differential games with continuous, switching, and impulse controls. Nonlinear Anal. **63**, 23–41 (2005)
138. Soner, H.M., Shreve, S.E., Cvitanić, J.: There is no non-trivial hedging portfolio for option pricing with transaction costs. Ann. Probab. **5**, 327–355 (1995)
139. Strook, D., Varadhan, S.: On the support of diffusion processes with applications to the strong maximum principle. In: Proceedings of the 6th Berkeley Symposium on Mathematical Statistics and probability, vol. 3. Probability Theory, pp. 333–359. University of California Press, Berkeley, CA (1972)
140. Strook, D., Varadhan, S.: Multidimensional Diffusion Processes. Springer, Berlin (1979)
141. Tarasov, V.E.: Fractional Dynamics: Applications of Fractional Calculus to Dynamics of Particles, Fields and Media. Springer, Berlin (2011)
142. Thiery, S.: Évaluation d'options "vanilles" et "digitales" dans le modèle de marché à intervalles. Ph.D. thesis, Université de Nice-Sophia Antipolis, France (2009)
143. Thiery, S., Bernhard, P., Olsder, G.: Robust control approach to digital option pricing: Synthesis approach. In: Bernhard, P., Gaitsgory, V., Pourtallier, O. (eds.) Advances in Dynamic Games and their Applications (also in 12th International Symposium on Dynamic Games and Applications, Sophia Antipolis, France, 2006). Annals of the International Society of Dynamic Games, vol. 10, pp. 293–310. Birkhaüser, Boston (2009)
144. Uchaikin, V.V., Zolotarev, V.M.: Chance and stability, Stable distributions and their applications. With a foreword by V. Yu. Korolev and Zolotarev. Modern Probability and Statistics. VSP, Utrecht, pp. xxii+570 (1999) ISBN: 90-6764-301-7
145. Wang: Scaling and long-range dependence in option pricing: pricing European options with transaction costs under the fractional Black–Scholes model. Phys. Abstr. **389**, 438–444 (2010)
146. Wilmott, P.: Derivatives: The Theory and Practice of Financial Engineering. Wiley, Chichester (1998)

147. Zabczyk, J.: Chance and decision: stochastic control in discrete time. Tech. rep., Scuola Normale de Pisa, Italy (1996)
148. Ziegler, A.: Incomplete information and heterogeneous beliefs in continuous-time finance. Springer Finance. Springer, Berlin (2003)
149. Ziegler, A.: A Game Theory Analysis of Options: Corporate Finance and Financial Intermediation in Continuous Time. Springer Finance. Springer, Berlin (2004)

Index

A
actualized value, 328
asset price path, 34, 95

B
binomial tree model, vi, 34
Black and Scholes, 17, 22, 38, 217, 294
  degenerate, 276
  fractional in time, 282
  generalized, 275
  nonlinear extension, 274
  with stochastic volatility, 281
Brownian motion, v, 3, 39, 294

C
Clio derivative, 334
cost range, 46, 48
  best-case cost, 50, 65
  interval, 47
  maximum profit, 65
  worst-case cost, 50, 65
Cox–Ross–Rubinstein, vi, 34, 43, 294
  two-color extension, 255
Cox-Ross-Rubinstein, 217
  multicolor extension, 257
current latitude, 66
cushion, 320

D
dynamic LLN, 288

F
fair price interval, 48, 50
  computation, 55, 56
floor, 320, 324

G
general position, 224

H
Hadamard product, 239
hedge, 100, 251
hedging, 99
  forward contract, 31
  futures contract, 31
  overhedging, 58
  under-hedging, 205
  underhedging, 59
hedging strategy, 45, 293, 301
  continuous, 46
  delta hedging, 34
  delta-hedging, 46, 66
  extreme delta hedging, 48
  nonanticipating, 46
  nonanticipative, 99
  path independent, 46
  stop-loss, 17, 20, 33, 46
  subgradient, 49

I
information structure
  closed-loop perfect state, 46
  feedback perfect state, 46
  one step delay, 105
innovation term, 38
instantaneous digital CDS, 285
  simultaneous defaults, 288
interval model, v, 95, 96, 249
  nonlinear, 263
  uniform, 15

P. Bernhard et al., *The Interval Market Model in Mathematical Finance*, Static & Dynamic Game Theory: Foundations & Applications, DOI 10.1007/978-0-8176-8388-7,

Isaacs equation
 continuous, 116
 discrete, 112
Isaacs' equation
 continuous, 119
 discrete, 157, 171

L
liquidating dividend, 329
lognormal distribution, v, 3, 39
lower expectation, 265

M
market model, 45, 95
 NGARCH, 314
Markowitz, 7

N
no arbitrage, 34, 101, 217

O
option, 18, 31, 94
 American, 37, 95, 104, 261
 barrier, 312
 Bermuda, 307
 call, 31, 107, 271, 293, 301
 capped style, 308
 colored, 249
 European, 18, 34, 94, 100, 249, 301
 maturity date, 250
 payoff function, 45
 put, 31, 107, 293
 rainbow, 249
  call on the maximum, 249
  delivering the best, 249
  multiple-strike, 249
  nonlinear jumps, 263
  portfolio, 250
  spread, 250
 real, 262

P
portfolio, 250, 293
 insurance, 319
 risk-free, 35
 self financed, 17, 19
positive completeness, 222
pre-Markov model, 270
price
 buyer's, 265
 fair, 46, 217
 hedging, 286
 seller, 93, 101
 seller's, 265
proportional jump factor, 34

R
realized hedge costs, 66, 205, 209
reduced Bellman operator
 American options, 261
 European options, 251
region of interest, 125, 175
risk-free interest rate, 19, 35, 95
risk-neutral probability, 221
robust control, 91, 251
robust cost constraint, 65

S
Samuelson, v, 18, 38
stress testing, 42
subdifferential, 49, 143
subgradient, 49
submodular function, 254

T
transaction costs, 32, 34, 93, 94, 258, 303
 fixed, 258, 303
 proportional, 98, 100, 258, 303
 reduced Bellman operator, 259

U
upper expectation, 265
utility
 fractional power, 6, 11
 logarithmic, 8, 14

V
Value-at-Risk, 41, 65
volatility, v, 38, 297, 331
 uncertain, 42

MIX
Papier aus verantwortungsvollen Quellen
Paper from responsible sources
FSC® C105338

If you have any concerns about our products,
you can contact us on
ProductSafety@springernature.com

In case Publisher is established outside the EU,
the EU authorized representative is:
Springer Nature Customer Service Center GmbH
Europaplatz 3, 69115 Heidelberg, Germany

Printed by Libri Plureos GmbH
in Hamburg, Germany